C0-CDD-528

WITHDRAWN

TOPICS IN DIFFERENTIAL GAMES

TOPICS IN DIFFERENTIAL GAMES

Editor

Austin BLAQUIÈRE

Professor of the University of Paris VII

1973

NORTH-HOLLAND PUBLISHING COMPANY - AMSTERDAM • LONDON
AMERICAN ELSEVIER PUBLISHING CO., INC. - NEW YORK

Library of Congress Catalog Card Number: 73-75528
ISBN North-Holland 07204 20806
ISBN American Elsevier 0444 10467 4

PUBLISHERS:

NORTH-HOLLAND PUBLISHING COMPANY – AMSTERDAM
NORTH-HOLLAND PUBLISHING COMPANY, LTD.–LONDON

SOLE DISTRIBUTORS FOR THE U.S.A. AND CANADA:

AMERICAN ELSEVIER PUBLISHING COMPANY, INC.
52 VANDERBILT AVENUE
NEW YORK, N.Y. 10017

PRINTED IN THE NETHERLANDS

LIST OF CONTRIBUTORS

Numbers in parenthesis indicate the pages on which the authors' contributions begin.

Austin BLAQUIERE (101, 151, 271)
Laboratoire d'Automatique Théorique,
Université de Paris 7, 2 Place Jussieu,
75005 Paris, France.

James CASE (377)
Mathematical Sciences,
The John Hopkins University,
Baltimore, Maryland 21218, U.S.A.

Pierre CAUSSIN (101, 151)
Laboratoire d'Automatique Théorique,
Université de Paris 7, 2 Place Jussieu,
75005 Paris, France.

Michael D. CILETTI (179)
Systems Control Inc.,
260 Sheridan Avenue,
Palo Alto, California 94306, U.S.A.

Rufus ISAACS (1)
Operations Research and Industrial Engineering,
The John Hopkins University,
Baltimore, Maryland 21218, U.S.A.

Libuska JURICEK (271, 311)
Laboratoire d'Automatique Théorique,
Université de Paris 7, 2 Place Jussieu,
75005 Paris, France.

George LEITMANN (345)
Mechanical Engineering,
University of California,
Berkeley, California 94720, U.S.A.

Boris PCHENITCHNY (45)
Institute of Cybernetics,
Academy of Sciences of the Ukrainian SSR,
109 Science Avenue, Kiev-28, USSR.

Martin SHUBIK (401)
Administrative Sciences,
Yale University, 2 Hillhouse Avenue,
New Haven, Connecticut 06520, U.S.A.

Harold STALFORD (345)
Department of the Navy,
Naval Research Laboratory,
Washington,D.C.20390, U.S.A.

Ward WHITT (401)
Administrative Sciences,
Yale University, 2 Hillhouse Avenue,
New Haven, Connecticut 06520, U.S.A.

Karl E. WIESE (271)
Laboratoire d'Automatique Théorique,
Université de Paris 7, 2 Place Jussieu,
75005 Paris, France.

PREFACE

In the past few years many works have been devoted to the theory of differential games. This book includes results which extend or generalize the theory and some of its applications. A thorough discussion and a classification of the problems and of their difficulties are presented by Professor R. Isaacs in his introductory chapter.

To the first level belong maximizing problems, including the theory of optimal control ; that is, one-player games. At a higher level comes the zero-sum two-player game with complete information. Whatever one player gains, his opponent loses, and both players possess complete knowledge of the current state.The two player zero-sum game with incomplete information is the next level. Here, either or both players may possess some essential knowledge of the current state not available to his opponent. To the next level of the hierarchy come two-player games which are not zero-sum. Not only the losses of one player not necessarily have to balance the gains of his opponent, but both players may loss or both may gain. Two kinds of non-zero sum games have been widely discussed in the literature, namely cooperative and non cooperative games. Associated with these two well shaped specimens are the concepts of Pareto and Nash equilibria, respectively. Between them one finds the whole gamut of partly cooperative games, with different kinds and different degrees of cooperation. The next rung of the hierarchy, and undoubtedly the most promising for the applications, might be games with an arbitrary number of players.

The papers published in this volume can be ideally grouped into two parts. The first part, to which the papers by Pchenitchny, Blaquiere and Caussin, and Ciletti belong, contains basic aspects of the theory of zero-sum two-player games, each emphasizing a particular aspect. In the second part some properties of non zero-sum (e.g. N-person) games are studied, with a view toward making the theory applicable to economic problems. Necessary and sufficient conditions for optimality of the strategies are given in the papers by Blaquiere Wiese and Juricek, and Stalford and Leitmann, respectively. Games with coalitions are investigated by Juricek. Economic examples are discussed. The last two chapters, by Case and by Shubik and Whitt, deal with economic applications.

Especial thanks go to Mrs Alice Roussel, who typed the manuscript and contributed to its edition.

January 1973 A. BLAQUIERE

CONTENTS

Several organizations have supported in part some of the research reported in this book, among them the Délégation Générale à la Recherche Scientifique et Technique (France), the Ford Foundation, the National Science Foundation and the Office of Naval Research (U.S.A.)

SOME FUNDAMENTALS OF DIFFERENTIAL GAMES

Rufus ISAACS[†]

A few years ago I wrote an article on differential games whose intent is stated in its own introduction. I take the liberty of reproducing it here as a starting point, for the notions therein can now be expanded in, I hope, a profitable way.

Throughout the following it will be referred to as "the article" and my book *Differential Games* (Wiley, 1965) as DG .

Section 4 contains some simple but I believe new problems, illustrating time-lag games. Some current aspects of the latter, which I find puzzling, appear in Section 3 .

† The John Hopkins University, Baltimore, Maryland, USA .

Differential Games : Their Scope, Nature, and Future[†]

Abstract. There is a profound distinction between classical mathematical analysis and game theory which comes into especial prominence with the advent of differential games. There is a hierarchy of theories of applied mathematics in which the classical theory is the bottom row. Thus, it is important in the inevitable pending developments of higher forms of game theory to be prepared for ideas and concepts which break with tradition. These general thoughts, not at present widely understood, are expounded here with some simple examples which already illustrate the novelties of future research.

1. *Introduction.*

Since the appearance of my book (Ref.1), I have felt increasing concern over a certain prevalent misunderstanding of the real nature of differential games. My own reflections, from the perspective that trails the setting down of all the complex technical details, conversations with others and their writings, and reactions to my lectures, all bolster this view. At present, there appear to be comparatively few people who have grappled in depth with the problems of differential games. Hence, there is a rather widespread failure to grasp the distinction from classical analysis.

This distinction is due to the presence of two players with contrary purposes. When each makes a decision, he must take into account his opponent's portending action toward the

---

† Reprinted from J. of Optimization Th. and App., Vol.3, N°5 (May 1969) .

opposite end, his opponent's similar wariness of the first player's actions, and so forth. This situation is basically different from that of much of classical mathematics, which, as I shall argue below, consists of one-player games.

Thus, my main thesis is that differential games lie in a stratum distinctively above the mathematics to which we are accustomed ; accordingly, our thinking in this field must transcend the habits inculcated by traditional training. Beyond are further strata which can be expected to harbor further novelties. I shall shortly attempt to arrange them in a rough hierarchy, although the emphasis will be on the distinction already mentioned.

2. *Classical Applied Mathematics*.

We shall be concerned with those problems whose answers are numbers or sets of numbers. Included in the latter are functions, so that differential-equation problems or problems of variational calculus are among the legions of examples. The science of such problems can be termed *classical applied mathematics*.

There are familiar ways to frame many of these problems as maximizing or minimizing problems. Thus, the domain of classical mechanics reduces to minimizing the Hamiltonian ; Fermat's principle of least time applies to optics ; finding the root of a polynomial can be done by minimizing its modulus. If we tolerate a bit of artifice, there is no doubt that all such problems can be so framed. Such problems are one-player games. Their matrix has a single column or row.

This viewpoint provides perspective on the relation of game theory to general applied mathematics and on the relation of differential games to general game theory. Thus, we can apprehend a basic and cardinal fact : generally, a problem of game theory is not merely a bit harder than its traditional one-player counterpart but many times as hard. In the simplest terms, the former has a full two-dimensional matrix ; the latter, a single-row matrix.

We can also perceive the limitations of traditional game theory, where the approach is through first writing the matrix. Should we adopt the same idea for classical problems, such as those mentioned above, the one-row matrix would simply be a list of candidates for the solution† with the ensuing *payoff* entered for each. The game theorist's dictum "find the maximum" would be as useless for obtaining the solution as the order "win the battle" would be for obtaining a military victory.

A great deal of game-theoretic exposition stops with finite matrices, which can be solved by well-known methods. But let us not forget how near they are to the threshold of puerile thinking, at least in principle. Their counterpart in traditional mathematics, a finite one-row matrix, would be the epitome of trivia ; it asks such a question as : "What is the maximum of 1, 5, and 10 ?"

Passing from these trivial questions of maximizing over a finite set to the substantial problems of classical applied mathematics (to maximize over infinite sets) is thus a giant step. We can now put the theory of differential games in its place. Differential games are the same step from finite matrix games as classical applied mathematics is from finite maximizing.

---

† Of course, this list would generally be infinite and could be written in principle only .

3. *The Hierarchy of Applied Mathematics.*

To the first level belong maximizing problems : as we have seen, they embrace virtually all classical problems of applied mathematics, that is, those with numerical answers.

At a higher level comes the zero-sum, two-player game with complete information. Such is a matter of pure conflict : whatever one player gains, his opponent loses. But, as noted in an earlier section, the interactions of the two players make these games extremely more complex than their counterparts with one player, that is, those of level one.

Between these levels belongs Danskin's max-min theory (Ref.2), which likewise rests on conflict between two players rather than on the striving of one. The distinction from game theory is that here one player selects his strategy after and with full knowledge of his opponent's strategy. The latter then seeks to maximize the minimum that he knows the former has chosen against him. If game theory can be applied to the waging of war, then max-min theory relates to the large-scale preparation for war.

Before going on with the hierarchy, we compare more closely the one-player and two-player games already mentioned.

4. *Classical Applied Mathematics and Two-Player, Zero-Sum Games.*

Let us note the difference already incurred in the concept of solution. There is no doubt about the meaning of the correct answer to a classical problem which bids us to maximize a single quantity. For the game problems on our present rung, what is meant by an answer is not immediately clear. However, the *mini-max* or *saddlepoint* definition, proposed by Von Neumann, is universally accepted. Besides this minimax of the payoffs, called the *Value*[†] of the game, the solution should include at least one optimal strategy for each player. The basic relation between these three entities, which is tantamount to their definitions, is as follows : If either player plays an optimal strategy, he is certain to reap a payoff equal to or better than the Value. If he plays a strategy that is not optimal, *there is a strategy for his opponent leading to a payoff worse than the Value* (for the first player).

Playing optimally, in the game-theoretic sense, thus guarantees a player a certain outcome which he cannot expect to improve, *provided his opponent acts rationally*. As this outcome, the Value, is the same for both players, it is unique and has won acceptance as a constituent of the mathematically defined solution.

The solution so defined appears conservative and herein lies a possible reason why it might not always be used. Note the italicized clauses above. For instance, if a military commander has reason to believe that his opponent will not act rationally (that is, will not play optimally), he may be able to do better than the Value if his belief turns out to be justified. But, to reap this gain, our commander must deviate from his optimal strategy ; and there will be a way for the opposition to exploit this defection by causing a payoff to our commander worse than the Value.

Examples of this kind of risk taking are not unknown in history. To a commander bent on such a risk, the game theorist can offer little quantitative sound advice, unless (and how likely is it in practice ?) he can reliably assign numerical probabilities to the opponent's deviation from optimality.

---

† We capitalize the word *Value* used in this context.

Virtually all of my own work on differential games and most of that of others has been on this rung. We can now perceive the place of these results in the broad mathematical scheme. This level of differential games is in the same relation to the usual matrix presentation of game theory as our *classical applied mathematics* is to maximizing over a small, finite set of numbers.

If one player in the preceding type of differential game is deprived of all volition, what remains is a particular type of maximizing problem (one-player game). The study of these problems is called control theory or, more properly, the theory of optimal control. If *control theory* replaces the phrase italicized in the preceding paragraph, the relationship given there becomes meticulously exact.

Finally, a pragmatic word must be given to control theorists about a pitfall which besets them when they first embark on differential games. They are accustomed to *open-loop control*, which means that the control variables are functions of the time rather than of the state variables, as is the case with a strategy. But the former does not suffice for two-player games. The interim actions of the opponent cannot be ignored. Imagine how a chess player would fare if, before a match, he were to decide his first, second, third,..., moves without regard to those of his opponent.

5. *Classical Applied Mathematics and Differential Games.*

We return to the important point made earlier : differential games involve an essentially higher and novel level of difficulty than *classical applied mathematics* (again, the reader may prefer to substitute the words *control theory*, thus gaining precision but losing scope). We employ three examples diversely typical of novel phenomena which arise only in the two-player case.

Example 1 : The Homicidal Chauffeur Game and A Rendezvous Problem. As the former game is presented at length in Ref.1, I suppose that the reader is familiar at least with its definition. On page 11 of Ref.1 is described, in intuitively plausible terms, the swerve maneuver : at an early stage of certain parties, the evader E commences by chasing the pursuer P ; these actions of course are later interchanged. Note that, in the diagram on page 13 of Ref.1, the path of E lies on two straight lines, while the path of P consists of several segments, all of which are either straight or of sharpest possible right or left curvature. Later (pages 298 and 300), we learn that this need not be so : for part of the play, each player can optimally traverse curved paths more complex in nature and presumably transcendental, corresponding to the equivocal curve in the reduced space.

For starting points in the region labeled with a question mark on page 301, there appears to be still a third phase of optimal play. This is not treated in Ref.1, nor, as far as I know, has it been quantitatively analyzed ; but we can draw some reasonably certain heuristic conclusions. Should E try the swerve from such a starting position, he would be in jeopardy of an almost immediate capture were P to turn right at the outset. Furthermore, it would be advantageous for P initially to force E away so as to enhance the separation of P and E. Then, P would gain in the swerve to follow, for, when E pursues P, he must do so at a greater distance ; and so P attains the sooner room for a turnabout leading to the final kill. Thus, we can expect the final optimal play to enjoy three phases ; first P swings to the right and E flees from him ; second, P veers left and E,switching direction, pursues him ; and third, P, now having adequate maneuvering room, turns right again, E flees again, and capture ensues.

Professor J. Breakwell and his associates at Stanford University (Ref.3) offer an alternative conjecture as to the still open phase of the solution. Starting at a reduced position in the region labeled with a question mark, P swings to the left with E in pursuit but taking care not to come too close to P . The path in the reduced space arrives tangent to the capture circle at some point, follows this circle for a while, and leaves tangent to this circle until E is directly behind P ; and only later does P swing to the right.

It is not easy to ascertain which, if either, of these two concepts is correct. Here is a typical instance of a main tenet of this paper : an utterly new phenomenon occurs unlike anything in classical, one-player analysis. But, in any case, there is still an additional phase in the solution with at least three segments in the evader's optimal path.

Note how this complexity would evaporate, should we turn the problem into a one-player game. Of the several ways to do so, the most appropriate is to leave the kinematics unchanged and let both players be minimizing. The pursuit game becomes a rendezvous problem ; both players now strive to minimize the time until "capture". The game is now of the one-player type in the sense that we can envisage a single mind controlling both craft with their mutual desideratum.

If S is the starting point of E and, under optimal play, he terminates at T, it is clear that his optimal path is the straight segment ST. For obviously such path minimizes the time to reach T. Any motivation, such as above, for another path is now void, for certainly E has no reason to fear a premature meeting with P (such would lead to a termination time smaller than the Value end, hence, be contradictory).

Note that, in each of these two games, the kinematic equations, the playing space, and the terminal surface are identical. The main equation of the latter differs from that of the former only in that the original max-min (over the control variables) is replaced by min-min. Thus, the solutions in the small (those having to do with the integration of differential equations) are very much alike. The distinction rests entirely on singular surfaces.

Example 2 : The Lady in the Lake†. The lady E is swimming at speed w_1 in a circular lake of radius R and center O. The pursuer P, a lascivious gentleman, remains on the shoreline on which he runs at speed w_2 considerably greater than w_1. As E wishes to reach the shore as far from P as possible, the payoff is the distance PE, either metrically or arcwise, at the instant the lady leaves the water.

Let us suppose that we set up the usual formal apparatus to solve this game. The standard procedure leads at first to well-behaved optimal paths constructed retrogressively from the shore (or, rather, its equivalent in a suitable reduced space) : E travels on certain straight lines ; P follows the shore in the expected direction ; there is an obvious dispersal surface, with an instantaneous mixed strategy, when P and E start from diametrically opposite points.

But, mysteriously, this solution soon breaks down. Formally, it involves the square root of a quantity which becomes negative as soon as the distance OE becomes less than $(w_1/w_2)R$. What has happened ?

---

† This game does not appear in Ref. 1 but has been inserted in both the French and Russian translations.

The answer is clear if we drop the formal analysis and consider the following policy open to the lady. Let K be a disk concentric with the lake and of radius $(w_1/w_2)R$. When E is within K, the lady can travel at a greater angular speed about O than P. Therefore, the lady can at any time swim to O and then reach the rim of K, always remaining diametrically opposite to P[†]. From this instant on, she utilizes the optimal straight paths found earlier.

Observe that, when E is in K, neither E nor P is under any compulsion at any instant, for relevance of the termination time to the payoff has not been postulated (but unbounded endurance of the swimmer tacitly has). Both players may loiter before E completes her play ; all strategies here are optimal.

This is true even beyond K . Let V_o be the Value when E utilizes the preceding ploy. Clearly, she can do so from *any* starting point. Thus, our earlier formal solution is valid only in that region of the playing space in which the computed Value is less than V_o. Outside this region, the Value is V_o and all strategies are optimal.

It is difficult to conceive any one-player version of this game for which the preceding situation would hold.

Example 3 : Bang-Bang-Bang Surfaces. In the earlier days of differential games, the simplest singular surface appeared to me to be the transition surface. Later, I learned that

† Strictly, she is not here using a strategy, for her control variable depends on that of P rather than on the state variables. But this is easily remedied if we are willing to settle for locations arbitrarily close to diametrically opposite when E reaches the edge of K .

control theorists used the same entity but under the more onomatopoeic title of *bang-bang control*.

The phenomenon occurs when the kinematic equations are linear in a control variable ϕ, which is constrained, say, $-1 \leq \phi \leq 1$. Then, almost everywhere in the playing space, the optimal $\bar{\phi}$ assumes one of its extreme values, except when there is a region in which all admissible ϕ are optimal. A *transition surface* is one crossed by optimal paths and at which $\bar{\phi}$ switches from one extreme to the other. These can usually be detected through the *switching function* A, a quantity which can be computed along each optimal path as its differential equations are integrated. When A = 0, the switching surface occurs. The details are well-known to control theorists ; it is from them that I borrow the name for A.

After Ref.1 was written, I found, to my surprise, that it is possible to have transition surfaces at which A does not vanish. I called this phenomenon, tentatively and jocularly, *bang-bang-bang control*. At such surface, a control variable of the opponent, *which appears nonlinearly in the kinematic equations*, also abruptly shifts its optimal value. Thus, I do not believe this phenomenon is possible in one-player cases. An example or two is all that is now known of this subject (see the Appendix) but a student of mine, Po-Lung Yu, has selected these surfaces as his dissertation topic (Ref.4).†

A ship P, with the curvature of its trajectory bounded, endeavors to maintain surveillance over a slower submarine E which moves with simple motion. Thus, the kinematics are as in the homicidal chauffeur game. But the payoff is the maximum of the

† His term for the loci of this new phenomenon is *double transition surface*. He has by now admirably clarified the whole question.

distance PE occuring during a partie, which P seeks to minimize. Our new surfaces arise here. At a state on one, P, who has a linear vectogram, shifts from sharp right to sharp left rudder under optimal play. At the same instant, E, who has a circular vectogram, also discontinuously changes his travel direction.

Implications of the Examples. In all three examples, the local or differential equation aspect of the solution entailed no essential novelty over one-player games other than the two sets of control variables, one maximizing and one minimizing . Thus, what is new can be said to reside in singular surfaces . Their role has already been discussed in Example 1, and Example 3 is about a new type of surface. Our statement also holds for Example 2, provided we grant the boundary of the region where $V = V_o$ the status of singular surface. Whether this is valid or not depends on the definition of singular surface.

I have not attempted a sharp definition of singular surface, because as yet there is no unified theory of these surfaces, nor do we know what strange new types are yet to appear. Could there be, for example, a bang-bang-bang counterpart of the universal surface ? This is a surface which is universal, yet fails to satisfy the analytic conditions for such, due to the opponent having some kind of discontinuous optimal behavior there.

My dictum is that the emphasis for two-player differential games with full information should be on singular surfaces. Through them will the theory be completed.

6. *The Hierarchy Resumed.*

The two-player zero-sum game with incomplete information is the next level. Here, either or both players may possess some essential knowledge of the current state not available to his opponent. Any book on basic game theory explains that now the optimal strategies usually are mixed or randomized. To deceive his opponent, to conceal his own intent, a player generally should make his decisions probabilistically. Solving the game consists of finding the best probability distribution of decisions for each player, his optimal mixed strategy, and the Value of the game, which now means the minimax of the expected value of the payoff. This discovery that, for finite games, solutions in this sense always exist is a major achievement of Von Neumann ; methods for finding the solutions are in the text books .

We can feel certain that mixed strategies will dominate differential games of incomplete information also. But very little is known of solution methods. All that I know is written in Chapter 12 of Ref.1 .

Let us note that, at this level, we have something radically different from anything on a lower level. We are confronted with a new potential domain of ideas. I suspect that it is ill understood by many working in allied fields such as control theory.

Stochastic problems are well known in this subject. But they are concerned with one-player affairs in which the data or current state variables are known only probabilistically. If the criterion is the expected value of the payoff, it is often possible to find optimal stochastic strategies. *But it is not necessary to employ them.* Pure strategies do as well. There is no opponent here from which to conceal our intentions.

I am convinced, and have been for many years, that differential games will ultimately be a vital military tool. Already, interest is keen in the USSR. When American authorities are sufficiently convinced of this military value, there will inevitably ensue a concentrated and frenetic effort to close the gap . The incomplete information case will be of paramount interest.

I urge a search now for the breakthroughs and new ideas that will render possible a useful theory.

To the next level of the hierarchy, we assign two-player games which are not zero-sum. Not only do the losses of one player not necessarily have to balance the gains of his opponent, but both players may lose or both may gain. Consequently, we are no longer faced with pure conflict ; these games entail cooperation as well.

At each level, we advance to an essentially new type of mathematical problem. We are now at a stage where the very meaning of a solution is far from clear. To formulate the problem fully, we must state (because cooperation is an element) whether or not the players are permitted to negociate. Also, do they do so explicitly or tacitly ? Are there means of enforcing agreements, if any, between players ? Is player A allowed to make a side payment to player B in order to induce B to play a strategy greatly beneficial to A but, according to the rules, only slightly so to B ?

Different answers to these questions engender different problems. I know of no unequivocal, uncontested definition of solution to any such problems. Various candidates, such as the equilibrium points of J. Nash (for an account of his ideas and list of references, see Ref.5) have been proposed and many seem to have justifiable merit in many cases. But counterexamples have been offered in which natural expectations of reasonableness seem to evanesce, with almost an aura of paradox.

The next rung of the hierarchy might be n-person games. Here, J. Case is breaking new ground (Ref.6). We have reached rarefied air now and I will no longer attempt to charp delineations.

Much of the research in game theory for the past several years has been devoted, to n-player games. Many promising and deep results have been obtained, but they are by many authors and scattered in many papers. I am no expert and shall not attempt a summary or an evaluation, except to say that here, too, there is no single generally accepted concept of a solution.

7. *Conclusions.*

I have tried to show that differential games, at their broadest, open to us a hierarchy of new types of applied mathematics ; among these, the classical is at the bottom level. Each rung demands essentially new concepts. I advocate efforts to climb this difficult ladder rather than to descend by adhering too closely to traditional thinking.

Appendix

Bang-Bang-Bang Surfaces in Differential Games.

The transition surfaces appear early in differential games as an apparently very simple type of singular surface. The idea has been extensively studied in the one-player case (control theory) under the name *bang-bang control.*

If the kinematic equations are linear in a control variable ϕ, which is bounded, then it is well known that the optimal $\bar{\phi}$ generally assumes one or the other of its extreme values according to the sign of a calculable quantity, the switching function A.

We demonstrate here what we believe is a new phenomenon. In a proper, two-player differential game, it is possible to have a transition surface, at which each player's optimal strategy is discontinuous even though A *does not change sign*. This transition surface does not appear possible in the one-player case.

Consider the differential game with playing space in the half-plane xy with $y \geqslant -1$. The terminal surface $\mathcal{C}$ is at $y = -1$. The payoff is terminal with H, the Value on $\mathcal{C}$, being given as $-x$. The kinematic equations are the following :

$$\dot{x} = \phi(-\frac{1}{2}\sqrt{3}y) + \sqrt{3}(2-\frac{1}{2}y) + 2\sqrt{3}\cos\psi$$

$$\dot{y} = \phi(3 + \frac{1}{2}y) + (\frac{1}{2}y - 1) + 2\sqrt{3}\sin\psi$$

with $-1 \leqslant \phi \leqslant 1$. Here, as usual, ϕ is minimizing while ψ is maximizing.

If we solve this game in the usual way, we find that

$$A = -\frac{1}{2}\sqrt{3}yV_x + (3 + \frac{1}{2}y)V_y$$

and the optimal ϕ is $\bar{\phi} = -\operatorname{sgn} A$. On and above $\mathcal{C}$, we find that

$$V_x = -1, \qquad V_y = -\sqrt{3}$$

so that $A = -3\sqrt{3}$ is negative and constant. The optimal paths are straight lines of inclination $-\pi/6$. On them, $y = 1 - 2\exp(-\tau)$, so that the paths approach the line $y = 1$. The optimal ψ is $\bar{\psi} = -2\pi/3$, also constant.

Yet, despite the constant negative A, there is a transition surface at $y = 0$. For, in the upper half-plane, vertical lines are semipermeable with the proper orientation. This is seen formally by noting that

$$\max_{\psi}\ \min_{\phi}\ (-\dot{x}) = 0$$

when $y \geqslant 0$. It is not hard to see that these vertical lines are optimal paths, and the original $-\pi/6$ lines must be discarded above the x-axis, although they fulfill the formal requirements

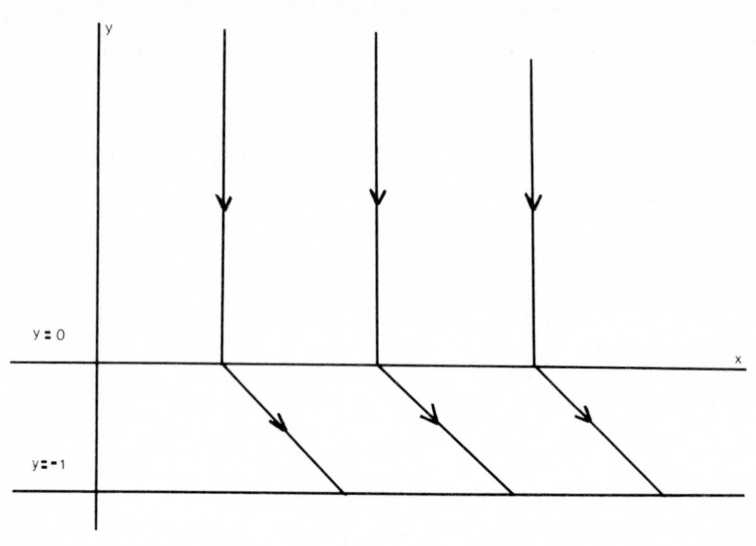

Fig.1

of a solution. The correct paths appear in Fig. 1. The optimal strategies in the upper half-plane are $\overline{\phi} = -1$, $\overline{\psi} = -\pi$, so that both have abruptly changed at the bang-bang-bang surface $y = 0$.

Note that we must adjoin the ukase to the above game that, should it not terminate, both players would suffer a drastic penalty. The submarine surveillance game show that these surfaces can be meaningful in realistic problems.

We have specifically proved that the above phenomenon cannot occur in one-player games of the same format ; that is, when the kinematic equations are of the form

$$\dot{x} = a\phi_1 + u + \cos\phi_2 , \qquad \dot{y} = b\phi_1 + v + \sin\phi_2$$

where a, b, u, v are given functions of x, y and both control variables are minimizing .

References.

1. ISAACS, R., *Differential Games*, John Wiley and Sons, New York, 1965.

2. DANSKIN, J., *The Theory of Max-Min*, Springer-Verlag New York, New York, 1967.

3. BREAKWELL, J.V., MITCHELL, A., and MERZ, T., Personnal Communication, 1967.

4. YU, P.L., *Transition Surfaces of a Class of Differential Games*, John Hopkins University, Department of Operations Research and Industrial Engineering, Ph. D. Thesis, 1969.

5. LUCE, R.D., and RAIFFA, H., *Games and Decisions*, John Wiley and Sons, New York, 1957.

6. CASE, J. H., *Equilibrium Points in N-Person Differential Games*, University of Michigan, Department of Industrial Engineering, Technical Report N°1967-1, 1967.

1. Terms, Symbols and Genesis

A galaxy of notation and vocabulary already pervades differential games. This is hardly a serious question unless diversion reaches the point at which communication is impaired. I don't believe this has as yet occured, but it may. Perhaps we ought to look now towards a basis of unification.

Primarily the vernacular split because of the two developments : differential games and control theory. The latter had the lion's share with a number of exclusive journals and meetings, even though it is the special case of one-player differential games.

When there is a choice of usages one criterion might be priority ; I do not know which subject came first. Differential games began in the late 1940's at the Rand Corporation. My audience there at the informal seminars on the subject included several who later wrote on control theory, but I know of a specific "borrowing" of ideas in only one case.

The first paper on differential games is my Rand Report P-257, *Games of Pursuit,* 17 November 1951. I had not yet coined "differential games" and the report contains some naive errors, but the germ of the subject is there. In particular it contains the *tenet of transition.*

Once I felt that here was the heart of the subject and cited it often in the early Rand seminars. Later I felt that it - like other mathematical hearts - was a mere truism. Thus, in DG it is mentioned only by title. This I regret. I had no idea that Pontryagin's principle and Bellman's maximal principle (a special case of the tenet, appearing a little later in the Rand seminars) would enjoy such widespread citation.

The tenet of transition, quoted from the 1951 report, runs:

> If the play proceeds from one position to a second and V is thought of as known at the second, then it is determined at the first by demanding that the players optimize (that is, make minimax) the increment of V during the transition.

Some details in my terminology I regret : I did not recognize the Hamiltonian form of the main equation until my work was well underway and then I was naive enough to underestimate its classic precedence. I wish DG in discussing the main equation, had used H where it sometimes used Q and a new letter replaced H in the terminal boundary condition.

My term *envelope barrier* foolishly overlooks its role in games of degree as well as of kind. John Breakwell's term *safe contact* is much better.

The name *switching surface* would have been ideal for what I called *transition surface*, but I believe that control theory has usurped the former term for something else. Po-Lung Yu (see the article) named his extension *double transition surfaces*, but it could as well be *double switching surfaces*.

I will not forcibly advocate a reform in nomenclature. But I do plead that future writers adhere to either the established control or differential game terminology unless there is a definite motive for change. It sometimes almost seems that research workers feel they are advancing the theory when they pointlessly alter existing terms.

Let us avoid an n-player tower of Babel.

2. A Major Problem for Research

My colleague, Professor James Case, has solved some excellent examples of differential games that are not zero-sum. He seeks Nash equilibrium points as a means ; his end is realistic economic situations. He is forced to use his experience to first conjecture a solution and then to prove it is one. This is because of an essential and surprising gap in the theory ; filling it would be as important a contribution as I can imagine.

Given a first order partial differential equation with the unknown function $V(x_1,\ldots,x_n)$, there is a well-known and beautiful technique for constructing the characteristic curves (or, more technically, strips) involving Monge cones and the like. It is elegantly expounded in Courant-Hilbert, *Methods of Mathematical Physics*, Volume II, Chapter 2 .

In zero-sum differential games, where V is the Value and x_j the state variables, the differential equation is the main equation. Its characteristics are the optimal paths. They are found more simply than in the above general case by merely differentiating the ME with respect to each x_j as explained in DG.

What is lacking is a characteristic theory of, say, a pair of simultaneous first-order partial differential equations in two unknown functions. Dr Case has inquired of authorities in the field and they do not seem to know of any means of constructing characteristic curves (or strips or some analogue).

In a non-zero-sum game with, say, two players, their two Values will be just the unknowns of such a pair. Let us suppose what appears the most general case : we have such a game and it has a solution (based on the Nash equilibrium criterion). Then there will be optimal paths over the playing space. These paths ought to be, if not characteristics of the main equation pair, something of a very similar ilk.

Thus it appears that some version of characteristic paths should exist generally for pairs of simultaneous partial differential equations or at least for that type which can appear as a pair of main equations in a non-zero-sum game. I deem a construction technique one of the most cogent problems for research.

3. Whence Mixed Strategies ?

Since writing the article I have encountered several papers which deal with two-player zero-sum differential games with a time lag. These are certainly cases with incomplete information. Their solutions may or may not have to entail mixed strategies. What I find amazing is that the authors, although they apparently find correct pure strategies, never even mention mixed ones. Is it that they have not heard of the latter ? Or have they been so bred on control theory as to deem them not worth a remark.

Now when pure strategy solutions are found the game is essentially solved and no more need be said. But small changes in the rules or the parameters may shift it into the domain of necessarily randomized mixed strategies. I illustrate with three related time lag examples where such strategies are indispensable.

a. *Aiming and Evasion*. The evader E moves with some simple kinematics, his sole objective being to avoid being hit by the gunner G who is firing at him. There is a time lag between G's sighting of E's position and the arrival of the bullet. For simplicity we suppose this lag is G's only drawback ; his aim at any chosen point is excellent.

Clearly G must predict E's location at the time of arrival of the bullet. As his sole objective, E must frustrate this prediction and so moves unpredictably, hence randomly, in accord with a mixed strategy. In a reasonably realistic model, G also should invoke a mixed strategy. For, if his predicted aim points were calculable by E, E need merely travel so as never to be there. Plainly mixed strategies are the quintessence of this game.

There are a number of papers on such problems. For a sketch of one version and bibliography, see DG, Chapter 12.

b. *Aiming and Evasion with a Destination.* In the preceding case E had no other concern than maneuvering to avoid being hit. Now let him - more realistically - have a destination. Once there he is safe and play is done.

If G can fire steadily, it is no longer clear that E is best off with a mixed strategy. The more he zigzags to frustrate G's aiming predictions, the longer his path and so his exposure to fire. Might not the most direct and speedy dart to his haven be best for E ?

The following section will solve some simple discrete models of this type of game. These solutions, to my knowledge have not been published before. I devised them for a course on game theory two years ago at The Johns Hopkins University.

We will learn that the answer to the preceding question is negative ; both players' optimal strategies are mixed. To foil G's prognostic aim, E should randomize his route ; to foil E's dodging pending target sites, G should randomize his aim points.

c. *Evasion and Pursuit.* We now turn to a class of games much studied by myself when the information is complete (see DG) and others : The gunner G is replaced by a pursuer P.

If P's knowledge of E's location suffers a time lag, does it pay P to exploit this by random zigzagging ?

In most cases, based on reality I should think not. His devious route would seem to penalize E more than his gain from the slight uncertainty he would induce on P's flight direction. No doubt we could construct exceptions : P is subject to some kind of severe kinematic restraint so that a slight error in his bearing angle causes an appreciable loss of payoff.

My intuition tells me that usually optimal strategies are pure. Such games seem to belong to those of the aforementioned papers concerned with only pure strategies. Thus a just defense of this concern might be argued. But there may well be limits to it and I would prefer some probing of them. Here is a case :

d. *Composite Pursuit and Aiming Games.* Here let us suppose the pursuer is also a gunner. His aim accuracy is increased greatly with less range so that it pays him to get close to E before firing. The time lag assumptions hold as in Example 1 (and likely 2).

Now, should E randomly zigzag to defeat aim even though he allows P (or G) too close thereby ? This question cannot be evaded in solving. The answer may well be that pure strategies are optimal over one phase of a partie and mixed over another.

4. Aiming and Evasion with a Destination

We shall present three versions of a game in which a gunner G fires steadily at a traveling evader E. As in the last section the gunner's aim is subject to a time lag. He must predict E's location and E seeks to frustrate G's estimates.

But E has a destination, called *home* ; once there he is immune to G's fire. If he heads there as fast and directly as possible he is subject to least exposure time to G. Should E randomize his path to foil aim or spurt home to attain least time and hence least fire ?

In Problems I and II, E is confined to a straight course to home, but he has the option of varying his speed. Problem III requires fixed speed but entails geometric path variation.

Very simple discrete models will be used in all cases. For problems such as these have a way of growing stupendously difficult with more versimilitude.

Problem I. The path is the line to home sketched and E occupies one of the numbered points. His variable speed is thus simulated: at each unit time he may move 0, 1, 2,...,k consecutive points in either direction. There is a one unit time delay in G's fire which he does after each move of E. The payoff is the number of hits (or correct guesses by G of E's next position) until E reaches home. Of course G strives to maximize and E to minimize this.

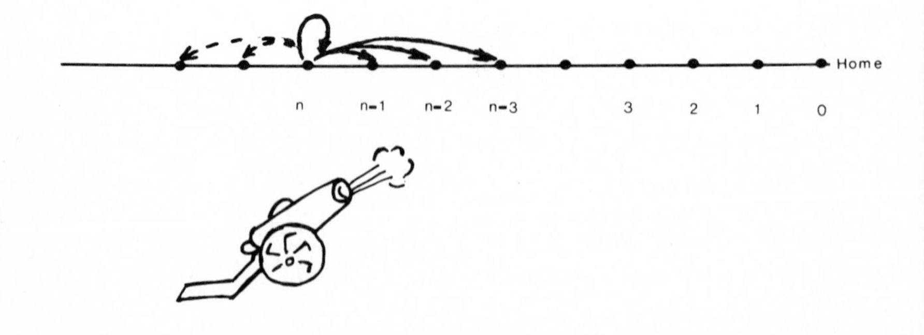

Fig. 2

For simplicity we shall take $k = 3$; more generality is obvious. We show first that E should never move away from home. (Thus if he is at n, he has the four effective choices shown by the solid arrows). Call our original game J and let J' differ only by the rule that E cannot move away from home (left, in the figure 2). When J is played let G use an optimal strategy of J'. Should E move left, although he is currently safe from a hit, he must ultimately move right again to terminate. As these right moves are either more numerous or else from more leftward points, E can only lose from his left choices. In fact his payoff will be less than the Value of J and therefore left moves cannot belong to his optimal strategies.

If E starts from 1, 2, or 3 clearly he should make the one jump taking him home which gives him the minimal payoff zero. Thus, a spurt home is best. We shall see that this is true in no other case.

Let V_n be the Value of the game - the expected number of hits under optimal play - if E starts from point n . If $n > 3$, E has the four immediate choices shown in the figure and G may choose to aim at any of the four destinations. If E moves j points to the right to point n-j, the Value becomes $1+V_{n-j}$ or V_{n-j} according as G aims at n-j or not. Thus we are led to the current game matrix.

$$\text{(1)} \qquad G \quad \overset{E}{\begin{matrix} 1+V_n & V_{n-1} & V_{n-2} & V_{n-3} \\ V_n & 1+V_{n-1} & V_{n-2} & V_{n-3} \\ V_n & V_{n-1} & 1+V_{n-2} & V_{n-3} \\ V_n & V_{n-1} & V_{n-2} & 1+V_{n-3} \end{matrix}}$$

in which G picks a row and E a column. The Value of this matrix game is, of course, V_n .

Proceeding from here is not hard. It turns out that E has these optimal strategies : the probability p_o of not moving (first matrix column) is subject only to $0 \leqslant p_o \leqslant \frac{1}{4}$ and the other three p_j are each equal to $\frac{1}{3}(1-p_o)$. Thus, aside from a certain arbitrariness of standing still, E should make (if $n > 3$) all his moves equiprobably. Note the independence of his distance from home. Thus, E's optimal strategy is as mixed as can be.

We can now find the Value of the matrix game (1) and obtain

$$V_n = \frac{1 + V_{n-1} + V_{n-2} + V_{n-3}}{3} \tag{2}$$

This linear difference equation with the initial conditions

$$V_1 = V_2 = V_3 = 0$$

determine the V_n . (For the present case of $k = 3$, I have found the following approximation for large n :

$$V_n \sim \frac{7}{18} + \frac{n}{6} .) \tag{3}$$

For G to aim j points to the right (at $n-j$) his optimal probability turns out to be

$$V_n - V_{n-j} . \tag{4}$$

Thus, if we put $j = 0$, we see that G never fires at the last seen location of E. Clearly these probabilities increase with j. It seems reasonable that he should so stress the longer moves of E. Thus G should not employ an equiprobable distribution ; further it changes with n . (From (3), it follows that this distribution becomes more linear with increasing n.)

We see now that E's arbitrary p_o merely signifies an insipid delay with no effect on the outcome. Thus E may as well always play

$$0, \frac{1}{3}, \frac{1}{3}, \frac{1}{3}.$$

Problem II. In the preceding game the payoff was the number of hits. For, say, military problems such has significace only when a large composite target is fired at by a small weapon where each hit does its small bit of damage. In Problem II we posit a more potent gun so that the payoff is the probability of a single hit. Were n large enough, E would lose if G made but one correct guess and such seems too lopsided to be interesting. Therefore we will assume a (small) probability ρ such that : if G picks his aim point correctly, the hit probability is ρ ; otherwise it is 0.

All other aspects are as in Problem I.

The principles are unchanged, but the formal mathematics now is harder. We will now demand that E advance on each move so that he has three choices. The game matrix (1) is now replaced by

$$\begin{matrix} \rho + (1-\rho)V_{n-1} & V_{n-2} & V_{n-3} \\ V_{n-1} & \rho + (1-\rho)V_{n-2} & V_{n-3} \\ V_{n-1} & V_{n-2} & \rho + (1-\rho)V_{n-3} \end{matrix}$$

The recurrence relation for V_n (compare (2)) is here nonlinear, but the interested reader may verify that

$$U_n = \frac{1}{1 - V_n}$$

satisfies

$$U_n = \frac{1}{3-\rho} \sum_{i=1}^{3} U_{n-i}$$

with initial conditions : $U_1 = U_2 = U_3 = 1$.

The optimal strategies are given by, if play starts from n:

E's probabilities : $U_{n-j} \Big| \sum_i U_{n-i}$

G's probabilities : $(1 - U_{n-j} \; U_n) \Big| \rho$

Here j (= 1, 2, or 3) has the same meaning as before.

Thus both players should employ mixed strategies but now they both depend on n.

Problem III. Now let E move in the plane with fixed speed but a free choice of direction (in the terms of DG, with simple motion). I cannot solve this problem ; it will be replaced by a somewhat convincing discrete version.

At each unit time E moves one segment of the triangular lattice shown. As before G has a unit time lag and we adopt the simpler payoff of Problem I.

At the six points adjacent to home, V = 0 as marked. Then working outward, we can find Values at other points successively, of course exploiting the obvious symmetry. For, reasoning as with Problem I, we can show that E never should move away from home.

I have not executed more than a sample of such steps. But they show that mixed strategies are best ; E does not always take the shortest route to home and G does not always aim exclusively at the most obvious next position of E.

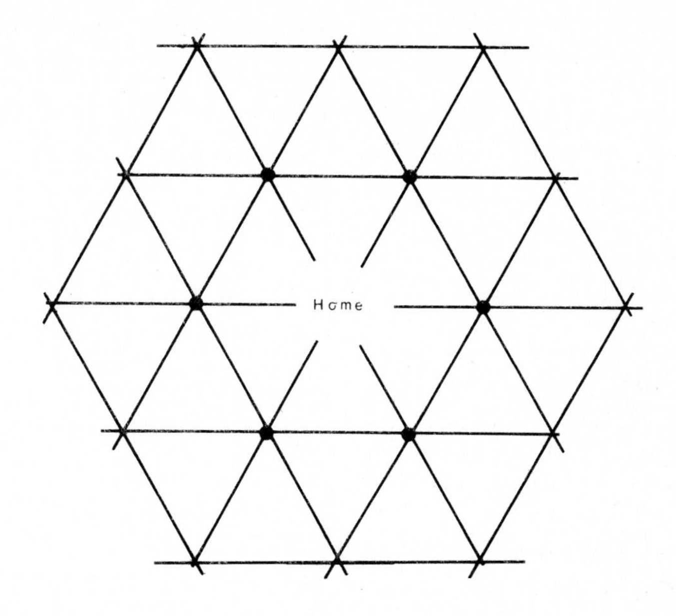

Fig. 3

Such results are persuasive that mixed strategies are optimal in the continuous case. I hope enough so that I may be spared the probably difficult rigorous proof.

5. Precise Proofs and Rigor

The article seeks to show some basic differences between classical applied mathematics and game theory. Perhaps proofs of results need some innovation too.

In all applied mathematics we simplify reality drastically so that all hinges on some finite set of conditions or axioms (in game theory they may be called "rules"). The result is termed a *model*. We should never underestimate the importance of models. All our analysis (and it could be argued, all our thinking of any sort) pertains to them. The human mind is incapable of anything else.

I have long has the hazy idea that in games with proper conflict there can exist a notion intermediate to reality and the usual mathematical model. I shall try to explain with the proviso that I know I may be talking nonsense.

Suppose we have a traditional problem which reduces to finding a function which maximizes some quantity. To apply precise reasoning we usually limit the candidate functions in the model to a class A defined by such adjectives as differentiable, measurable, etc. Without such it is generally impossible to apply the tools of analysis.

Now let us pass to a competitive differential game. Each of the several players picks a function - say, his strategy - with the object of maximizing his payoff. We formulate a model in which each player's choice is limited to A and solve.

What if *one* player picks his function outside of A ?

I can envisage the indignant reader : - you formulated a model, did you not ? Only functions in A are allowed. Maximizing over A is your problem and you must stick with it !

Let us return to our recalcitrant player. Say that his calculated payoff, even under optimal play, is discouragingly niggardly. Out of sheer perversity he picks a strategy violating A so that play cannot be completed and no one scores.

In the real situation usually he cannot act so. For example, let him be piloting a moving craft and one of his option is to move some control lever. Its displacement will be a function of the state variables or time. Whether the operation is manually or by practicable machine, reality will not permit very weird functions - unmeasurable or nowhere differentiable, say. In short, A, if well chosen, will be binding.

In the mathematical model all players are confined to A by rule. Thus, if we wish to consider a player violating A, we have to imagine an intermediate situation between reality and the conventional mathematical model. Here the competitors have the power to deviate from the accustomed premises. They can break the rules. Such is not unknown in this world.

Such thoughts - admittedly vague and perhaps futile - were disturbing when I was first trying to fabricate differential games. A rigorous theory in which there were non functional restrictions on the strategies at all (no class A) would dispel the quandary. (Note that constraints on strategy functions in the form of upper and lower bounds were allowed. Thus, in our recent example, reality would put bounds on the possible displacement of the lever. The same bounds would fit the mathematical model nicely and there is no difficulty).

The suggestion of my friend, Professor Samuel Karlin, was ingenious and welcome. Karlin strategies (K-strategies for short) are defined in DG : each player picks an *arbitrary* function and an unbounded increasing sequence $\{t_n\}$ of time values. In play he can make strategy changes only at the t_n[†].

My explanation in DG in unclear as I have learned from students and its lack of impact on other contributors to the field. The main idea is replacing the usual existence theorem by the verification theorem. I abandoned attempts at the former as I explored more examples. The array of new phenomena seemed unending ; essentially novel and unprecedented features marked the most innocuously appearing problems. Most existence attempts I have seen by others were either innocently shallow or candid enough to acknowledge that all possibilities were not covered.

The principle behind use of the verification theorem, which I think not well understood, is :

> First, obtain an alleged formal solution by techniques such as in DG. Then let the players be restricted to Karlin strategies and apply the verification theorem which tells us: let, say, $\bar{\varphi}(x)$ be the alleged optimal strategy for the minimizing player P and let $\varepsilon > 0$ be given. Then P can find a sequence $\{t_n\}$ which he uses, with $\bar{\varphi}$, to form a K-strategy such that, if the opponent plays any K-strategy, P will obtain a payoff $< V + \varepsilon$. Here V is the found alleged Value. Similarly the maximizing player can guarantee a payoff $> V - \varepsilon$.

---

† An objection, like my preceding one might be raised : why could not a perverse player make continuous changes ? I am satisfied because it certainly looks as if anything he does could be arbitrarily closely approximated by a K-strategy if $\{t_n\}$ is sufficiently fine. I confess this argument a bit weak, but will drop the matter here.

To me this seems exactly what is needed, at least for two-player zero-sum games with full information. The realistic and theoretic are nicely wed ; the difficulties aforementioned are nicely dispelled. I am astounded that others do not share my view. Most likely the exposition in DG is at fault ; the idea is simply not presented clearly.

For example, Avner Friedman, in his recent excellent book, also titled *Differential Games* (Wiley - Interscience, 1971) omits all mention. He repeats a number of my problems, gets the same solutions, but bolsters them with rigorous proofs which seem to me superfluous. Either he omits mention of the source of these problems or implies that my treatment was heuristic . I do not acknowledge the latter.

I shall sketch the treatment of one in the next section, not only to clarify the preceding matter but to point out some interesting oversights committed by both myself and Friedman.

6. The Most-Least-Time Game

As the brachistochrone problem is the historic cornerstone of the variational calculus, I thought it apt to open my essential examples in DG with a natural game version. To the Greek etymology of its name "least time" I wished to insert a leading "most" as my innovation was the second, maximizing player. What I should have coined was macrobrachistochrone ; my classic ignorance is shameful† .

---

† Friedman uses my erroneous term. To quote him : "This game is called the *dolichobrachistochrone game*." The classic status he gives it I find flattering but embarassing. "Dolicho", I understand, applied to "large" (or "most") in the spatial, not the temporal, sense.

But I made a genuine error in the solution. As was discovered by Dr John Breakwell, whose insight seems uncanny, I was wrong about the game of kind aspect : there is another distinct part of the solution below the line (let us call it L) where $y = w^2$ in Figure 5.2.2 of DG. I falsely stated (See DG, Example 8.6.4) that here E can prevent termination and hence attain unboundedly large payoffs. Breakwell shows that the correct barrier B is comprised of a curve (below L) and the part of L lying to right of the point of tangency of this curve and L . Some optimal paths below L reach the terminal manifold C, here the y-axis. Once the descriptive point arrives on C it remains on it and rises. Breakwell calls this *safe contact*.

There seems an absurdity here : how can x be on the terminal surface C with yet no termination ? One remedy is to alter the rules : termination means penetration of C rather than just reaching it. Such ideas are discussed in DG in the latter parts of Sections 8.2 and 8.3, for safe contact seems a game of degree version of what I called an envelope barrier in games of kind.

From now on let us consider only states above L . Friedman does so and here we are on common ground. Let us now see how this mutual conclusion can follow from the verification theorem. If the reader turns to Theorem 4.4.1 in DG, he will see four hypotheses. M.I. Zelikin, editor of the Russian translation of DG, correctly points out there should be one more :

> 0. If $x \in E'$, then so does the arc of the alleged optimal path extending from x to C.

We begin by choosing a region E' which lies above L, left

of the parabolic transition surface TS (See DG, Figure 5.2.2)[†] and satisfies 0. Then hypotheses 1, 2, and 3 of the verification theorem are satisfied, for the $V(x)$ actually constructed in detail in DG (Section 5.2) (and also in Friedman's book) fulfills them.

It remains to treat 4 : the $V(x)$ found is unique. Rather than invoke a general theorem on partial differential equations, we shall show explicitly how this may be done for the present game. The RPE are integrated as is done in DG. The integrals obtained, using the initial conditions, are unique. This follows either from the uniqueness theorem for ordinary differential equations or else from the step by step integration in DG, for each step, by standard textbook integration, leads unequivocally to the next. We may note also here that the optimizing φ and ψ of the ME, are also unique. Thus the optimal paths are given by x and y expressed as unique functions of s and τ which are (5.2.5) and (5.2.4) of DG. The Jacobian can be calculated and be found not to vanish when $s > w^2$ and $0 < \tau < \pi\sqrt{s}$ (compare (5.2.5) of DG). The former implies paths that start above L ; the latter, from (5.2.5), assures that the paths reach the TS. Thus in E', we can solve uniquely for τ and s in terms of x and y. (Or this conclusion can be reached geometrically by studying the behaviour of the cycloidal paths.)

The main equation can be written

$$V_x \dot{x} + V_y \dot{y} - 1 = 0$$

or $\dot{V}$ = the τ - derivative along a path = 1.
Consequently $V = \tau$ and so is unique.

† The theorem applies only to games with terminal payoff. As explained in DG, Theorem 2.4.1, our game can be converted to this type by adjoining a third state variable. We leave the simple modifications in our discussion to the reader.

As (alleged) optimal strategies $\bar{\varphi}(x)$ and $\bar{\psi}(x)$ likewise follow uniquely, we can apply the verification theorem to E' . This means that, from any starting point x therein, a player can, by using a sufficiently fine K-strategy attain, against any opposing K-strategy, the Value V(x) with a penalty less than a prescribed ε. As every point above L and between C and the TS belongs to some E', the above holds for all these points.

The next step is go beyond the TS. Now the specific integration of the RPE to the right of the TS, though elementary in principle, seems arduous to do. Neither I nor Friedman carried it out. I did show there can be no further sign changes of the switching function A so that all that is lacking is a proof of :

> (*) The region R to the right of the TS and above L is covered completely and uniquely by the alleged optimal paths.

This is certainly true for some right half-neighborhood R' of the TS and very likely true (and provable) for R as stated.

We apply the same reasoning as before to R' (or R if (*) is true) with the TS now playing the role of C with the V on it (H in the notation of DG) being the V(x) already discussed above. This second application of the verification theorem extends the preceding conclusions to starting points in R' (or R, which means everything above L and right of C).

I trust I have improved the clarity here over DG. If not, I plead the reader strive on his own for comprehension. The concept is worth the trouble despite its obduracy to lucid explication.

7. Conclusions On Rigor

The macrobrachistochrone problem has served as a sample of what I assert to a rigorous proof of the correctness of a formal solution to a differential game. Similar reasoning holds rather generally but not completely. The elisions occur because I did not bother in DG to cover all types of singular surfaces. Universal surfaces give no trouble ; see Theorem 7.4.2 and sequel in DG. Dispersal surfaces are manageable too, although Example 4.4.3 is hardly a model of clarity. Equivocal surfaces, envelope barriers and safe contact I deferred investigating, hoping others would take them up.

The macrobrachistochrone also serves when we recall Breakwell's contribution, as an example of the novelties which seem forever to pop up in differential games and which have no classical counterparts, (as discussed in the article). It is a nice instance of the pitfalls of applying classical existence techniques.

8. Are Games Always Solvable ?

We live in a milieu of faith in reason. The American heritage especially teaches that all problems are solvable. In our era this outlook was no doubt born with the brilliant success of Newton's precise analysis of the behaviour of the solar system and the subsequent success of this new mathematics on what seemed like everything else. The faith in reason was distilled into a basic philosophy in eighteenth century France and, despite Kant, is still largely with us.

The first precise refutation came with Gödel's famous proof of the existance of unprovable statements. Since then a handful of instances, such as the continuum hypothesis, have become specifically known. In physics there were parallels : quantum mechanics showed that the exact location of a particle was unknowable.

It is my point here - admittedly now but ignorant speculation - that we should at least be prepared for something of the kind in game theory. I revert to the hierarchy described in the article. Seeking a maximum in the classic problems at least gives a well defined goal. The saddle point in simple games is accepted as a desideratum, but still a bit open to some doubts. On n-person or non-zero-sum games the literature contains much discussion of the nature of a solution and various alternative definitions. Should our minds be open to the possibility that there is no inherently right one and that some of the problems we labor over so assiduously simply have no answers?

I do not mean that an entire attractive branch of mathematics should be discarded. Rather that some of our efforts might well be directed to the solvability question. This question is likely to be difficult ; witness the few extant examples in the earlier paragraph and how hard they were to reach.

But we need not plumb the depths : solvability can be in doubt on a much more practical level. I have spent most of my working years solving practical mathematical problems and often have been beset with qualms. I have done papers, sometimes with pride in their mathematical elegance, but often with scepticism of their utility. Sometimes, as I now reflect on the past, I think I might have been wiser if my output had been an attempt to prove (if I could) that the problem was, in some practical sense, insoluble. A few mundane classes of such dilemnas are widespread enough to be specifically mentioned.

As stated in the article, most (and possible all) classical problems are optimizing. There is a basic and evident fact of which I sometimes suspect many people are not aware : optimizing can only mean maximizing one quantity. A basic question that confronts us throughout our personal as well as technical lives is : "What quantity ?"

As a typical case, take the design of optimal military aircraft. Criteria of merit may be range, speed, bomb load, rate of climb and many more. One can be enhanced only at the expense of others. What shall be maximized ? An obvious reply is to attach a weight to each criterion and use a linear combination. But what weights ? This is a question of judgment. Then why not use qualitative judgment on the original problem directly ?

Such a quandary permeates many human decisions. Yet I have seen research proposals blandly demanding an optimum on questions equally ambiguous.

Game theory can suffer the quandary in magnified form, for each competitor is limited to maximizing one quantity, which we call his payoff. Perhaps our familiarity with recreational games, where the payoff is defined by the rules, tends to mislead us. Thus we can easily be too confident when our game is a model of some realistic human conflict. Goals are not given us and are sometimes baffling to delineate.

One broad class of such occurs in games which extend over an indefinite future. For example, in a business game, where the players are competing firms, each seeks to maximize his profit. Over what period ? We choose one - say T. We ought not be surprised if a painstaking analysis give us a ludicrous answer which demands that each firm liquidates its plant just prior to T. Obviously this correctly maximizes cash at T, but it is hard to imagine an executive rejoicing at our recommendation.

There are many other varied instances of this paradox of the indefinite future. A possible remedy is to use a weighted integral of the entire future

$$\int_0^\infty p(t)\ w(t)\ dt\ .$$

Here p(t) is the rate of gaining payoff and the weighting function w(t) decreases fast enough to attain convergence. But how is w(t) to be chosen? What decrement of value should attach to the remote future? Again we are forced to a judgment and our solution is of little use unless it is largely independent of this judgment.

A milder form of quandary beset me at the outset of differential games. The first problems were of pursuit and evasion with military applications in mind, such as missile vs.aircarft or torpedo vs.ship. The dominant question, of course, was the game of kind : could P capture E ? If he could, the resulting game of degree demanded that both players act with optimal efficiency. How define such? I chose time of capture as payoff with P minimizing and E maximizing. But I felt uneasy. If a torpedo could certainly hit a ship, did a few seconds delay really matter? But since then many others have taken up pursuit games and they freely use capture time as payoff, apparently without the qualms that beset me.

I am not advocating scrapping game theory because of difficulties of the last two ilks. While at their worst they might strip the solution of realistic meaning, at their best they leave but a harmless margin of indeterminacy. But I am urging attention to such matters and a healthy doubt as to the practical solvability and use of game theory.

Part . I

Zero-Sum Differential Games

ε-STRATEGIES IN DIFFERENTIAL GAMES

B.N. PCHENITCHNY†

In the past few years many works have been devoted to the theory of differential games (1),(2),(4),(6). A basic difference between them lies in the definition of strategies.

The contents of the present chapter are based on earlier works (7)-(9). We shall discuss the concept of ε-strategies, defined in (7). In the strategy space thus defined we shall be interested in two-player zero-sum games and we shall prove the existence of a value for such games. As we shall see, the state space can be separated into four regions. Each of them is a set of initial states for which the odds are in the favor of one of the two players. This approach leads to a new derivation of Isaacs equation, for the boundary between regions from which termination of the game is favourable to one or the other player.

In this brief introduction it may be worth-while to speak in favour of the ε-strategies. The main strictures upon them

---

† Institute of Cybernetics Academy of Sciences of the Ukrainian SSR, Kiev, U.S.S.R.

Translated from the Russian by M. Granger and A. Blaquiere.

lie in the fact that, when making use of ε-strategies, one is led to assuming that one of the players has full knowledge of the control of the opposite player on some future time interval. Though this time interval is supposed to be small, this assumption is not satisfactory from a practical point of view. Yet it could be legitimated from a practical point of view also, but we don't consider this matter as an important one. The ε-strategies are to be regarded as a mathematical generalization of the classical concept, namely the one of strategies as functions of the state. This is legitimated by the fact that, when a differential game has a solution with strategies which are functions of the state, one gets a deeper insight into the structure of the game, and one can obtain existence theorems, by using ε-strategies. The very question of knowing when the ε-strategies can be transformed into the classical ones is a subsidiary one, which can be considered independently of the problem of existence of a solution of the game. Furthermore, one can already state that in a region where the value of the game is a differentiable function of the state, the ε-strategies can be replaced by the classical ones.

1. Problem Statement, ε-Strategies

We shall consider a differential game governed by the differential equation

$$\dot{z} = f(z,u,v) \tag{1}$$

where z is an n-dimensional vector with components z^i, $i = 1,\ldots,n$; $f(z,u,v)$ is an n-dimensional vector function of z, u and v, where *control variables* u and v are r- and s-dimensional vectors, belonging to the sets U and V, respectively. As usual, $\dot{z}$ is the time derivative of z .

Target θ is a given set of points in E^n .

From now on, we shall assume that

A. target θ is closed ;

B. function f(z,u,v) is continuous with respect to all of its arguments, and continuously differentiable with respect to z ;

C U and V are compact sets ;

D. the set f(z,U,v,) is convex for all z and v ;

E. there exists a constant C such that
$|\langle z,f(z,u,v)\rangle| \leq C(1 + \|z\|^2)$
where we denote by $\langle x,y\rangle$ the scalar product of vectors x and y, and

$$\|z\| = \sqrt{\sum_{i=1}^{n} (z^i)^2}$$

Players P and E make their decisions through choosing the controls u and v, respectively. P endeavors to "steer" the state of the game to θ ; if he succeeds we shall say that he is a *winner*. E seeks to prevent this event. In the classical statement of the problem the choices of u and v are governed by functions of z, namely u(z) and v(z), respectively. The reason why we shall not define the strategies in this way is clear. Among the arguments, there is the fact that only a small number of results have been obtained as yet, from this point of view. Also, note that all the authors dealing with the proof of existence theorems [2],[5],[7], have felt necessary to give up stating the problem in this way.

Now let us define ε-*strategies*.

Definition 1. An ε-strategy for player E, namely Γ_E is defined by attaching (i) a number ε(z), ε(z) > 0 ; and (ii) a measurable function of t with range in V ; namely $v(t) \equiv \Gamma_E(t;z), 0 \leqslant t \leqslant \varepsilon(z)$, to each point $z \in E^n$.

Definition 2. For given $\varepsilon > 0$ and function $v(t)$, $0 \leqslant t \leqslant \varepsilon$ an ε-strategy for player P, namely Γ_P, is defined by attaching a measurable function of t with range in V ; namely, $u(t) \equiv \Gamma_P(t;\varepsilon,v(\cdot),z)$, $0 \leqslant t \leqslant \varepsilon$, to each point $z \in E^n$.

Now for given strategy pair (Γ_P,Γ_E), we shall define a trajectory of system (1), emanating from given point z_o at time $t = 0$.

Such a trajectory can be roughly defined as follows. At the begining of a play, player E chooses the number $\varepsilon_o = \varepsilon(z_o)$ and the control function $v(t) \equiv \Gamma_E(t;z_o)$, $0 \leqslant t \leqslant \varepsilon(z_o)$, and player P chooses the control function $u(t) \equiv \Gamma_P(t;\varepsilon_o,v(\cdot),z_o)$, $0 \leqslant t \leqslant \varepsilon(z_o)$. By integrating system (1) from the initial point z_o, with $u = u(t)$ and $v = v(t)$, on the time interval ε_o , one obtains a piece of trajectory $z(t)$, $0 \leqslant t \leqslant \varepsilon_o$. Then, by taking $z(\varepsilon_o)$ as a new starting point, one can integrate further equation (1) on a new interval of time, and so on.

However, this definition of a trajectory is unsatisfactory since one may obtain a finite interval of time by piecing together the $\varepsilon(z)$. In that case the trajectory will be defined on a finite interval of time only.

For that reason we shall define a trajectory as follows.

Definition 3. We shall say that $z(t)$ is a trajectory defined on $[0, t_o)$, or $[0, t_o]$, emanating from point z_o, if $z(t)$ is an absolutely continuous function of t, $z(0) = z_o$, and there exists a set $T \subset [0, t_o)$, resp. $[0, t_o]$, such that

a) $0 \in T$, and if $\tau \in T$ and $\varepsilon(z(\tau)) + \tau < t_o$

$(\varepsilon(z(\tau)) + \tau \leqslant t_o$ for $[0, t_o])$, then $\tau + \varepsilon(z(\tau)) \in T$ and the interval $(\tau, \tau+\varepsilon(z(\tau)))$ has no point in T ;

b) the set $T \cup \{t_o\}$ is closed .

c) Let $\tau_o = \sup \{ \tau : \tau \in T \}$. On every interval $[\tau, \tau']$, $\tau' = \tau + \varepsilon(z(\tau)) < t_o$, and on the interval $[\tau_o , t_o)$, if $\tau_o \neq t_o$, $z(t)$ fulfils the differential equation

$$(2) \qquad \begin{aligned} \dot{z} &= f(z,u(t),v(t)) \\ v(t) &= \Gamma_E(t-\tau;\ z(\tau)) \\ u(t) &= \Gamma_P(t-\tau;\ \varepsilon(z(\tau)),v(\cdot),z(\tau)) \end{aligned}$$

almost everywhere ; a similar condition holds for a closed interval $[0, t_o]$;

d) in the case of a closed interval $[0, t_o]$, if $\tau_o = t_o$, then $t_o \in T$.

Theorem 1. *Given an initial point* z_o *and strategies* Γ_P *and* Γ_E *there is a unique trajectory* $z(t) \equiv z(t;z_o,\Gamma_P,\Gamma_E)$ *defined on the whole interval* $[0, \infty)$

Proof. On the time interval $[0, \varepsilon(z_o)]$ there is a unique trajectory, which can be constructed by similar arguments as in the rough discussion above.

Now let us prove that, if a trajectory has been defined on some half-open interval $[0,t_o)$, then it can be extended to the closed interval $[0,t_o]$.

Consider a trajectory defined on $[0,t_o)$. From Definition 3 there exists a set T . Two cases need be considered, namely $\tau_o < t_o$ and $\tau_o = t_o$.

If $\tau_o = \sup\{\tau : \tau \in T\} < t_o$, then $\tau_o + \varepsilon(z(\tau_o)) \geqslant t_o$, and, on the time interval $[\tau_o, \tau_o + \varepsilon(z(\tau_o))]$, there is a unique trajectory defined by Eq.(2), in which we put $\tau = \tau_o$.

If $\tau_o = t_o$, then

$$z(t_o) = \lim_{t \to t_o - 0} z(t)$$

According to Assumptions A-E above, this limit always exists. It follows that the trajectory is defined on the closed interval $[0, t_o]$.

If we replace T by $T \cup \{\tau_o + \varepsilon(z(\tau_o))\}$ in the first case, and by $T \cup \{t_o\}$ in the second one, then one can see easily that all the assumptions in Definition 3 are satisfied. Now consider a trajectory on the interval $[0, t_o]$. If $\tau_o < t_o$ then, according to Conditions a and b in Definition 3, $\tau_o \in T$ and $\tau_o + \varepsilon(z(\tau_o)) > t_o$. By letting z(t) coincide with the solution of Eq.(2), on the interval $[\tau_o, \tau_o + \varepsilon(z(\tau_o))]$, the trajectory is extended to the interval $[0, \tau_o + \varepsilon(z(\tau_o))]$. If $\tau_o = t_o$ then, according to Condition d in Definition 3, $t_o \in T$ and again z(t) can be defined on the time interval $[t_o, t_o + \varepsilon(z(t_o))]$ by Eq.(2) with $\tau = t_o$.

Hence, in both cases the trajectory is uniquely extended to a larger interval, while replacing the set T by the new one $T \cup \{\tau_o + \varepsilon(z(\tau_o))\}$.

The uniqueness of the construction of the trajectory follows from the uniqueness of the extension, owing to Definition 3, for one can see easily that if z'(t) and z"(t) are two trajectories which coincide on some interval, then because of the uniqueness of the extension, they must coincide on a larger interval .

Hence Theorem 1 is proved.

Now, if the trajectory $z(t;z_o,\Gamma_P,\Gamma_E)$ intersects the closed set θ, let $I(z_o,\Gamma_P,\Gamma_E)$ be the first time at which θ is reached. If the trajectory does not intersect θ, then let $I(z_o,\Gamma_P,\Gamma_E) = +\infty$. Hence the value of the function $I(z_o,\Gamma_P,\Gamma_E)$ is defined for all initial points z_o and for all strategies Γ_P and Γ_E .

$I(z_o,\Gamma_P,\Gamma_E)$ is called the *cost function.*

In the following, we shall be interested in a new problem, in connection with the question :

does there exist a *value* in a differential game with ε-strategies ; that is, does

$V^-(z_o) = V^+(z_o)$, where

$$V^-(z_o) = \sup_{\Gamma_E} \inf_{\Gamma_P} I(z_o,\Gamma_P,\Gamma_E)$$

and

$$V^+(z_o) = \inf_{\Gamma_P} \sup_{\Gamma_E} I(z_o,\Gamma_P,\Gamma_E)$$

2. Semi-Group of Set to Set Mappings of a Game

In this paragraph we shall be interested in a semi-group of set to set mappings, which entirely determines the solution of a differential game.

Definition 4. Let T_ε , $\varepsilon \geqslant 0$, be a mapping that associates with any subset X of E^n the set of points $T_\varepsilon(X)$ in E^n . By definition, $T_\varepsilon(X)$ is the set of points $z \in E^n$ such that, for any measurable control $v(t)$, $t \geqslant 0$, $v(t) \in V$, there exists a measurable control

$u(t)$, $u(t) \in U$, such that the solution of Eq.(1), with $u = u(t)$ and $v = v(t)$ and the initial condition $z(0) = z$, reaches the set X no later than time ε.

Some properties of T_ε are given below.

Property 1.

a) $T_\varepsilon(X) \subset T_{\varepsilon'}(X)$ if $\varepsilon' \geqslant \varepsilon$;

b) $T_\varepsilon(X) \subset T_\varepsilon(X')$ if $X' \supset X$

c) $T_0(X) = X$ $T_\varepsilon(X) \supset X$

Property 2.

$$T_{\varepsilon_1} T_{\varepsilon_2}(X) \subset T_{\varepsilon_1+\varepsilon_2}(X)$$

These properties are direct consequences of the definition of T_ε.

Property 3.

If X is closed, then $T_\varepsilon(X)$ is closed.

It follows that if $z_i \in T_{\varepsilon_i}(X)$, and $\varepsilon_i \to \varepsilon_0$ and $z_i \to z_0$ as $i \to \infty$, then $z_0 \in T_{\varepsilon_0}(X)$.

Property 4.

Given any family of sets X_α , $\alpha \in A$, we have

$$\bigcap_{\alpha \in A} T_\varepsilon(X_\alpha) \supset T_\varepsilon\left(\bigcap_{\alpha \in A} X_\alpha\right)$$

Property 5.

For a decreasing family of closed sets X_i, $i = 1,2,\ldots$, i.e. $X_{i+1} \subset X_i$, we have

$$\bigcap_{i=1}^{\infty} T_\varepsilon(X_i) = T_\varepsilon\left(\bigcap_{i=1}^{\infty} X_i\right)$$

Property 4 is a direct consequence of Property 1.b.

The proofs of Properties 3 and 5 are similar ; accordingly, here we shall give the proof of Property 5 only.

Taking account of Property 4, we have but to prove that

$$\bigcap_{i=1}^{\infty} T_\varepsilon(X_i) \subset T_\varepsilon\left(\bigcap_{i=1}^{\infty} X_i\right)$$

Let $z_o \in T_\varepsilon(X_i)$ for all i. Given a control v(t) there exists a control $u^i(t)$ such that the corresponding trajectory of system (1), namely $z^i(t)$, emanating from z_o , reaches the set X_i no later than time ε. Let

$$z^i(\delta_i) \in X_i \ , \qquad \delta_i \leqslant \varepsilon$$

According to Refs. [10] and [11], and to Assumptions A-E, there exists a sequence of trajectories $z^i(t)$ that uniformly converges towards a trajectory $z^o(t)$ of system (1), generated by controls $u^o(t)$ and v(t). We can suppose as well, without restricting the generality of the arguments, that $z^i(t)$ converges towards $z^o(t)$ and $\delta_i \to \delta_o$ as $i \to \infty$. Since $X_{i+1} \subset X$, and since the sets X_i are closed, we have

$$z^m(\delta_m) \in X_i \ , \qquad m \geqslant i$$

and

$$z^o(\delta_o) = \lim_{m \to \infty} z^m(\delta_m) \in X_i$$

That is

$$z^o(\delta_o) \in \bigcap_{i=1}^{\infty} X_i \qquad \delta_o \leqslant \varepsilon$$

and accordingly $z_o \in T_\varepsilon(\bigcap_{i=1}^{\infty} X_i)$ which concludes the proof.

Definition 5. By a *rational subdivision* ω we shall mean any finite sequence of rational numbers τ_i, $i=0,1,\dots,m$, with $\tau_o = 0$ and $\tau_i \leqslant \tau_{i+1}$. We shall let $|\omega| = \tau_m$.

Definition 6. We shall say that the rational subdivision ω' is *finer than* ω, which we shall denote by $\omega' \succ \omega$, if $|\omega'| \leqslant |\omega|$ and if each τ_i such that $\tau_i \leqslant |\omega'|$ coincides with a number τ'_j of the rational subdivision ω'.

Thus we introduce a partial ordering on the set of all the rational subdivisions.

We shall let

$$T_\omega(X) = T_{\delta_m} T_{\delta_{m-1}} \ldots T_{\delta_1}(X)$$

$$\delta_i = \tau_i - \tau_{i-1} \quad , \; i = 1,\ldots,m$$

Lemma 1. *If* $\omega' \succ \omega$, *then* $T_{\omega'}(X) \subset T_\omega(X)$.

This lemma is a direct consequence of Properties 1 and 2 of the mapping T_ε.

Definition 7. $\tilde{T}_t(X) = \bigcap_{|\omega|>t} T_\omega(X)$

The mapping $\tilde{T}_t$ plays an important role in the theory. As we shall see later, it determines the solution of the differential game.

Lemma 2.

a) $\tilde{T}_{t'}(X) \supset \tilde{T}_t(X)$ *if* $t' \geqslant t$;

b) $\tilde{T}_t(X') \supset \tilde{T}_t(X)$ *if* $X' \supset X$;

c) $\tilde{T}_o(X) = X$;

d) *If* X *is closed, then* $\tilde{T}_t(X)$ *is closed.*

The proof of Lemma 2 follows at once from Properties 1 and 3 of the mapping T_ε.

For instance, $\tilde{T}_t(X)$ is closed if X is closed because $T_\omega(X)$ is closed as follows from Property 3, and $\tilde{T}_t(X)$ is an intersection of closed sets $T_\omega(X)$.

Theorem 2. $\tilde{T}_{t_1+t_2}(X) = \tilde{T}_{t_1}\tilde{T}_{t_2}(X)$ *for* $t_1, t_2 \geqslant 0$ *and* X *closed.*

Proof. We shall denote by ω_σ the rational subdivisions for which there exist ω_1 and ω_2 such that

$$|\omega_\sigma| > t_1+t_2$$

$$T_{\omega_1}T_{\omega_2}(X) = T_{\omega_\sigma}(X) \tag{3}$$

$$|\omega_1| > t_1 \quad , \qquad |\omega_2| > t_2$$

Let t_1 and t_2 be given. If $|\omega| > t_1+t_2$, $\omega = \{\tau_o , \tau_1 ,\ldots, \tau_m \}$, then there exists a rational number τ such that $\tau > t_2$,$\tau_m - \tau > t_1$.

Let

$$k = \max \{ i : \tau_i \leqslant \tau \}$$

and

$$\omega_1 = \{ 0 , \tau_{k+1} - \tau,\ldots, \tau_m - \tau \}$$

$$\omega_2 = \{\tau_o, \tau_1 ,\ldots, \tau_k, \tau \}$$

$$\omega_\sigma = \{\tau_o, \tau_1 ,\ldots, \tau_k, \tau, \tau_{k+1},\ldots,\tau_m \}$$

Indeed, we have

$$|\omega_1| = \tau_m - \tau > t_1$$

$$|\omega_2| = \tau > t_2$$

Hence, $\omega_\sigma \prec \omega$ and

$$T_{\omega_1} T_{\omega_2}(X) = T_{\omega_\sigma}(X) \subset T_\omega(X)$$

Thus, with any subdivision ω, $|\omega| > t_1+t_2$, one can associate a subdivision finer than ω_σ. Moreover, with any two subdivisions ω_1 and ω_2, $|\omega_1| > t_1$, $|\omega_2| > t_2$, one can associate the subdivision

$$\omega_\sigma = \{\tau_0^2, \tau_1^2, \ldots \tau_{m_2}^2, \tau_1^1+\tau_{m_2}^2, \ldots \tau_{m_1}^1 + \tau_{m_2}^2\}$$

such that (3) is satisfied.

Accordingly

$$\tilde{T}_{t_1+t_2}(X) = \bigcap_{|\omega|>t_1+t_2} T_\omega(X) = \bigcap_{|\omega_\sigma|>t_1+t_2} T_{\omega_\sigma}(X)$$

Indeed, if $z_0 \in T_\omega(X)$ for any ω such that $|\omega| > t_1+t_2$, then $z_0 \in T_{\omega_\sigma}(X)$, since the subdivision ω_σ is contained in ω. Moreover, if $z_0 \in T_{\omega_\sigma}(X)$ for any ω_σ then, for given ω, there exists ω_σ such that $\omega_\sigma \prec \omega$, and accordingly

$$z_0 \in T_{\omega_\sigma}(X) \subset T_\omega(X)$$

Now from Property 4 of the mapping T_ε we have

$$\tilde{T}_{t_1+t_2}(X) = \bigcap_{|\omega_\sigma|>t_1+t_2} T_{\omega_\sigma}(X)$$

$$= \bigcap_{|\omega_1|>t_1} \bigcap_{|\omega_2|>t_2} T_{\omega_1} T_{\omega_2}(X) \supset \bigcap_{|\omega_1|>t_1} T_{\omega_1}\Big(\bigcap_{|\omega_2|>t_2} T_{\omega_2}(X)\Big)$$

$$= \tilde{T}_{t_1} \tilde{T}_{t_2}(X)$$

that is

(4) $$\widetilde{T}_{t_1+t_2}(X) \supset \widetilde{T}_{t_1}\widetilde{T}_{t_2}(X)$$

Now, let us prove that

(4)' $$\widetilde{T}_{t_1+t_2}(X) \subset \widetilde{T}_{t_1}\widetilde{T}_{t_2}(X)$$

Since any subdivision ω such that $|\omega| > t_2$ is made of a finite family of rational numbers, then the set of all such subdivisions is denumerable. Let us denote these subdivisions by $\overline{\omega}_1, \overline{\omega}_2, \ldots$

Thus $\overline{\omega}_k$, $k = 1,\ldots$ is the set of all the subdivisions ω satisfying $|\omega| > t_2$.

Now, let us construct a subdivision ω_k as follows. Let the points τ_i^k of this subdivision be the points of the subdivisions $\overline{\omega}_j$, $j \leqslant k$, satisfying the condition $\overline{\tau}_i^j \leqslant \min\limits_{1\leqslant j\leqslant k} |\overline{\omega}_j|$.

Then clearly we have

$$|\omega_k| = \min_{1\leqslant j\leqslant k} |\overline{\omega}_j| > t_2$$

$$\omega_k \ni \overline{\omega}_j \qquad j = 1,\ldots,k$$

$$\omega_{k+1} \ni \omega_k$$

Let us show that

(5) $$\widetilde{T}_{t_2}(X) = \bigcap_{k=1}^{\infty} T_{\omega_k}(X)$$

i.e. that $\widetilde{T}_{t_2}(X)$ is the intersection of the nesting sets $T_{\omega_k}(X)$.

In fact, since ω_k, $k = 1,\ldots$ is a portion of the set of all the subdivisions ω, we have :

$$\tilde{T}_{t_2}(X) = \bigcap_{|\omega|>t_2} T_\omega(X) \subset \bigcap_{k=1}^{\infty} T_{\omega_k}(X)$$

Moreover, let $z_o \in T_{\omega_k}(X)$ for any k.

For given ω there exists a number j such that $\omega = \bar{\omega}_j$ and accordingly $\omega_j \succ \bar{\omega}_j = \omega$. That means that $z_o \in T_{\omega_j}(X) \subset T_\omega(X)$, from which follows that

$$\tilde{T}_{t_2}(X) \supset \bigcap_{k=1}^{\infty} T_{\omega_k}(X)$$

Hence (5) is established.

Now, by making use of Property 5 of the mapping T_ε we obtain

$$\tilde{T}_{t_1}\tilde{T}_{t_2}(X) = \bigcap_{|\omega_1'|>t_1} T_{\omega_1'}\Big(\bigcap_{|\omega_2'|>t_2} T_{\omega_2'}(X)\Big)$$

$$= \bigcap_{|\omega_1'|>t_1} T_{\omega_1'}\Big(\bigcap_{k=1}^{\infty} T_{\omega_k'}(X)\Big)$$

$$= \bigcap_{|\omega_1'|>t_1} \bigcap_{k=1}^{\infty} T_{\omega_1'}T_{\omega_k}(X) \supset \bigcap_{|\omega_\sigma|>t_1+t_2} T_{\omega_\sigma}(X) = \tilde{T}_{t_1+t_2}(X)$$

Hence (4)' is established. By comparing (4) and (4)' Theorem 2 is proved.

Corollary. $\tilde{T}_t(X) \subset T_\varepsilon\tilde{T}_{t-\varepsilon}(X)$ *for* $0 \leqslant \varepsilon \leqslant t$.

Proof. From Theorem 2 we have

$$\tilde{T}_t(X) = \tilde{T}_\varepsilon\tilde{T}_{t-\varepsilon}(X) \subset T_\varepsilon\tilde{T}_{t-\varepsilon}(X)$$

as $T_\varepsilon(Y) \supset \tilde{T}_\varepsilon(Y)$. Indeed, from the definition of $\tilde{T}_\varepsilon$ we have

$$\tilde{T}_\varepsilon(Y) \subset T_\gamma(Y), \qquad \gamma > \varepsilon$$

since one can consider a subdivision ω consisting of two points $\tau_o = 0$ and $\tau_1 = \gamma$. For fixed γ and ε, $\gamma > \varepsilon$, we obtain the expected result from Property 3 of the mapping T_ε.

3. Optimal Strategy Γ_P^*. A Basic Theorem

In this paragraph, we shall be interested in the set of all initial points from which the game can be terminated in finite time. As we shall see, this set can be obtained through the set to set mapping $\tilde{T}_t$, defined in paragraph 2. First of all we shall define an optimal strategy for player P, namely Γ_P^*

Let

$$t_o = \inf \{ t : z_o \notin \tilde{T}_t(\theta) \}$$

and let us prove that the inf, in the right hand side, is reached.

Suppose that $z_o \notin \tilde{T}_{t_o}(\theta)$. Then there exists a rational subdivision ω_o such that $|\omega_o| > t_o$,

$$z_o \notin T_{\omega_o}(\theta)$$

But

$$|\omega_o| > t_o + \delta$$

for δ sufficiently small, $\delta > 0$; and hence $z_o \notin \tilde{T}_{t_o+\delta}(\theta)$ which contradicts the definition of t_o; Accordingly

$$z_o \in \tilde{T}_{t(z_o)}(\theta)$$

We shall let

$$t_o = \min \{ t : z_o \in \tilde{T}_t(\theta) \} \tag{6}$$

If $z_o \notin \tilde{T}_t(\theta)$ for $t \geqslant 0$, we shall let $t(z_o) = +\infty$.

Now let z_o be such that $t(z_o) < +\infty$, and let $\varepsilon > 0$ and the control $v(t) \in V$ on $[0, \varepsilon]$ be given. We shall associate with z_o, ε and $v(\cdot)$ a control $u(t)$ on the interval $[0, \varepsilon]$

In the following discussion we shall use a variable z^o , which is a function of time t such that

$$\dot{z}^o = 1$$

If $z^o(0) = 0$ then $z^o(t) = t$.

We shall add the equation that governs z^o to Eqs.(1) .

Now we shall define strategies Γ_P^o and Γ_E^o for the augmented system.

Let us attach a number $\varepsilon(z,z^o) > 0$, a control $\bar{v}(t)$ and a control $u(t)$ defined on the same interval $[0,\varepsilon(z,z^o)]$, to each point (z,z^o).

If (z,z^o) is such that $t(z) = +\infty$ or $z^o \notin [0,\varepsilon]$ we let $\varepsilon(z,z^o)$, $\bar{v}(t)$ and $u(t)$ be arbitrary.

If (z,z^o) is such that $t(z) < +\infty$ and $0 \leqslant z^o \leqslant \varepsilon$ we consider two cases, namely

First case $\varepsilon - z^o \geqslant t(z)$

In that case we let $\varepsilon(z,z^o) = \varepsilon - z^o$, and $\bar{v}(t) = v(z^o + t)$

We have

$$z \in \tilde{T}_{t(z)}(\theta) \subset T_{t(z)}(\theta)$$

and, from the definition of T_ε , it follows that there exists a control $u(t)$ such that the trajectory $z(t)$ of system (1) generated by $u(t)$ and $\bar{v}(t)$, emanating from z, reaches θ no later than time $t(z)$. By definition Γ_P^o is the control $u(t)$ thus defined, and Γ_E^o is the control $\bar{v}(t)$.

Second case $\varepsilon - z^o < t(z)$

Again we let $\overline{v}(t) = v(z^o + t)$, $t \geqslant 0$.

Since, from Theorem 2, we have

$$z \in \widetilde{T}_{t(z)}(\theta) \subset T_{\varepsilon - z^o} \widetilde{T}_{t(z)-(\varepsilon - z^o)}(\theta)$$

again, from the definition of $T_{\varepsilon - z^o}$, it follows that there exists a control u(t) such that the trajectory z(t) of system (1), generated by u(t) and $\overline{v}(t)$, emanating from z, reaches the set

$$\widetilde{T}_{t(z)-(\varepsilon - z^o)}(\theta)$$

no later than time $\delta \leqslant \varepsilon - z^o$. Let $\varepsilon(z,z^o) = \delta$, and let u(t) be the control defined above. Again we shall let Γ_P^o and Γ_E^o be the controls u(t) and $\overline{v}(t)$, respectively.

From the definitions of Γ_P^o and Γ_E^o , it follows that, in the first case, the trajectory z(t) of system (1), emanating from z, reaches θ no later than time $\varepsilon(z,z^o)$, and in the second case the following condition is satisfied.

$$z(\varepsilon(z,z^o)) = z(\delta) \in \widetilde{T}_{t(z)-(\varepsilon - z^o)}(\theta)$$

and hence

$$t(z(\varepsilon(z,z^o))) \leqslant t(z)-(\varepsilon - z^o) \leqslant \leqslant t(z)-\delta \leqslant t(z)-\varepsilon(z,z^o) \tag{7}$$

Now we shall define Γ_P^* . For given z_o, $\varepsilon > 0$ and control v(t) on [0, ε], let us construct the trajectory z(t), emanating from z_o , $z^o = 0$, generated by Γ_P^o and Γ_E^o . Since $z^o(0) = 0$ we have z(t) = t . Let *T* be the set depicted in Definition 3, corresponding to the trajectory z(t) on the interval [0, ε].

With each $\tau \in T$ there is associated an interval $[\tau,\tau')$, $\tau' = \tau + \varepsilon(z(\tau),\tau)$, such that $T \cap [\tau,\tau') = \tau$. Of course $[0,\varepsilon]$ is covered by such intervals. From Definition 3, the trajectory $z(t)$ satisfies Eqs.(2) on the interval $[\tau,\tau')$; that is, there exists a solution of system (1), on the interval $[\tau,\tau')$, generated by controls $u(t)$ and $v^o(t)$ given by Eqs.(2) with Γ_P and Γ_E replaced by Γ_P^o and Γ_E^o, respectively.

For instance we have, on the interval $[\tau,\tau')$

$$v^o(t) = \Gamma_E^o(t-\tau;z(\tau)) = \overline{v}(t-\tau)$$

$$= v(z^o(\tau)+(t-\tau)) = v(\tau + (t-\tau)) = v(t)$$

The control $u(t)$ can be computed in an analogous fashion by setting $\Gamma_P = \Gamma_P^o$ in Eqs.(2), since it is defined on each interval $[\tau,\tau')$, its definition can be extended to the entire interval $[0,\varepsilon]$.

For given z_o, $\varepsilon > 0$ and control $v(t)$ on $[0,\varepsilon]$, we shall let Γ_P^* be the control thus obtained.

Lemma 3. *If* $u(t)$, $t \in [0,\varepsilon]$, *is the control associated with* z_o, ε *and* $v(t)$ *through the strategy* Γ_P^*, *then the trajectory* $z(t)$ *generated by the controls* $u(t)$ *and* $v(t)$, *emanating from* z_o, *has one of the following properties*

- *either it reaches* θ *at a time* δ, $\delta \leqslant \varepsilon$
- *or the following inequality is satisfied*

$$t(z(\varepsilon)) \leqslant t(z_o)-\varepsilon \tag{8}$$

Proof. From the discussion above, it appears that the trajectory $z(t)$, emanating from z_o, generated by the control $v(t)$ and the strategy Γ_P^*, on the interval $[0,\varepsilon]$, is the same as the one emanating from z_o , generated by the strategies Γ_P^o and Γ_E^o .

Let T be the set depicted in Definition 3, and suppose that $z(t) \notin \theta$ for $t \in [0,\varepsilon]$. Let us show that

$$(9) \qquad t(z(\tau)) \leqslant t(z_o) - \tau, \quad \tau \in T$$

For $\tau = 0$, condition (9) is satisfied.

Let α_o be the upper bound of the numbers τ, $\tau \in T$, for which condition (9) is satisfied, and suppose that $\alpha_o < \varepsilon$. Then from condition b in Definition 3 we have $\alpha_o \in T$, and there exists a sequence of numbers $\tau_i \in T$, $\tau_i < \tau_{i+1}$, $\tau_i \to \alpha_o$ as $i \to \infty$, for which condition (9) is satisfied. Indeed

$$z(\tau_i) \in \tilde{T}_{t(z_o)-\tau_i}(\theta) \subset T_{\alpha_o - \tau_i}\, \tilde{T}_{t(z_o)-\alpha_o}(\theta)$$

Since $z(\tau_i) \to z(\alpha_o)$ as $i \to \infty$, in view of the property 3 of the mapping T_ε we obtain

$$z(\alpha_o) \in \tilde{T}_{t(z_o)-\alpha_o}(\theta)$$

that is

$$(10) \qquad t(z(\alpha_o)) \leqslant t(z_o) - \alpha_o$$

Hence the upper bound α_o is reached.

Now consider $z(\alpha_o)$ as a new initial point for the trajectory $z(t)$ generated by Γ_P^o and Γ_E^o . Two cases need be considered, namely

a) θ is reached on the interval $[\alpha_o, \varepsilon]$; or

b) at point $\alpha_o + \varepsilon(z(\alpha_o),\alpha_o)$, $\varepsilon(z(\alpha_o),\alpha_o) \leqslant \varepsilon - \alpha_o$, condition (7) is satisfied, that is

$$(11) \qquad t(z(\alpha_o + \varepsilon(z(\alpha_o),\alpha_o))) \leqslant t(z(\alpha_o)) - \varepsilon(z(\alpha_o),\alpha_o)$$

But according to an assumption above, $z(t)$ does not reach θ. Hence we have $\alpha = \alpha_o + \varepsilon(z(\alpha_o),\alpha_o) \in T$, according to Definition 3, and

$$(12) \qquad t(z(\alpha)) \leqslant t(z(\alpha_o)) - \varepsilon(z(\alpha_o),\alpha_o) \leqslant$$

$$t(z_o) - \alpha_o - (\varepsilon - \alpha_o) = t(z_o) - \varepsilon \leqslant t(z_o) - \alpha_o$$

in view of conditions (10) and (11).

But that contradicts the fact that α_o is the upper boundary of the numbers τ, $\tau \in T$, such that condition (9) is satisfied and $\alpha_o < \varepsilon$. Indeed $\alpha > \alpha_o$, $\alpha \in T$ and condition (9) is satisfied, as follows from (12).

Hence we have $\alpha_o = \varepsilon$, and condition (8) follows from (10).

Hence Lemma 3 is established.

Theorem 3. *If* $t(z_o) < +\infty$, *then*

(i) *If the strategy of player* P *is* Γ_P^* *then, for a play begining at point* z_o, P *is sure to win, that is, to "steer" the state of the game to* θ, *in finite time* $t(z_o)$, *no matter what the strategy* Γ_E *of player* E *is ; and*

(ii) *for* $\alpha > 0$, α *arbitrarily small, there exists a strategy* Γ_E^α *such that the trajectory generated by* Γ_P *and* Γ_E^α *, emanating from* z_o, *does not reach* θ *sooner that* $t(z_o) - \alpha$, *no matter what the strategy* Γ_P *of player* P *is ; that is,*

$$I(z_o , \Gamma_P , \Gamma_E^\alpha) \geqslant t(z_o) - \alpha \qquad \forall\, \Gamma_P$$

If $t(z_o) = +\infty$, *then for any* α, $\alpha > 0$, *there exists a strategy* Γ_E^α *such that*

$$I(z_o , \Gamma_P , \Gamma_E^\alpha) \geqslant \alpha \qquad \forall\, \Gamma_P$$

Proof. Let $t(z_o) < +\infty$, and suppose that the strategy of player P is Γ_P^* ; then the trajectory generated by Γ_P^* and Γ_E, emanating from z_o , reaches θ at a time belonging to the interval $[0, t(z_o)]$. Indeed, suppose that the converse is true. Then, from Lemma 3 we have

$$t(z(\tau')) \leqslant t(z(\tau)) - (\tau'-\tau), \qquad \tau \in T \tag{13}$$

where T is the set depicted in Definition 3, corresponding to the trajectory $z(t)$ on the interval $[0, t(z_o)]$, and τ' is the point of T next to the right of τ ; that is, $\tau' = \tau + \varepsilon(z(\tau))$. By similar arguments as in the proof of Lemma 3, we obtain

$$t(z(\tau)) \leqslant t(z_o) - \tau, \qquad \tau \in T \tag{14}$$

Now let

$$\tau_o = \sup \{ \tau : \tau \in T \}$$

Two cases need be considered, namely

a) $\tau_o = t(z_o)$. Then from Definition 3 we have $\tau_o = t(z_o) \in T$ and from (14) we obtain

$$t(z(\tau_o)) \leqslant t(z_o) - \tau_o = 0$$

Since $t(z)$ is non-negative, we see that

$$t(z(\tau_o)) = 0$$

that is,

$$z(\tau_o) \in \widetilde{T}_o(\theta) = \theta$$

But that contradicts the assumption according to which the trajectory $z(t)$ does not reach θ in time $t(z_o)$

b) $\tau_o < t(z_o)$. Then T has no point in the time interval $[\tau_o, t(z_o)]$. In other words, we have $\tau_o \in T$ and

$\varepsilon(z(\tau_o)) + \tau_o > t(z_o)$. Then

$$(15) \qquad t(z(\tau_o)) \leqslant t(z_o) - \tau_o \leqslant \varepsilon(z(\tau_o))$$

Now let us return to the definition of Γ_P^*. Remember that the trajectory $z(t)$ generated by Γ_P^* and Γ_E, emanating from $z(\tau_o)$, is the same as the one generated by Γ_P^o and Γ_E from that point. Accordingly, the trajectory $z(t)$ reaches θ at a time belonging to $[\tau_o, t(z(\tau_o)) + \tau_o]$. But in view of (14) we have

$$\tau_o + t(z(\tau_o)) \leqslant \tau_o + (t(z_o) - \tau_o) = t(z_o)$$

Accordingly the trajectory $z(t)$ reaches θ at a time belonging to $[\tau_o, t(z_o)]$, which contradicts one of our assumptions, namely the one according to which $z(t) \notin \theta$, $t \in [0, t(z_o)]$

Hence the first part of Theorem 3 is established.

Now,suppose that $t(z_o) < +\infty$, $\alpha > 0$. Then, from the definition of $t(z_o)$ we have

$$z_o \notin \tilde{T}_{t(z_o)-\alpha}(\theta)$$

From the definition of $\tilde{T}_t$, there exists a rational subdivision ω_o such that

$$|\omega_o| > t(z_o) - \alpha$$

$$z_o \notin T_{\omega_o}(\theta)$$

Suppose that ω_o is defined by points $\tau_o, \tau_1, \ldots, \tau_m$, and consider the numbers $\delta_i = \tau_i - \tau_{i-1}$, $i = 1,\ldots,m$. Since they are rational numbers, there exists a rational number δ such that

$$\delta_i = k_i \, \delta \,, \quad i = 1,\ldots,m$$

where the numbers k_i, $i = 1,\ldots,m$, are integers. Then, since

$\tau_o = 0$, we have $\tau_i = n_i\delta$ where the numbers n_i are integers also. Let us consider the subdivision $\omega = \{ 0, \delta, 2\delta, \ldots N\delta \}$, $N = n_m$. Clearly we have $\omega \ni \omega_o$ $|\omega| = |\omega_o|$.

Hence

$$N\delta = |\omega| > t(z_o) - \alpha$$

$$z_o \notin T_\omega(\theta)$$

Let $A_i = T_\delta A_{i-1}$, $i = 1,\ldots,N$ and $A_o = \theta$.

Clearly we have

$$A_N = T_\omega(\theta)$$

$$z_o \notin A_N$$

Moreover, from the properties of T_δ it follows that

$$A_N \supset A_{N-1} \supset \ldots \supset A_1 \supset \theta$$

Now let us define a strategy Γ_E^α as follows. For $z \notin A_N$ let $\varepsilon(z) = \delta$. Since $A_N = T_\delta A_{N-1}$ we have $z \notin T\ A_{N-1}$. Accordingly, from the definition of T_δ , there exists a control $v(t), 0 \leqslant t \leqslant \delta$, such that for any $u(t)$ the trajectory $z(t)$ of system (1), generated by controls $u(t)$ and $v(t)$, emanating from z, does not reach A_{N-1} .

Likewise if $z \in A_{i+1}$ but $z \notin A_i = T_\delta A_{i-1}$, there exists a control $v(t)$, $0 \leqslant t \leqslant \delta$, such that, no matter what the control $u(t)$ is, the trajectory emanating from z, generated by $u(t)$ and $v(t)$, does not reach A_{i-1} during the interval of time $[0,\delta]$.

By definition Γ_E^α is the strategy of player E which associates with each point z the corresponding control $v(t)$ thus defined.

Now suppose that Γ_P is arbitrary and let us prove by recurrence that

$$I(z;\Gamma_P,\Gamma_E^\alpha) > i\delta \quad \text{if} \quad z \notin A_i$$

For $i = 0$ we have $A_o = \theta$ and the assertion is true. Let us suppose that the assertion has been proved for $i \leqslant j$ and $z \notin A_{j+1}$.

Let k be the subscript such that $z \in A_{k+1}$ and $z \notin A_k$. Clearly we have $k \geqslant j+1$. Then, by the definition of Γ_E^α, the trajectory $z(t)$ emanating from z, generated by Γ_P and Γ_E^α, does not intersect A_{k-1} on the time interval $[0,\delta]$; that is,

$$z(t) \notin A_{k-1} \quad , \qquad 0 \leqslant t \leqslant \delta$$

but $k-1 \geqslant j$, accordingly

$$z(t) \notin A_j \quad , \qquad 0 \leqslant t \leqslant \delta$$

since $A_{k-1} \supset A_j \supset \theta$.

For our assumption above, according to which

$$I(z(\delta); \Gamma_P , \Gamma_E^\alpha) > j\delta$$

we have

$$I(z ; \Gamma_P , \Gamma_E^\alpha) = \delta + I(z(\delta); \Gamma_P, \Gamma_E^\alpha) \geqslant \delta + j\delta = (j+1)\delta$$

which concludes the proof.

It follows that, if $z_o \notin A_N$, then

$$I(z_o; \Gamma_P , \Gamma_E^\alpha) \geqslant N\delta = |\omega_o| > t(z_o) - \alpha$$

Hence the second part of Theorem 3 is established.

The third part of Theorem 3 can be easily proved by invoking similar arguments.

Theorem 4. *A differential game with ε-strategies has a value, and this value is* $t(z)$ *; that is,*

$$\inf_{\Gamma_P} \sup_{\Gamma_E} I(z; \Gamma_P , \Gamma_E) = \sup_{\Gamma_E} \inf_{\Gamma_P} I(z; \Gamma_P , \Gamma_E) = t(z)$$

Proof. Let us suppose that $t(z) < +\infty$. The case where $t(z) = +\infty$ can be treated by invoking similar arguments.

From Theorem 3 we have

$$I(z\ ;\ \Gamma_P^* , \Gamma_E) \leqslant t(z)$$

Hence

$$V^+(z) = \inf_{\Gamma_P} \sup_{\Gamma_E} I(z\ ;\ \Gamma_P , \Gamma_E) \leqslant t(z) \tag{16}$$

On the other hand we have

$$I(z\ ;\ \Gamma_P , \Gamma_E^\alpha) \geqslant t(z) - \alpha$$

Hence

$$V^-(z) = \sup_{\Gamma_E} \inf_{\Gamma_P} I(z\ ;\ \Gamma_P, \Gamma_E) \geqslant t(z) - \alpha$$

and since α, $\alpha > 0$, is arbitrary

$$\sup_{\Gamma_E} \inf_{\Gamma_P} I(z\ ;\ \Gamma_P , \Gamma_E) \geqslant t(z)$$

But $V^-(z) \leqslant V^+(z)$, and so we have

$$t(z) \leqslant V^-(z) \leqslant V^+(z) \leqslant t(z)$$

that is

$$V^-(z) = V^+(z) = t(z)$$

which concludes the proof.

4. A Decomposition of the State Space, Isaacs-Bellman Equation

In the previous paragraph we have proved an existence theorem for the value of the game, from which one can derive a number of results. For any point $z \in E^n$ let us list four situations, that players P and E can meet, namely

1) $t(z) < +\infty$. Then there exists a strategy of player P, namely Γ_P^*, which guarantees termination of the game no later than time $t(z)$. We shall let $\Omega_P^* = \{ z : t(z) < +\infty \}$

2) $t(z) = +\infty$ and there exists a strategy Γ_P^{**} such that

$$I(z, \Gamma_P^{**}, \Gamma_E) < +\infty \qquad \forall \Gamma_E$$

We shall denote by Ω_P^{inf} the set of all such points.

3) $t(z) = +\infty$ and there exists a strategy Γ_E^* such that

$$I(z, \Gamma_P, \Gamma_E^*) = +\infty \qquad \forall \Gamma_P$$

We shall denote by Ω_E the set of all such points z.

4) $t(z) = +\infty$ and with every Γ_P one can associate a strategy Γ_E such that

$$I(z, \Gamma_P, \Gamma_E) = +\infty$$

and with every Γ_E one can associate a strategy Γ_P such that

$$I(z, \Gamma_P, \Gamma_E) < +\infty$$

We shall denote by Ω_o the set of all such points z.

As we see, in cases 1-3 one of the two players can win no matter what the strategy of his opponent is. In case 4 the situation is different ; that is, the fact that one or the other player is informed of the strategy of his opponent has an effect on his possibility of winning.

There is no doubt about the fact that Ω_P^* and Ω_E are non-empty sets, in concrete examples. In what follows we shall discuss an example in which Ω_P^{inf} is non-empty. Unfortunately we have no example for which $\Omega_o \neq \phi$, however one can show that Ω_o is not always empty.

Example. Let us consider a system whose behavior is governed by the equations

$$\begin{aligned} \dot{x} &= f(x) + v \\ \dot{y} &= u \\ |u| \leqslant 1 \,, &\quad |v| \leqslant 1 \end{aligned} \tag{17}$$

where x, y, v $\in$ R . Target θ is composed of the point x = 0, y = 0 and of the half ray y = 0, x $\geqslant$ 1.

f(x) is a continuous function with the following properties

a) $f(x) + 1 < 0, \quad x \neq 1, \quad f(1) + 1 = 0$

b) $\displaystyle\int_{x_o}^{0} \frac{dx}{f(x) + 1} \to + \infty \quad \text{if} \quad x_o \to 1, \quad x_o < 1$

Players P and E control coordinates y and x, respectively. Obviously, player P must let y tend to zero as fast as possible ; that is, his optimal strategy is

$$\begin{aligned} u &= - \operatorname{sign} y & & y \neq 0 \\ u &= 0 & \text{if} \quad & y = 0 \end{aligned}$$

It follows from Eqs. (17) that

$$y(t) = y_o - (\operatorname{sgn} y)t , \qquad t \leqslant |y_o|$$
$$y(t) = 0 \qquad t \leqslant |y_o|$$

Since

$$f(x) + v \leqslant f(x) + 1 \leqslant 0$$

x is a non increasing function of t . One can see that if $y(t_o) = 0$ and $x(t_o) \geqslant 0$ at time $t_o = |y_o|$, then the game is a terminating one and P wins. Indeed, the point $x = 0$, $y = 0$ belongs to θ, and

(i) if $x(t_o) < 1$, then $x(t)$ is a strictly decreasing function of t, for $t \geqslant t_o$, and $y(t) = 0$; and

(ii) if $x(t_o) \geqslant 1$, then the game terminates at time t_o .

Now consider an initial point $x_o = 1$, $y_o \neq 0$; and suppose that the following condition is satisfied, namely

$$\int_1^0 \frac{dx}{f(x) - 1} > |y_o| \tag{18}$$

Let us show that if the strategy of P is the one stated above, then the trajectory reaches θ in finite time. Let $v(t)$ be the control of player E.

If $\bar{t} < |y_o|$ and $v(t) = 1$ for $0 \leqslant t \leqslant \bar{t}$, then $x(t) = 1$ for $0 \leqslant t \leqslant \bar{t}$, and $y(t)$ is a strictly decreasing function of t . Its value at time $\bar{t}$ is $y_o - (\operatorname{sgn} y_o)\bar{t}$.

If $\bar{t} \geqslant |y_o|$, then the game terminates at time $t_o = |y_o|$ since $x(\bar{t}) = 1$ and $y(\bar{t}) = 0$.

We see that if $v(t) = +1$ on some time interval greater than $|y_o|$, then the game is a terminating one and P wins. If $v(t) = +1$ on some time interval smaller that $|y_o|$, then the earlier initial point can be replaced by the end point of the trajectory and the problem is unchanged since $x(\bar{t}) = 1$ and $y(\bar{t}) \neq 0$. For that reason, we shall let $v(t) < 1$ at the initial time. Then $x(t)$ is a strictly decreasing function of t since we have $f(1) + v(0) < f(1) + 1 = 0$ at the initial time, and $f(x) + v(t) \leqslant f(x) + 1 < 0$ later on as $x < 1$.

From (17) we get

$$t = \int_0^t \frac{\dot{x}(t)dt}{f(x(t))+v(t)} \geqslant \int_0^t \frac{\dot{x}(t)dt}{f(x(t))-1} = \int_1^{x(t)} \frac{dx}{f(x)-1}$$

It follows that $x(t) = 0$ at a time t such that

$$t \geqslant \int_1^0 \frac{dx}{f(x)-1} > |y_o|$$

and hence, at time $\bar{t} = |y_o|$ for which $y(\bar{t}) = 0$ we have $x(\bar{t}) \geqslant 0$.

From the discussion above, it follows that the trajectory reaches the point $x = 0$, $y = 0$, in the set θ, and P wins.

Now let us compute the value of the game at point $x_o = 1$, y_o, for which condition (18) is satisfied. Let us define strategy Γ_E^α as follows.

If $x_o = 1$, then $\varepsilon(x_o, y_o) = \alpha$, $v(t) = -1$; and if $x_o < 1$, then $v(t) = +1$ and $\varepsilon(x_o, y_o)$ is arbitrary.

Let us compute the transfer time to the target, when the strategies of P and E are Γ_P and Γ_E^α , respectively.

On the time interval $[0,\alpha]$, we have $v(t) = -1$ and $x(t)$

is the solution of

$$\dot{x} = f(x) - 1 \qquad \text{with} \quad x_o = 1$$

It follows that

$$\frac{dx}{f(x)-1} = dt \tag{19}$$

and

$$\int_1^{x(\alpha)} \frac{dx}{f(x)-1} = \alpha$$

If α is small, then $x(\alpha) < 1$ and $x(\alpha) \to 1$ as $\alpha \to 0$.

At time α, player E changes his control into $v(t) = +1$. Accordingly, from (17), the time $t(\alpha)$ for which $x(t) = 0$ is given by

$$\int_{x(\alpha)}^{0} \frac{dx}{f(x)+1} = t(\alpha) - \alpha$$

that is,

$$t(\alpha) = \alpha + \int_{x(\alpha)}^{0} \frac{dx}{f(x)+1}$$

Now, if $\alpha \to 0$ then $x(\alpha) \to 1$ and accordingly, in view of an earlier assumption, we have $t(\alpha) \to +\infty$. It follows that the value of the game is $+\infty$, and the initial point $x_o = 1$, y_o, belongs to the set Ω_P^{inf} .

Now let us pay attention to the boundaries of the sets Ω_P^* , Ω_P^{inf}, Ω_E and Ω_o . We shall say that the common boundary of the sets X and Y in E^n , namely $\overline{X} \cap \overline{Y}$, is *smooth* at point $z_o \in \overline{X} \cap \overline{Y}$ if there exists a continuously differentiable function $\varphi(z)$ and an open ball S_r with center z_o and radius r, such that

(i) $\varphi(z_o) = 0$; and

(ii) $\nabla\varphi(z) \neq 0$ at point z_o , where $\nabla\varphi(z)$ is the gradient of the function $\varphi(z)$; and

(iii) if $z \in S_r$ then

$$\varphi(z) < 0 \Rightarrow z \in X \text{ ; and}$$
$$\varphi(z) > 0 \Rightarrow z \in Y$$

If $\overline{X} \cap \overline{Y}$ is smooth at point z_o, we shall say that z_o is a *regular point* of $\overline{X} \cap \overline{Y}$.

In what follows, if z_o is a regular point of the common boundary of Ω_P^* and Ω_P^{inf} , we shall let

$$\left.\begin{array}{l} \varphi(z) < 0 \Rightarrow z \in \Omega_P^* \\ \varphi(z) > 0 \Rightarrow z \in \Omega_P^{inf} \end{array}\right\} \text{ for } z \in S_r$$

Theorem 5. *If, for any vector* $p \in E^n$ *and any point* z, *the following condition is satisfied, namely*

$$\min_{u\in U} \max_{v\in V} < p,f(z,u,v) > = \max_{v\in V} \min_{u\in U} < p,f(z,u,v) > \tag{20}$$

then at any regular point z_o *of* $\overline{X} \cap \overline{Y}$, $X, Y \in \{\Omega_P^*, \Omega_P^{inf}, \Omega_E, \Omega_o\}$, $z_o \notin \theta$, *the following condition is satisfied, namely*

$$\min_{u\in U} \max_{v\in V} < \nabla\varphi(z_o), f(z_o,u,v) > = 0 \tag{21}$$

with the provisio that, if z_o *is a regular point of the common boundary of* Ω_P^* *and* Ω_P^{inf}, *one must assume that* $\varphi(\overline{z}) \leqslant - \alpha < 0$, $\overline{z} \in S_r$, *implies that* $t(\overline{z}) \leqslant T(\alpha) < + \infty$.

Proof. We shall suppose that z_o is a regular point of the common boundary of Ω_P^* and Ω_P^{inf}, $z_o \notin \theta$. The other cases can be treated by invoking similar arguments.

Let

$$W(z) = \min_{u \in U} \max_{v \in V} < \nabla\varphi(z), f(z,u,v) >$$

and let $u_o \in U$ and $v_o \in V$ be such that

$$\max_{v \in V} < \nabla\varphi(z_o), f(z_o,u_o,v) > = W(z_o)$$

$$\min_{u \in U} < \nabla\varphi(z_o), f(z_o,u,v_o) > = W(z_o)$$

We intend to prove that $W(z_o) = 0$, and so we shall suppose that $W(z_o) \neq 0$.

Let us suppose first that $W(z_o) > 0$.

Let S_{r^*} be an open ball with center z_o and radius $r^* < r$, and suppose that r^* is sufficiently small so that we have, by continuity

$$\min_{u \in U} < \nabla\varphi(z), f(z,u,v_o) > \geqslant \frac{1}{2} W(z_o), \qquad z \in S_{r^*}$$

Then, consider the open ball $S_{\frac{1}{2} r^*}$. From the assumptions concerning U, V and f(z,u,v), and from Gronwall's lemma (13), one can prove that there exists an interval of time of length T^* such that the trajectory z(t), emanating from $z \in S_{\frac{1}{2} r^*}$, generated by arbitrary controls u(t) and v(t), remains in S_{r^*} during that interval of time.

Now let $\bar{z}$ be such that

$$0 > \varphi(\bar{z}) \geqslant -\frac{1}{4} W(z_o) T^*$$

$$\bar{z} \in S_{\frac{1}{2} r^*}$$

Indeed $\bar{z} \in \Omega_P^*$. Accordingly there exists a strategy Γ_P^* such that the trajectory z(t) reaches θ no later than $t(\bar{z}) < +\infty$.

Let us define Γ_E in such a way that, at point $\bar{z}$, $\varepsilon = T^*$ and $v(t) = v_o$, $0 \leqslant t \leqslant T^*$; and consider the trajectory z(t) generated by Γ_P^* and Γ_E on the interval $[0, T^*]$, emanating from $\bar{z}$.

We have

$$\varphi(z(t) - \varphi(\bar{z}) = \int_0^t < \nabla\varphi(z(t)), f(z(t), u(t), v_o) > dt \geqslant \frac{1}{2} W(z_o) t$$

since the trajectory remains in S_{r^*} on $[0, T^*]$.

It follows that

$$\varphi(z(T^*)) \geqslant \varphi(\bar{z}) + \frac{1}{2} W(z_o) T^* \geqslant$$

$$- \frac{1}{4} W(z_o) T^* + \frac{1}{2} W(z_o) T^* = \frac{1}{4} W(z_o) T^* > 0$$

Hence $z(T^*) \in \Omega_P^{inf}$ and, accordingly, there exists a strategy Γ_E^* such that the trajectory emanating from $z(T^*)$, generated by Γ_P^* and Γ_E^* reaches θ later than $t(\bar{z})$. But that contradicts the fact that the trajectory emanating from $\bar{z}$, generated by Γ_P^* and an arbitrary Γ_E, reaches θ no later than $t(\bar{z})$.

We thus conclude that $W(z_o)$ is non positive.

Now let us suppose that $W(z_o) < 0$.

Let S_{r^*} be such that

$$\max_{v \in V} < \nabla\varphi(z), f(z,u,v) > \leqslant \frac{1}{2} W(z_o), \qquad z \in S_{r^*} \tag{22}$$

As above, there exists an interval of time of length T^* such that the trajectory z(t), emanating from $z \in S_{\frac{1}{2} r^*}$, generated by arbitrary controls u(t) and v(t), remains in S_{r*} during that interval of time .

Let $\overline{z}$ be such that

$$\varphi(\overline{z}) > 0$$

$$\overline{z} \in S_{\frac{1}{2} r^*}$$

$$\varphi(\overline{z}) \leqslant - \frac{1}{4} W(z_o)T^*$$

Indeed $\overline{z} \in \Omega_P^{inf}$.

Accordingly, for any $\beta > 0$, there exists a strategy Γ_E^β such that the trajectory z(t), emanating from $\overline{z}$, generated by an arbitrary Γ_P and Γ_E^β , reaches θ no sooner than time β.

Let us define Γ_P as follows.

For $z \in \Omega_P^*$, $\varphi(z) \leqslant \frac{1}{4} W(z_o)T^*$ let Γ_P coincide with Γ_P^* . For $z \in S_{r*}$ $\varphi(z) > \frac{1}{4} W(z_o)T^*$, let Γ_P be such that, for given $\varepsilon > 0$ and v(t) on $[0, \varepsilon]$, $u(t) = u_o$.

Now consider the trajectory z(t), emanating from $\overline{z}$, generated by the strategy Γ_P defined above and an arbitrary Γ_E . It remains in S_{r*} for $t \in [0, T^*]$ and, as long as $\varphi(z(t)) > \frac{1}{4} W(z_o)T^*$ we have $u(t) = u_o$, so that according to (22) the following condition is satisfied, namely

$$\varphi(z(t)) - \varphi(\overline{z}) = \int_0^t < \nabla\varphi(z(t)),f(z(t),u_o,v(t)) > dt \leqslant \frac{1}{2} W(z_o)t$$

It follows that

$$\text{(23)} \quad \varphi(z(t)) \leqslant \varphi(\overline{z}) + \frac{1}{2} W(z_o)t \leqslant - \frac{1}{4} W(z_o)T^* + \frac{1}{2} W(z_o)t = \frac{1}{2} W(z_o)(t - \frac{1}{2} T^*)$$

We see that $\varphi(z(t)) \leqslant \frac{1}{4} W(z_o)T^*$ for $t \geqslant T^*$. Accordingly there exists a time $\bar{t} \in [0, T^*]$ such that

$$\varphi(z(t)) > \frac{1}{4} W(z_o)T^* \text{ for } t \in [0, \bar{t}) \quad ; \text{ and}$$

$$\varphi(z(\bar{t})) = \frac{1}{4} W(z_o)T^*$$

Indeed, $z(\bar{t}) \in \Omega_P^*$ since $\varphi(z(\bar{t})) < 0$. From that point $\Gamma_P = \Gamma_P^*$ and the trajectory reaches θ in a time that does not exceed $t(z(\bar{t}))$..

Accordingly θ is reached no later than $\bar{t} + t(z(\bar{t}))$.

But $\bar{t} \leqslant T^*$, and $t(z(\bar{t}))$ is bounded by $T(\alpha_o)$, $\alpha_o = -\frac{1}{4} W(z_o)T^*$, according to an assumption of Theorem 5. It follows that θ is reached no later than $T^* + T(\alpha_o)$. This upper bound of the terminal time does not depend on Γ_E, which contradicts the fact that $\bar{z} \in \Omega_P^{inf}$.

Hence Theorem 5 is proved.

Corollary. *If the value of the game* $t(z)$ *is a smooth function on a neighborhood of point* z_o, *then*

$$\min_{u \in U} \max_{v \in V} < \nabla t(z), f(z,u,v) > + 1 = 0$$

on that neighborhood.

Proof. Suppose that $t(z)$ is a smooth function on a neighborhood of point z_o, and consider the set $\tilde{\theta}$, in the augmented space E^{n+1}, such that

$$(z,z^o) \in \tilde{\theta} \text{ if } z \in \theta, \quad z^o \leqslant t_o, \quad t_o = t(z_o).$$

Let $\tilde{\theta}$ be the target of a game whose state in E^{n+1} is governed by Eq.(1) and

$$\frac{dz^o}{dt} = 1$$

Let

(24) $$\varphi(z,z^o) = z^o + t(z) - t_o$$

Since $z^o(t) = z^o + t$, one can see easily that the game defined above can be terminated in finite time, from the point (z,z^o), if and only if

$$\varphi(z,z^o) \leqslant 0$$

Indeed, in that case we have $t(z) \leqslant t_o - z^o$, and the trajectory $z(t)$ can reach θ no later than $t(z)$. If $\bar{t}$ is the terminal time, then $\bar{t} \leqslant t(z) \leqslant t_o - z^o$, and hence $z^o(\bar{t}) = z^o + \bar{t} \leqslant t_o$.

In other words, the common boundary of Ω_P^* and Ω_E is defined by

$$\varphi(z,z^o) = 0$$

From Theorem 5 we obtain

$$\min_{u \in U} \max_{v \in V} < \nabla_z \varphi(z,z^o), f(z,u,v) > + \frac{\partial \varphi(z,z^o)}{\partial z^o} = 0$$

At last, taking account of (24), we obtain the expected result.

5. Games of Fixed Duration

By a *game of fixed duration* we mean a game in which

(i) there is given a prescribed time T_o ; and

(ii) a trajectory $z(t)$ cannot reach the closed target at a time other than T_o .

The cost function is defined by

$$I(z, \Gamma_P , \Gamma_E) = \begin{cases} 0 & \text{if} \quad z(T_o) \in \theta \\ +\infty & \text{if} \quad z(T_o) \notin \theta \end{cases}$$

where $z(t)$ is the trajectory emanating from z, generated by the strategies Γ_P and Γ_E .

Here Γ_P and Γ_E are defined as in Paragraph 1, with the provisio that the number ε ($\varepsilon > 0$) and the control $v(t)$, $t \in [t_o, t_o+\varepsilon]$, associated with the point z through Γ_E , may depend on t_o . Likewise, the control $u(t)$, $t \in [t_o, t_o+\varepsilon]$ associated with z, ε and $v(t)$ through Γ_P, may depend on t_o . Clearly, the definition 3 of Paragraph 1 is now modified.

The properties of the set T remain unchanged and if $\tau \in T$ then, on the interval $[\tau, \tau + \varepsilon(z(\tau),\tau)]$, $z(t)$ obeys

$$\dot{z} = f(z,u(t),v(t)) \tag{25}$$

where $v(t)$ is the control associated with the point $z(\tau)$ and the time τ through the strategy Γ_E, and $u(t)$ is the control associated with $z(\tau)$, τ, ε and $v(t)$ through Γ_P .

By introducing the new variable z_o, governed by

$$\frac{dz^o}{dt} = 1$$

and the set $\tilde{\theta} \subset E^{n+1}$, $\tilde{\theta} \triangleq \{(z,z^o) : z \in \theta,\ z^o = T_o\}$, a game of fixed duration is reduced to a game of the same type as the ones studied in the earlier paragraphs. However it will be useful to return to the constructions of Paragraphs 2 and 3, with slight changes.

Definition 7. Let $X \subset E^n$. $P_\varepsilon(X)$ is the set of points z such that, for any measurable control $v(t) \in V$, $t \in [0,\varepsilon]$, there exists a measurable control $u(t) \in U$ such that the trajectory $z(t)$, emanating from z, generated by $u(t)$ and $v(t)$, reaches X at time ε ; that is, $z(\varepsilon) \in X$.

One can prove easily that P_ε has properties similar to the ones of T_ε (see Paragraph 2) ; for instance

1. $P_\varepsilon(X') \supset P_\varepsilon(X)$ if $X' \supset X$;

2. $P_o(X) = X$

3. $P_{\varepsilon_1} P_{\varepsilon_2}(X) \subset P_{\varepsilon_1+\varepsilon_2}(X)$

However $\varepsilon' > \varepsilon$ does *not* imply that $P_{\varepsilon'}(X) \supset P_\varepsilon(X)$.

Now we shall say that ω_t is a rational subdivision of the interval $[0,t]$ if ω_t is a sequence of numbers τ_i, $i = 0,\dots,m$, $\tau_i \leqslant \tau_{i+1}$, $\tau_o = 0$, $\tau_m = t$, and the numbers τ_i, $i < m$, are rational.

We shall let

$$P_{\omega_t}(X) = P_{\delta_m} \dots P_{\delta_1}(X)$$

where $\delta_i = \tau_i - \tau_{i-1}$, $i = 1,\dots,m$

Definition 8. $\tilde{P}_t(X) = \bigcap_{\omega_t} P_{\omega_t}(X)$

By similar arguments as in Paragraph 2, one can prove that $\tilde{P}_t$ has the following properties.

1. $\tilde{P}_o(X) = X$

2. If X is closed, then $\tilde{P}_t(X)$ is closed.

3. If X is closed, then

$$\tilde{P}_{t_1}\tilde{P}_{t_2}(X) = \tilde{P}_{t_1+t_2}(X), \quad t_1,t_2 \geqslant 0$$

4. If X is closed, then

$$\tilde{P}_t(X) \subset P_\varepsilon \tilde{P}_{t-\varepsilon}(X) \qquad 0 < \varepsilon < t .$$

Now let us define Γ_P^* as follows.

Let z be a given point and t_o be the present time, and

$z \in \tilde{P}_{T_o - t_o}(\theta)$. Let there be given ε and $v(t)$ on $[t_o, t_o + \varepsilon]$. Two cases need be considered, namely

Case 1 $\quad \varepsilon \geqslant T_o - t_o$

Then

$$z \in \tilde{P}_{T_o - t_o}(\theta) \subset P_{T_o - t_o}(\theta)$$

From the definition of P_t it follows that there exists a control $u(t)$ such that the trajectory $z(t)$ of system (25), with $z(t_o) = z$, reaches target θ at time $t_o + (T_o - t_o) = T_o$.

Γ_P^* is the strategy that associates with z, t_o, ε and $v(t)$ this control $u(t)$.

Case 2 $\quad \varepsilon < T_o - t_o$

Then

$$z \in \tilde{P}_{T_o - t_o}(\theta) \subset P_\varepsilon \tilde{P}_{T_o - t_o - \varepsilon}(\theta)$$

Accordingly there exists a control $u(t)$, $t \in [t_o, t_o + \varepsilon]$, such that the trajectory $z(t)$ of system (25), with $z(t_o) = z$, reaches the set $\tilde{P}_{T_o - t_o - \varepsilon}(\theta)$ at time $t_o + \varepsilon$; that is,

$$z(t_o + \varepsilon) \in \tilde{P}_{T_o - t_o - \varepsilon}(\theta) \tag{26}$$

Theorem 6. *If* $z_o \in \tilde{P}_{T_o}(\theta)$, *then*

$$I(z_o, \Gamma_P^*, \Gamma_E) = 0 \qquad \forall \Gamma_E$$

If $z_o \notin \tilde{P}_{T_o}(\theta)$, *then there exists a strategy* Γ_E^* *such that*

$$I(z_o, \Gamma_P, \Gamma_E^*) = +\infty \qquad \forall \Gamma_P$$

We shall suppose that the trajectories that emanate from z_o leave z_o at time $t_o = 0$; that is, $z(0) = z_o$.

Proof. Let $z_o \in \tilde{P}_{T_o}(\theta)$. From the definition of Γ_P^* it follows that, if condition (26) is satisfied, we have

$$z(\tau) \in \tilde{P}_{T_o-\tau}(\theta) \tag{27}$$

for $\tau \in T$, where T is the sequence of times that occurs in the definition of the trajectory z(t) (Definition 3).

The proof is similar to the one of Theorem 3.

From (27) and the definition of Γ_P^* one easily deduces that

$$z(T_o) \in \tilde{P}_{T_o-T_o}(\theta) = \theta$$

that is,

$$I(z_o , \Gamma_P^* , \Gamma_E) = 0$$

Now, if $z_o \notin \tilde{P}_{T_o}(\theta)$ there exists a subdivision $\omega_{T_o}^o$ such that

$$z_o \notin P_{\omega_{T_o}^o}(\theta)$$

that is,

$$z_o \notin P_{\delta_m^o} \dots P_{\delta_1^o}(\theta)$$

where $\delta_i^o = \tau_i^o - \tau_{i-1}^o$, and the times τ_i^o are those of the subdivision $\omega_{T_o}^o$.

We shall define the strategy Γ_E^* as follows.

If

$$z \notin P_{\delta_i^o} \dots P_{\delta_1^o}(\theta) \tag{28}$$

Γ_E^* is the strategy that associates with z, $T_o-\tau_i$, $\varepsilon = \delta_i^o$, the control v(t) ; and the control v(t), $t \in [T_o-\tau_i, T_o-\tau_{i-1}]$,is chosen in such a way that the trajectory of system (25), emanating from z at time $t_i = T_o - \tau_i$, generated by an arbitrary control u(t) and v(t), do not reach the set $P_{\delta^o_{i-1}} \ldots P_{\delta^o_1}(\theta)$ at time $t_{i-1} = T_o - \tau_{i-1}$ [the set θ if i = 1] .

From the definition of $P_{\delta^o_i}$ and from (28) it follows that such a control v(t) exists.

Now one can see easily that the set T associated with the trajectory emanating from z_o, generated by an arbitrary strategy Γ_P and Γ^*_E , is made of the points $t_m = T_o - \tau_m = 0$, t_{m-1} ,... $t_o = T_o - \tau_o = T_o$. Then, from the construction of the control v(t) associated with the strategy Γ^*_E , we have

$$z(t_i) \notin P_{\delta^o_i} \ldots P_{\delta^o_1}(\theta)$$

and, as a special case

$$z(T_o) \notin \theta$$

which concludes the proof of Theorem 6.

Theorem 7. *A game of fixed duration, with ε-strategies, has a saddle-point.*

Proof. If $z_o \in \widetilde{P}_{T_o}(\theta)$, we shall let Γ^o_E be arbitrary, and $\Gamma^o_P = \Gamma^*_P$. Then, from Theorem 6 we have

$$I(z_o, \Gamma^o_P, \Gamma_E) \leqslant I(z_o, \Gamma^o_P, \Gamma^o_E) \leqslant I(z_o, \Gamma_P, \Gamma^o_E) \tag{29}$$

and furthermore $I(z_o, \Gamma^o_P, \Gamma^o_E) = 0$.

If $z_o \notin \widetilde{P}_{T_o}(\theta)$, we shall let Γ^o_P be arbitrary, and $\Gamma^o_E = \Gamma^*_E$, where Γ_E^* is the strategy constructed in the proof of Theorem 6. Then, clearly, (29) is satisfied but $I(z_o, \Gamma_P^o, \Gamma_E^o) = +\infty$.

Hence, the game has a value in both cases. This value is 0 in the first case and $+\infty$ in the second one .

Now, let us suppose that the system (1) is linear; that is,

$$\dot{z} = Az + u - v \,, \quad u \in U, \quad v \in V \tag{30}$$

where U and V are compact convex sets in E^n, and A is a $n \times n$ matrix. If u(t) and v(t) are measurable functions on $[t_o, t_1]$ with ranges in U and V, respectively, the solution of system (30), with the initial condition $z(t_o) = z_o$, is

$$z(t) = \Phi(t-t_o)z_o + \int_{t_o}^{t} \Phi(t-\tau)\,[\,u(\tau)-v(\tau)\,]\,d\tau \tag{31}$$

where the matrix $\Phi(t)$ satisfies

$$\begin{aligned} &\dot{\Phi}(t) = A\,\Phi(t) \\ &\Phi(0) = I \end{aligned} \tag{32}$$

where I is the identity matrix.

Also, let us note that

$$\Phi(t) = e^{tA} \equiv I + \frac{1}{1!}\,tA + \dots + \frac{1}{k!}\,t^k A^k + \dots$$

Indeed we have

$$\Phi(t_1)\,\Phi(t_2) = \Phi(t_1 + t_2)$$

$$\Phi^{-1}(t) = \Phi(-t)$$

Now let X be a convex set. We shall pay attention to the set $P_\varepsilon(X)$.

Let us recall that any closed convex set Y can be completely depicted by the function

$$(33) \qquad W_Y(\psi) = \sup_{x \in Y} < x, \psi >$$

that is, $x \in Y$ if and only if

$$(34) \qquad < x, \psi > \leqslant W_Y(\psi) \qquad \forall \psi$$

Indeed, if $x \in Y$ the inequality (34) is trivial. Now suppose that (34) is satisfied for some x and for all ψ . If $x \notin Y$, it follows from the separation theorem for convex sets that there exists a non-zero vector ψ_o, and a number $\varepsilon > 0$, such that

$$< x, \psi_o > -\varepsilon \geqslant < y, \psi_o > , \quad y \in Y$$

but that contradicts (34) and proves the property.

$W_Y(\psi)$ is called the *support function* of Y . We see that, instead of working with convex sets, one can work with their support functions as well.

Let $z \in P_\varepsilon(X)$, where X is a closed convex set. From the linearity of Eq. (30) it follows that $P_\varepsilon(X)$ is convex. Indeed, since $z \in P_\varepsilon(X)$, no matter what the control v(t) is, $t \in [\,0,\varepsilon\,]$ $v(t) \in V$, there exists a control u(t) such that

$$z(\varepsilon) = \Phi(\varepsilon)z + \int_0^\varepsilon \Phi(\varepsilon-\tau)u(\tau)d\tau - \int_0^\varepsilon \Phi(\varepsilon-\tau)v(\tau)d\tau \in X$$

or, by taking account of the properties of the matrix $\Phi(\varepsilon)$

$$(35) \qquad z + \int_0^\varepsilon \Phi(-\tau)u(\tau)d\tau - \int_0^\varepsilon \Phi(-\tau)v(\tau)d\tau = \Phi(-\varepsilon)X$$

Let $U(\varepsilon)$ be the set of points y such that

$$y = \int_0^\varepsilon \Phi(-\tau)u(\tau)d\tau$$

for some measurable control $u(\tau)$, $u(\tau) \in U$, and define $V(\varepsilon)$ in an analogous fashion. Then (35) is equivalent to

$$z - \int_0^\varepsilon \Phi(-\tau)u(\tau)d\tau \in \Phi(-\varepsilon)X - U(\varepsilon) \tag{36}$$

but since the latter formula holds for all $v(\tau)$ (and $z \in P_\varepsilon(X)$), (36) rewrites

$$z \in \bigcap_{y \in V(\varepsilon)} [\Phi(-\varepsilon)X - U(\varepsilon) + y]$$

Thus we have proved that

$$P_\varepsilon(X) = \bigcap_{y \in V(\varepsilon)} [\Phi(-\varepsilon)X - U(\varepsilon) + y] \tag{37}$$

It follows directly that $P_\varepsilon(X)$ is convex if X is convex. Moreover, from the results of Ref.(12) it follows that $U(\varepsilon)$ is a compact set. Accordingly $P_\varepsilon(X)$ is closed if X is closed.

Let us compute the support function $W(\psi, \varepsilon, X, y)$ of the set $\Phi(-\varepsilon)X - U(\varepsilon) + y$. One can see easily that

$$W(\psi,\varepsilon, X, y) = \sup_{x \in X} < \psi, \Phi(-\varepsilon)x > +$$

$$\sup_{w \in U(\varepsilon)} < \psi, -w > + < \psi, y > =$$

$$W_X(\Phi^*(-\varepsilon)\psi) + \int_0^\varepsilon \sup_{u \in U} < -\Phi^*(-\tau)\psi, u > d\tau + < \psi, y >$$

Let

$$W_U(\psi) = \sup_{u \in U} < \psi, u >$$

From the arguments above it follows that

$$z \in \Phi(-\varepsilon)X - U(\varepsilon) + y$$

if, and only if

$$< \psi, z > \leqslant W_X(\Phi^*(-\varepsilon)\psi) + \int_0^\varepsilon W_U(-\Phi^*(-\tau)\psi)d\tau + < \psi, y >$$

But if $z \in P_\varepsilon(X)$, then $z \in \Phi(-\varepsilon)X - U(\varepsilon) + y$ for all $y \in V(\varepsilon)$. Accordingly $z \in P_\varepsilon(X)$ if and only if

$$< \psi, z > \leqslant W_X(\Phi^*(-\varepsilon)\psi) +$$

$$\int_0^\varepsilon W_U(- \Phi^*(-\tau)\psi)d\tau + \min_{y \in V(\varepsilon)} < \psi, y > =$$

$$W_X(\Phi^*(-\varepsilon)\psi) + \int_0^\varepsilon W_U(- \Phi^*(-\tau)\psi)d\tau - \int_0^\varepsilon W_V(- \Phi^*(-\tau)\psi)d$$

Thus we have proved that $z \in P_\varepsilon(X)$ if and only if

$$< \psi , z > \leqslant W_X(\Phi^*(-\varepsilon)\psi) + \int_0^\varepsilon [W_U(- \Phi^*(-\tau)\psi) - W_V(- \Phi^*(-\tau)\psi)] d\tau \tag{38}$$

Unfortunately the function in the right hand side of (38) is not, in general, the support function of the set $P_\varepsilon(X)$. This is because the support function of a convex set is a positively homogeneous convex function of ψ, as a consequence of its definition. The right hand side of (38) may not satisfy this condition. One can easily verify that it is positively homogeneous, but it is not always convex.

Let

$$\varphi(\psi,\varepsilon) = \int_0^\varepsilon [W_U(- \Phi^*(-\tau)\psi) - W_V(- \Phi^*(-\tau)\psi)] d\tau$$

We have proved the following

Lemma 4. *If* X *is a closed convex set in* E^n, *then* $z \in P_\varepsilon(X)$ *if and only if*

$$< \psi, z) \leqslant W_X(\Phi^*(-\varepsilon)\psi) + \varphi(\psi,\varepsilon), \qquad \forall \psi$$

Lemma 5. *Let* $W(\psi)$ *be a positively homogeneous closed convex function of* ψ, *which is not equal to* $+\infty$ *everywhere, and*

$$X = \{ x : \langle \psi, x \rangle \leqslant W(\psi), \quad \forall \psi \}$$

Then

$$W_X(\psi) \equiv \sup_{x \in X} \langle \psi, x \rangle = W(\psi)$$

We shall not give the proof of this lemma since it is based on the concept of conjugate convex functions. The exposition of this matter would go out of the scope of this chapter. The interested reader may refer to [14].

Theorem 8. *Let* X *be a closed convex set and suppose that, for any* $\varepsilon \in [0, T_0]$

$$W_X(\Phi^*(-\varepsilon)\psi) + \varphi(\psi, \varepsilon)$$

is a convex function of ψ. *Then*

1. $P_{\varepsilon_1} P_{\varepsilon_2}(X) = P_{\varepsilon_1+\varepsilon_2}(X), \quad 0 \leqslant \varepsilon_1, \varepsilon_2, \quad \varepsilon_1+\varepsilon_2 \leqslant T_0$;

2. $\tilde{P}_t(X) = P_t(X), \quad 0 \leqslant t \leqslant T_0$;

3. *the set* $\tilde{P}_t(X)$ *is depicted by*

$$\langle \psi, z \rangle \leqslant W_X(\Phi^*(-t)\psi) + \varphi(\psi, t), \qquad \forall \psi$$

Proof. From Lemmas 4 and 5, together with the assumptions of Theorem 8, we obtain

$$W_{P_\varepsilon(X)}(\psi) = W_X(\Phi^*(-\varepsilon)\psi) + \varphi(\psi, \varepsilon) \tag{39}$$

and since $\Phi(-\varepsilon_1)\,\Phi(-\varepsilon_2) = \Phi(-\varepsilon_1 - \varepsilon_2)$, and

$$\varphi(\Phi^*(-\varepsilon_1)\psi, \varepsilon_2) + \varphi(\psi, \varepsilon_1) = \varphi(\psi, \varepsilon_1 + \varepsilon_2)$$

as follows from the definition of $\varphi(\psi,\varepsilon)$, we have

$$W_{P_{\varepsilon_1} P_{\varepsilon_2}(X)}(\psi) = W_{P_{\varepsilon_2}(X)}(\Phi^*(-\varepsilon_1)\psi) + \varphi(\psi,\varepsilon_1) =$$

$$W_X(\Phi^*(-\varepsilon_2)\Phi^*(-\varepsilon_1)\psi) + \varphi(\Phi^*(-\varepsilon_1)\psi,\varepsilon_2) + \varphi(\psi,\varepsilon_1) =$$

$$W_X(\Phi^*(-(\varepsilon_1+\varepsilon_2))\psi) + \varphi(\psi,\varepsilon_1+\varepsilon_2) = W_{P_{\varepsilon_1+\varepsilon_2}(X)}(\psi)$$

Hence, the support functions of the closed convex sets $P_{\varepsilon_1} P_{\varepsilon_2}(X)$ and $P_{\varepsilon_1+\varepsilon_2}(X)$ are equal. It follows that these sets coincide, so that the first assertion of Theorem 8 is proved .

Now let ω_t be an arbitrary subdivision of $[0, t]$. From earlier arguments we have

$$P_{\omega_t}(X) = P_{\delta_m} \cdots P_{\delta_1}(X) = P_{\delta_m+\cdots+\delta_1}(X) = P_t(X)$$

and accordingly

$$\widetilde{P}_t(X) = \bigcap_{\omega_t} P_{\omega_t}(X) = \bigcap_{\omega_t} P_t(X) = P_t(X)$$

Hence the second assertion of Theorem 8 is proved.

At last the proof of the third assertion of Theorem 8 is a direct consequence of Lemma 4 with $\widetilde{P}_t(X) = P_t(X)$.

Theorem 9. *Let there be given a differential game of fixed duration* T_o *, governed by* Eq.(30), *with closed convex target* θ *. Suppose that, for any* $t \in [0, T_o]$,

$$W_\theta(\Phi^*(-t)\psi) + \varphi(\psi,t)$$

is a convex function of ψ *. Then, for a play begining at point* $z(0) = z_o$ *, if player* P *plays optimally,* P *is sure to win no matter what the strategy of* E *is, if and only if*

$$(40) \qquad \langle \psi, z_o \rangle \leqslant W_\theta(\Phi^*(-T_o)\psi) + \varphi(\psi,T_o), \quad \forall \psi$$

The proof of Theorem 9 is a direct consequence of Theorems 8 and 6.

Indeed, from Theorem 6, it follows that if P plays optimally, then P is sure to win, no matter what the strategy of E is, if and only if $z_o \in \tilde{P}_{T_o}(\theta)$; and from Theorem 8, together with the assumptions of Theorem 9, $\tilde{P}_{T_o}(\theta)$ is depicted by (40).

6. Games in the Presence of State Constraints

Again let us consider a game with state equations

$$\dot{z} = f(z,u,v) \tag{41}$$

and closed target set θ . Moreover, let there be given a closed set N . Player P endeavors to steer the state of the game to θ along a trajectory which lie entirely in N . If he succeeds he is a winner. If he does not succeed, for instance if the trajectory leaves the set N, he is a loser.

Without restricting the generality of our arguments, we shall suppose that $\theta \subseteq N$, since in any case θ can be replaced by $\theta \cap N$.

Now let us define the cost of a trajectory z(t) emanating from z_o , generated by the strategies Γ_P and Γ_E . If z(t) reaches θ , let t_1 be such that

$$z(t_1) \in \theta$$

$$z(t) \notin \theta \qquad t < t_1$$

If further, $z(t) \in N$ for all t in $[0, t_1]$, we shall let

$$I(z_o, \Gamma_P, \Gamma_E) = t_1$$

In any other case, we shall let

$$I(z_o, \Gamma_P, \Gamma_E) = +\infty$$

We intend to prove that a game in the presence of state constraints has a value.

Let us introduce a new mapping S_ε such that the transform of a set $X \subset N$ is

$$S_\varepsilon(X) = N \cap T_\varepsilon(X) \tag{42}$$

Then, for any rational subdivision ω, let us define $S_\omega(X)$, namely

$$S_\omega(X) = S_{\delta_m} S_{\delta_{m-1}} \cdots S_{\delta_1}(X) \tag{43}$$

and $\tilde{S}_t(X)$, namely

$$\tilde{S}_t(X) = \bigcap_{|\omega|>t} S_\omega(X)$$

From these definitions it follows that

$$\tilde{S}_t(X) \subset N, \quad \tilde{S}_o(X) = X \;, \quad \text{if } X \subset N \;.$$

One can easily verify that the mapping S_ε has the same properties as the mapping T_ε . Accordingly, $\tilde{S}_t$ has the same properties as $\tilde{T}_t$. Among these properties we have

$$\begin{aligned} &\tilde{S}_o(X) = X \;; \\ &\tilde{S}_t(X) \subset \tilde{S}_{t'}(X) \quad \text{if} \quad t' \geqslant t \;; \\ &\tilde{S}_{t_1}\tilde{S}_{t_2}(X) = \tilde{S}_{t_1+t_2}(X) \quad ; \\ &\tilde{S}_t(X) \subset S_\varepsilon \tilde{S}_{t-\varepsilon}(X) \;, \end{aligned} \tag{44}$$

if X is a closed subset of N .

Let

$$\begin{aligned} &t(z) = \min \{ \; t : z \in \tilde{S}_t(\theta) \; \} \\ &t(z) = +\infty \quad \text{if} \quad z \notin \tilde{S}_t(\theta), \qquad t \geqslant 0 \end{aligned} \tag{45}$$

Now let us construct the optimal strategy Γ_P^*.

Let z_o, ε and the admissible control $v(t)$, $t \in [0, \varepsilon]$ be given. Moreover let $t(z_o) < +\infty$.

Let

$$\varepsilon_o = \begin{cases} \varepsilon & \text{if} \quad t(z_o) \geqslant \varepsilon \\ t(z_o) & \text{if} \quad t(z_o) < \varepsilon \end{cases}$$

$$\varepsilon_m = \frac{\varepsilon_o}{m}$$

Then

$$z_o \in \widetilde{S}_{t(z_o)}(\theta) \subset S_\varepsilon \, \widetilde{S}_{t(z_o)-\varepsilon_m}(\theta)$$

From the definition of S_ε we have $z_o \in N$, and there exists a control $u(t)$, $t \in [0, \varepsilon_m]$, such that the trajectory $z^m(t)$ of system (41) on the interval $[0, \varepsilon_m]$, emanating from z_o, generated by $u(t)$ and $v(t)$, reaches $\widetilde{S}_{t(z_o)-\varepsilon_m}(\theta)$ at a time δ_m^1; that is,

$$z^m(\delta_m^1) \in \widetilde{S}_{t(z_o)-\varepsilon_m}(\theta) \tag{46}$$

The trajectory can be extended over an interval of length $\delta_m^1 + \delta_m^2$. Thus

$$z^m(\tau_m^2) \in \widetilde{S}_{t(z_o)-2\varepsilon_m}(\theta)$$

$$\tau_m^2 = \delta_m^1 + \delta_m^2 \ , \quad \delta_m^1 \leqslant \varepsilon_m \ , \quad \delta_m^2 \leqslant \varepsilon_m$$

and from (46) we have

$$t(z^m(\delta_m^1)) \leqslant t(z_o) - \varepsilon_m$$

In general, by repeating these arguments, one can extend the trajectory $z^m(t)$ over $[0, \tau_m^m]$, and we have

$$z^m(\tau_m^i) \in \tilde{S}_{t(z_o)-i\varepsilon_m}(\theta)$$

$$\tau_m^i = \delta_m^1 + \cdots + \delta_m^i \tag{47}$$

$$0 < \delta_m^i \leqslant \varepsilon_m \quad ; \quad i = 0, 1, \ldots, m$$

Now let $m \to +\infty$. We obtain a sequence of trajectories $z^m(t)$, $t \in [0, t_m]$, $t_m = \tau_m^m$. In view of the assumptions A-E of Paragraph 1, one can consider a sub-sequence of trajectories $z^m(t)$ which tend to a trajectory $z^o(t)$ of Eq.(41), emanating from z_o, generated by the control $v(t)$ and some control $u^o(t)$. Without restricting the generality of our arguments, we shall suppose that $z^m(t) \to z^o(t)$ and $t_m \to t_o$.Since $t_m = \tau_m^m \leqslant m\varepsilon_m = \varepsilon_o$, we have

$$t_o \leqslant \varepsilon_o$$

Moreover, from (47) we have

$$z^m(t_m) \in \tilde{S}_{t(z_o)-\varepsilon_o}(\theta)$$

and since $\tilde{S}_t(\theta)$ is closed and $z^m(t_m) \to z^o(t_o)$

$$z^o(t_o) \in \tilde{S}_{t(z_o)-\varepsilon_o}(\theta) \tag{48}$$

Let us prove that $z^o(t) \in N$ for all $t \in [0, t_o]$. With this end in view, let us suppose that the converse is true; that is, that there exists $t' < t_o$ such that

$$z^o(t') \notin N$$

Since N is a closed set, its complement is open and, accordingly, not only $z^o(t') \notin N$ but there exists an open ball with center $z^o(t')$ and radius r whose intersection with N is empty .

Since the points τ_m^i lie in $[0, t_m]$, and the interval between any two successive points does not exceed ε_m , then for a fixed t' there exists a $\tau_m^{i_m}$ such that

$$\left| t' - \tau_m^{i_m} \right| \leqslant \varepsilon_m$$

Consider the point $z^m(\tau_m^{i_m})$. Since the trajectory $z^m(t)$ tends to $z^o(t)$ and

$$\tau_m^{i_m} \to t'$$

then $z^m(\tau_m^{i_m}) \to z^o(t')$. Accordingly $z^m(\tau_m^{i_m}) \notin N$ for m sufficiently large. But from (47) we have

$$z^m(\tau_m^i) \in \tilde{S}_{t(z_o)-i\varepsilon_m}(\theta)$$

which implies that $z^m(\tau_m^{i_m}) \in N$, since

$$\tilde{S}_t(X) \subset N$$

for any t and X, as follows from the definition of $\tilde{S}_t(X)$. Hence we arrive at a contradiction, and accordingly $z^o(t) \in N$, $t \in [0, t_o]$.

Now, let us remark that if $\varepsilon > t(z_o)$ then $\varepsilon_o = t(z_o)$ and, from (48) it follows that

$$z^o(t_o) \in \tilde{S}_o(\theta) = \theta$$

At last let us summarize the results obtained in this paragraph :

If $t(z_o) < +\infty$, then for given z_o , ε and $v(t)$ one can construct a control $u^o(t)$ on $[0,t_o]$ such that the corresponding trajectory $z^o(t)$ either reaches θ, or satisfies condition (48) say

$$z^o(t_o) \in \tilde{S}_{t(z_o)-\varepsilon}(\theta)$$

and accordingly $z^o(t) \in N$, $t \in [0,t_o]$. From then on, one can repeat word for word all the arguments of Paragraph 3, and extend the control $u^o(t)$ over the whole interval $[0,\varepsilon]$. The strategy Γ_P^* thus constructed has all the properties of the one studied in Paragraph 3 and, in addition, the corresponding trajectory does not leave the set N .

Now we can state a basic theorem, namely

Theorem 10. *A game in the presence of state constraints has a value, and this value is* t(z), *where the function* t(z) *is defined by* (45). *In other words*

$$\inf_{\Gamma_P} \sup_{\Gamma_E} I(z, \Gamma_P, \Gamma_E) = \sup_{\Gamma_E} \inf_{\Gamma_P} I(z, \Gamma_P, \Gamma_E)$$

and, accordingly, there exists an optimal strategy for player P.

Moreover, for given $\alpha < t(z)$ *there exists a strategy* Γ_E^α *such that*

$$I(z, \Gamma_P, \Gamma_E^\alpha) \geqslant \alpha$$

The proof of Theorem 10 repeats word for word the arguments given in the proofs of Theorems 3 and 4 of Paragraph 3, with the only difference that the properties of the mapping $\tilde{T}_t$ are to be replaced by the properties of the mapping $\tilde{S}_t$.

Note from the Editor

The translation has been as faithful as is possible within the constraints of approximately good English usage† and in keeping with the differences in the style and in the technical terminologies in English and Russian.

---

† the Editor makes apologies for the unavoidable french flavour of the translation.

References

(1) R. ISAACS, *Differential Games* (Wiley, 1965) Russian Translation in Izd-Vo "MIR", M., 1967

(2) A. FRIEDMAN, On the Definition of Differential Games and the Existence of Value and Saddle Points, J. Differential Equations 7, (1970) 69-71

(3) A. FRIEDMAN, Existence of Value and of Saddle Points for Differential Games of Pursuit and Evasion, J. Differential Equations 7, (1970) 92-110

(4) L.S. PONTRYAGIN, On the Theory of Differential Games, Uspekhi Matem. Nauk 21, N°4 (1966) 219-274

(5) L.S. PONTRYAGIN, On Linear Differential Games, Doklady Akad. Nauk S.S.S.R. 175, N°4 (1967) 764-766

(6) N.N. KRASOVSKII, *Problems on Pursuit Games*, Izd-Vo "Nauka", M., 1970

(7) B.N. PCHENITCHNY, Structure of Differential Games, Doklady Akad. Nauk S.S.S.R. 184, N°2 (1969) 285-287

(8) B.N. PCHENITCHNY, Leçon sur les jeux différentiels, I.R.I.A., Contrôle optimal et jeux différentiels, cahier 4 (mars 1971) 146-226

(9) B.N. PCHENITCHNY and M.I. SAGHAIDAK, On Differential Games of Fixed Duration, Kiberbetika, N°2 (1970) 54-63

(10) A.F. FILIPPOV, On Some Questions about Optimal Control, Vestnik MGU, Seria Matem., Mekh., Astr., N°2 (1969)

(11) G.S. GOODMAN, On a Theorem of Scorza-Dragoni and its application to Optimal Control in *Mathematical Theory of Control*, Academic Press Inc., 1967, 222-233

(12) L.W. NEUSTADT, The Existence of Optimal Control in the Absence of Convexity, J. Math. Anal. Appl. N°7 (1963) 110-117

(13) S. LEFSCHETZ, *Differential Equations Geometric Theory*, Wiley - Interscience, New York, 1963

(14) A.D. IOFFE and V.M. TIKHOMIROV, Duality of Convex Functions and Extremal Problems, Uspekhi Matem. Nauk 23, N°6 (1968) 51-116

FURTHER GEOMETRIC ASPECTS OF DIFFERENTIAL GAMES

A. BLAQUIERE and P. CAUSSIN*

1. State Equations and Strategies

The aim of this chapter is to discuss further geometric aspects of Differential Games (DG) with a view to making the theory applicable to a class of problems in which the *state of the game,* at any instant, is a point in a Banach space. DG with memory and DG with time lag belong to this class.

In the following we shall consider only the case of two players J_P and J_E whose interests are in conflict, or at least different. We shall assume that the state of the game, at any instant t is a point $z = (t,x) \in G \subseteq E$, where G is an open subset of E, $E \triangleq R \times N$, and N is a Banach space.

Let U and V be two Banach spaces, and let F be a mapping from $G \times U \times V$ into E, namely

$$F : G \times U \times V \to E$$
$$(z, \mu, \nu) \mapsto F(z, \mu, \nu)$$

By a *transfer* of the state from initial state $z^i = (t_i, x^i)$ to end state $z^j = (t_j, x^j)$, $t_j > t_i$, we shall mean a mapping $\tilde{z}$

$$\tilde{z} : [t_i, t_j] \to G$$
$$t \longmapsto z = \tilde{z}(t)$$

Such that $\tilde{z}(t_i) = z^i$, and $\tilde{z}(t_j) = z^j$.

* Laboratoire d'Automatique Théorique, Université de Paris 7, France.

For given mappings $\tilde{\mu}$ and $\tilde{\nu}$

$$\tilde{\mu} : [t_i, t_j] \rightarrow U$$
$$t \longmapsto \mu = \tilde{\mu}(t)$$
$$\tilde{\nu} : [t_i, t_j] \rightarrow V$$
$$t \longmapsto \nu = \tilde{\nu}(t)$$

We shall say that a transfer $\tilde{z} : [t_i, t_j] \to G$ is a *solution* of

$$\dot{z} = F(z, \tilde{\mu}(t), \tilde{\nu}(t)) \tag{1}$$

if

(i) $\tilde{z}$ is continuous on $[t_i, t_j]$;

(ii) $\tilde{z}$ is differentiable almost everywhere ; that is, it is differentiable for all $t \in [t_i, t_j]$, except on a subset of $[t_i, t_j]$ at most denumerable ;

(iii) the equality

$$d\tilde{z}(t)/dt = F(\tilde{z}(t), \tilde{\mu}(t), \tilde{\nu}(t))$$

is satisfied almost everywhere.

We shall consider *strategies* for players J_P and J_E to be functions of z, $\varphi : z \mapsto \varphi(z)$ and $\psi : z \mapsto \psi(z)$, respectively, defined on an open subset Z of G .

We shall let

$$\varphi(z) \in K_P \subseteq U$$
$$\psi(z) \in K_E \subseteq V$$

for all $z \in Z$. K_P and K_E are *constraint sets*. We shall assume that they do not depend on the state z .

We shall denote by S_P and S_E the sets of strategies for players J_P and J_E, respectively, and we shall call members (φ,ψ) of $S_P \times S_E$ *strategy pairs* .

Assumption 1. For any $\varphi', \varphi'' \in S_P$ and $\psi', \psi'' \in S_E$, and for any $z^j = (t_j, x^j) \in Z$, the functions $\varphi''' : z \mapsto \varphi'''(z)$ and $\psi''' : z \mapsto \psi'''(z)$, $z \in Z$, defined by

$$\left.\begin{array}{l}\varphi'''(z) = \varphi'(z)\\ \psi'''(z) = \psi'(z)\end{array}\right\} \text{ for all } z = (t,x) \text{ in } Z \text{ such that } t \leqslant t_j$$

$$\left.\begin{array}{l}\varphi'''(z) = \varphi''(z)\\ \psi'''(z) = \psi''(z)\end{array}\right\} \text{ for all } z = (t,x) \text{ in } Z \text{ such that } t > t_j$$

are strategies ; that is $\varphi''' \in S_P$ and $\psi''' \in S_E$.

For a given strategy pair (φ, ψ), the equation that governs the motion of the state is

$$\dot{z} = F(z, \varphi(z), \psi(z)) \tag{2}$$

We shall say that a transfer $\tilde{z} : [t_i, t_j] \to G$ is a solution of equation (2) if

(i)' $\tilde{z}$ is continuous on $[t_i, t_j]$;

(ii)' $\tilde{z}(t) \in Z$ for all $t \in [t_i, t_j)$;

(iii)' $\tilde{z}$ is differentiable almost everywhere ;

(iv)' the equality
$d\tilde{z}(t)/dt = F(\tilde{z}(t), \varphi(\tilde{z}(t)), \psi(\tilde{z}(t)))$
is satisfied almost everywhere .

Assumption 2. F is of class C^1 on $G \times U \times V$.

We shall denote by $F_z(z, \mu, \nu;\cdot)$, $F_\mu(z, \mu, \nu;\cdot)$ and $F_\nu(z, \mu, \nu;\cdot)$ the Frechet derivatives of F with respect to z, μ, and ν , respectively, at point $(z, \mu, \nu) \in G \times U \times V$.

Note that, since $z = (t,x)$, we have

$$F(z, \mu, \nu) = (1, f(z, \mu, \nu))$$

where $f : G \times U \times V \longrightarrow N$

2. Qualitative and Quantitative Games

Roughly speaking, a *qualitative game* (or game of kind) is a game in which player J_P endeavors to "steer" the state of the system to a given subset of G, say to his *target* θ_P. The other player J_E endeavors to "steer" the state to another given subset of G, say θ_E .

In qualitative games we shall say that a player is a *winner* if he has transferred the state to his own target, despite the efforts of his opponent. Of course, if $\theta_P \cap \theta_E \neq \phi$, both players may win .

We may encounter two situations. Either both targets are non-empty, or one of the targets is empty. The latter case can arise in a game of avoidance. For instance, suppose $\theta_P \neq \phi$ and $\theta_E = \phi$; then player J_P endeavors to "steer" the state to θ_P , and J_E seeks to prevent this eventuality .

A *quantitative game* (or game of degree) is a game in which (i) there is given a single target $\theta = \theta_P = \theta_E$; and (ii) there is given a functional which assigns a unique real number to each transfer effected by a strategy pair. Such a functional is called a *performance index*, and the number which it assigns to a transfer is called the value of the functional, or the *cost* .

In zero-sum, two player games, the intention of one of the players is to maximize the cost of transfer from an initial state to the target, while the intention of the other is to minimize it . In such games there is no winner, since the outcome of the game is a compromise between the players .

3. Play, Terminating Play, Playable Strategy Pair

We shall let $\theta \triangleq \theta_P \cup \theta_E$, and we shall assume that θ belongs to the boundary ∂Z of Z, and that it is defined by a single equation $m(z) = 0$, that is

$$z \in \theta \Rightarrow m(z) = 0$$

where the function $m : \Delta \to R$ is defined on a neighborhood Δ of θ in G.

Assumption 3. m is of class C^1, and $m_z(z;\cdot) \neq 0$ on Δ, where $m_z(z;\cdot)$ is the Frechet derivative of m with respect to z, at point $z \in \Delta$.

For any given strategy pair (φ,ψ), the set $\{z : z = \tilde{z}(t),\ t \in [t_i,t_j]\}$ where $\tilde{z}$ is a solution of equation (2) such that $z(t_j) \in Z \cup \theta$, will be called a *play* or a *path in* $Z \cup \theta$ *generated by* (φ,ψ).

A *terminating play* (or a *terminating path*) is a play (or a path) whose end state $z^j = \tilde{z}(t_j)$ belongs to θ. It *terminates* at $t = t_j$.

We shall say that strategy pair (φ,ψ) is *playable at state* $z \in Z$ if it generates a terminating play from z. The set of all strategy pairs which are playable at z will be denoted by $I(z)$. If $\theta_P \neq \phi$ [$\theta_E \neq \phi$], we shall say that strategy pair (φ,ψ) is *strongly playable for* J_P [J_E] *at state* $z \in Z$ if every path generated by (φ,ψ) and emanating from z is a portion of a terminating path ending in $\theta_P-\theta_E$ [$\theta_E-\theta_P$].

If, in a game of avoidance, $\theta_P \neq \phi$ and $\theta_E \neq \phi$, (φ,ψ) will be said to be strongly playable for J_E at state $z \in Z$ if no path generated by (φ,ψ) and emanating from z is a terminating path ending in θ_P.

We shall let $I_P(z)$ and $I_E(z)$ denote the sets of all strongly playable pairs for J_P and J_E, respectively, at state z .

4. Optimal Strategies in a Qualitative Game, Sets of the Game.

In a qualitative game, we shall say that $\varphi^* \in S_P$ is an *optimal strategy for player* J_P *at state* $z \in Z$, if (φ^*, ψ) is strongly playable for J_P at that state for every $\psi \in S_E$; that is

$$(\varphi^*, \psi) \in I_P(z) \qquad \forall\, \psi \in S_E$$

Likewise we shall say that $\psi^* \in S_E$ is an *optimal strategy for player* J_E *at state* $z \in Z$, if (φ, ψ^*) is strongly playable for J_E at that state, for every $\varphi \in S_P$; that is

$$(\varphi, \psi^*) \in I_E(z) \qquad \forall\, \varphi \in S_P$$

Let us define two subsets of Z, namely

$$\mathcal{S}_P \triangleq \{z : \text{there exists an optimal strategy for } J_P \text{ at state } z\}$$

$$\mathcal{S}_E \triangleq \{z : \text{there exists an optimal strategy for } J_E \text{ at state } z\}$$

Sets $\mathcal{S}_P^* \triangleq \mathcal{S}_P \cup (\theta_P - \theta_E)$ and $\mathcal{S}_E^* \triangleq \mathcal{S}_E \cup (\theta_E - \theta_P)$ are called the *sets of the game.*

Various aspects of qualitative games are discussed in Refs. (1) and (2). We shall not discuss such games any further here, but rather turn to quantitative games.

5. Optimality in a Quantitative Game

Now, as we have said before we shall introduce a performance index ; that is, a rule that assigns a unique real number called the cost to every play .

Let f_o be a given mapping from $G \times U \times V$ into R .

Assumption 4. f_o is of class C^1 on $G \times U \times V$.

We shall denote by $f_{oz}(z, \mu, \nu ; \cdot)$, $f_{o\mu}(z, \mu, \nu ; \cdot)$ and $f_{o\nu}(z, \mu, \nu ; \cdot)$ the Frechet derivatives of f_o with respect to z, μ, and ν, respectively, at point $(z, \mu, \nu) \in G \times U \times V$.

Assumption 5. For all $z^i \in Z$, for all $\varphi \in S_P$, for all $\psi \in S_E$, and for all paths π^{ij} in $Z \cup \theta$, emanating from z^i , generated by (φ, ψ), given by $\tilde{z} : t \mapsto \tilde{z} = z(t)$, $t \in [t_i , t_j]$, the integral

$$\int_{t_i}^{t_j} f_o(\tilde{z}(t), \varphi(\tilde{z}(t)), \psi(\tilde{z}(t)))dt$$

is defined .

Then we shall consider an integral performance index such that

$$V(z^i,z^j ; \varphi,\psi,\pi^{ij}) \triangleq \int_{t_i}^{t_j} f_o(\tilde{z}(t), \varphi(\tilde{z}(t)), \psi(\tilde{z}(t)))dt$$

The cost of a terminating play transferring the state from z^i to a state $z^f \in \theta$ along a path π generated by $(\varphi,\psi) \in I(z^i)$ will be denoted by

$$V(z^i, \theta; \varphi,\psi) \triangleq V(z^i, z^f ; \varphi, \psi ,\pi)$$

Since, for given z^i and $(\varphi,\psi) \in I(z^i)$, the terminating path π need not be unique, $V(z^i , \theta ; \varphi , \psi)$ need not be unique.

Furthermore, we let

$$V(z^i, \theta; \varphi,\psi) = 0 \qquad \forall z^i \in \theta , \forall (\varphi,\psi) \in S_P \times S_E$$

We shall say that a strategy pair (φ^*,ψ^*) is *optimal at* *state* z^i if

(i) $(\varphi^*, \psi^*) \in I(z^i)$;

(ii) $V(z^i, \theta; \varphi^*, \psi^*)$ is defined ; if so, it will be termed the *value of the game* at state z^i , and denoted by
$V^*(z^i) \triangleq V(z^i, \theta ; \varphi^*, \psi^*)$

(iii) the *saddle-point condition*
$$V(z^i,\theta;\varphi^*,\psi) \leqslant V(z^i,\theta;\varphi^*,\psi^*) \leqslant V(z^i,\theta;\varphi,\psi^*)$$
is satisfied for all $(\varphi^*,\psi) \in I(z^i)$, for all $(\varphi,\psi^*) \in I(z^i)$, and for all terminating plays generated from z^i by (φ^*,ψ) and (φ,ψ^*) .

Let $I^*(z^i)$ denote the set of all strategy pairs optimal at z^i.

We shall say that (φ^*, ψ^*) is *optimal on* Z if
$$(\varphi^*, \psi^*) \in I^*(z) \qquad \forall\, z \in Z$$

Assumption 6. There exists a strategy pair (φ^*, ψ^*) optimal on Z.

A path in $Z \cup \theta$ generated from initial state $z^i \in Z$ by a strategy pair (φ^*, ψ)[(φ, ψ^*)] will be termed a P-path [E-path]. A path in $Z \cup \theta$ generated from $z^i \in Z$ by (φ^*, ψ^*) will be termed a PE-path .

A terminating PE-path will be termed an *optimal-path.* A terminating P-path [E-path] will be termed a P-*optimal path* [E - *optimal path*] .

We shall use sub-scripts P, E, PE for P-paths, E-paths, and PE-paths, respectively. An optimal path will be also denoted by π^* .

6. Paths in Augmented State Space

Let us introduce a real variable y_o, and consider the *augmented state space* $R \times \mathcal{E}$ of points $\zeta \triangleq (y_o, z)$.

Let us define a *path* $\Pi^{ij}(C)$ *in* $Z \cup \Theta$, where $Z \triangleq R \times \mathcal{Z}$ and $\Theta \triangleq R \times \theta$, by

$$\Pi^{ij}(C) \triangleq \{\zeta^k = (y_o^k, z^k) : y_o^k + V(z^k, z^j; \varphi, \psi, \pi^{kj}) = C, \quad \pi^{kj} \subseteq \pi^{ij}\}$$

where π^{ij} is a path in $\mathcal{Z} \cup \theta$ generated by (φ, ψ) from initial state $z^i \in \mathcal{Z}$, and C is a real constant parameter .

Likewise, let us define a P-*path in* $Z \cup \Theta$ by

$$\Pi_P^{ij}(C) \triangleq \{\zeta^k = (y_o^k, z^k) : y_o^k + V(z^k, z^j; \varphi^*, \psi, \pi^{kj}) = C, \quad \pi^{kj} \subseteq \pi_P^{ij}\}$$

and E-*path in* $Z \cup \Theta$ by

$$\Pi_E^{ij}(C) \triangleq \{\zeta^k = (y_o^k, z^k) : y_o^k + V(z^k, z^j, \varphi, \psi^*, \pi^{kj}) = C, \quad \pi^{kj} \subseteq \pi_E^{ij}\}$$

and a PE-*path in* $Z \cup \Theta$ by

$$\Pi_{PE}^{ij}(C) \triangleq \{\zeta^k = (y_o^k, z^k) : y_o^k + V(z^k, z^j, \varphi^*, \psi^*, \pi^{kj}) = C, \quad \pi^{kj} \subseteq \pi_{PE}^{ij}\}$$

By varying the value of the parameter C one generates a one-parameter family of paths $\{\Pi^{ij}(C)\}[\{\Pi_P^{ij}(C)\}\ \{\Pi_E^{ij}(C)\}$, and $\{\Pi_{PE}^{ij}(C)\}]$ in $Z \cup \Theta$.

P, E and PE-paths in $Z \cup \Theta$ whose end states belong to Θ are termed P-*optimal*, E-*optimal* and *optimal paths in* $Z \cup \Theta$, respectively, denoted by $\Pi_P(C)$, $\Pi_E(C)$ and $\Pi^*(C)$, respectively.

If path $\Pi^{ij}(C)$, for given C, and its projection π^{ij} on $Z \cup \theta$, are represented by $\tilde{\zeta} : t \mapsto \zeta = \tilde{\zeta}(t)$ and $\tilde{z} : t \mapsto z = \tilde{z}(t)$, $t \in [t_i, t_j]$, respectively, we have

$$y_o^j - y_o^k = \int_{t_k}^{t_j} f_o(\tilde{z}(t), \varphi(\tilde{z}(t), \psi(\tilde{z}(t)))dt$$

or

$$\dot{y}_o = f_o(z, \varphi(z), \psi(z)) \tag{3}$$

7. Game and Isovalue Surfaces

Since $V^*(z)$ is defined on $Z \cup \theta$, we can define a *game surface* $\Sigma(C)$ by

$$\Sigma(C) \triangleq \{\zeta = (y_o, z) : y_o + V^*(z) = C\} \tag{4}$$

where C is a real constant parameter.

The intersection of $\Sigma(C)$ with $Z \cup \theta$ will be called an *isovalue surface*,

$$S(C) \triangleq \{z : V^*(z) = C\} \tag{5}$$

Note that $S(C)$ may be empty.

The surface $S(C)$ is the locus of all initial states for which the value of the game is the same, namely C, hence the name isovalue surface.

As the value of parameter C is varied, equations (4) and (5) define one-parameter families of surfaces, namely $\{\Sigma(C)\}$ and $\{S(C)\}$.

For each given C, equation (4) defines a single-sheeted surface in $Z \cup \Theta$; that is, $\Sigma(C)$ is a set of points which are in one-to-one correspondance with the points of $Z \cup \theta$.

Consider now two game surfaces $\Sigma(C_1)$ and $\Sigma(C_2)$, corresponding to two different parameter values C_1 and C_2 respectively . Let y_{o1} and y_{o2} denote the values of y_o on $\Sigma(C_1)$ and $\Sigma(C_2)$ respectively, for the same state z . Then it follows from (4) that

$$y_{o1} - y_{o2} = C_1 - C_2 \quad \forall\, z \in Z \cup \theta$$

Consequently the members of the one-parameter family $\{\Sigma(C)\}$ may be deduced from one another by translation parallel to the y_o-axis. Furthermore these surfaces are ordered along the y_o-axis in the same way as the parameter value C ; that is, the "higher" a surface in the family the greater is the parameter value C . Thus one and only one surface $\Sigma(C)$ passes through a given point ζ in $Z \cup \Theta$.

Now let us introduce some more nomenclature. A given game surface $\Sigma(C)$ separates the set $Z \cup \Theta$ in two disjoint sets ; that is, sets which have no point in common . We shall denote these sets by $A/\Sigma(C)$ ("above" $\Sigma(C)$) and $B/\Sigma(C)$ ("below" $\Sigma(C)$), respectively. For a game surface $\Sigma(C)$ corresponding to parameter value C, we have

$$A/\Sigma(C) \triangleq \{ \zeta : y_o > C - V^*(z) \} \tag{6}$$

$$'\Sigma(C) \triangleq \{ \zeta : y_o < C - V^*(z) \} \tag{7}$$

A point $\zeta \in A/\Sigma(C)$ will be called an A-*point* relative to $\Sigma(C)$, and a point $\zeta \in B/\Sigma(C)$ a B-*point* relative to $\Sigma(C)$.

8. A Fundamental Property of Game Surfaces

A fundamental property of game surfaces is embodied in

Theorem 1. (a) *No point of a* P-*path in* $Z \cup \Theta$, *emanating from* $\zeta^i \in \Sigma(C)$ *is an* A-*point relative to* $\Sigma(C)$.

(b) *No point of an* E-*path in* $Z \cup \Theta$, *emanating from* $\zeta^i \in \Sigma(C)$, *is a* B-*point relative to* $\Sigma(C)$

(c) *All points of a* PE-*path in* $Z \cup \Theta$, *emanating from* $\zeta^i \in \Sigma(C)$, *belong to* $\Sigma(C)$.

Consider a P-path in $Z \cup \Theta$, generated by (φ^*, ψ), emanating from $\zeta^i = (y_o^i, z^i) \in \Sigma(C)$ and let $\zeta^k = (y_o^k, z^k)$ be any one of its points.

Let π_P^{ij} be the corresponding path in $Z \cup \theta$, and let $\pi_P^{ik} \subset \pi_P^{ij}$ be the portion of π_P^{ij} ending at z^k.

Since $z^k \in Z \cup \theta$, there exists a path π_{PE}^{kf} emanating from z^k, generated by (φ^*, ψ^*), that reaches θ at point z^f.

In view of Assumption 1, there exists a strategy pair (φ^*, ψ') that generates the path $\pi_P^{if} = \pi_P^{ik} \cup \pi_{PE}^{kf}$. The associated cost is

$$V(z^i, \theta ; \varphi^*, \psi') = V(z^i, z^f ; \varphi^*, \psi', \pi_P^{if}) \tag{8}$$

In view of the saddle-point condition we have

$$V(z^i, \theta ; \varphi^*, \psi') \leqslant V(z^i, \theta ; \varphi^*, \psi^*) \triangleq V^*(z^i) \tag{9}$$

From the additivity property of cost we have

$$\begin{aligned} &V(z^i, z^f ; \varphi^*, \psi', \pi_P^{if}) \\ &= V(z^i, z^k ; \varphi^*, \psi, \pi_P^{ik}) + V(z^k, z^f ; \varphi^*, \psi^*, \pi_{PE}^{kf}) \end{aligned} \tag{10}$$

with

(11) $$V(z^k, z^f ; \varphi^*, \psi^*, \pi_{PE}^{kf}) \triangleq V^*(z^k)$$

Relations (8)-(11) result in

(12) $$V(z^i, z^k : \varphi^*, \psi, \pi_P^{ik}) \leqslant V^*(z^i) - V^*(z^k)$$

From the definition of a P-path in $Z \cup \Theta$, it follows that

$$y_o^k - y_o^i = V(z^i, z^k ; \varphi^*, \psi, \pi_P^{ik})$$

and hence, by (12) we have

(13) $$y_o^k - y_o^i \leqslant V^*(z^i) - V^*(z^k)$$

At last, it follows from relation (13) together with the definition of $A/\Sigma(C)$ that ζ^k is not an A-point relative to $\Sigma(C)$. Hence, condition (a) of Theorem 1 is established. Condition (b) can be proved in analogous fashion, and condition (c) is a direct consequence of (a) and (b) .

9. Decomposition of a Game Surface

We shall say that $\{Z_1, Z_2, \dots Z_K\}$ is a *décomposition of* Z if

(i) Z_σ , $\sigma = 1,2,\dots K < \infty$ is a domain in E

(ii) $Z_k \cap Z_\ell = \emptyset$, $k \neq \ell$

(iii) $Z_\sigma \subseteq Z$, $\sigma = 1,2,\dots K$

(iv) $Z \subset \bigcup_{\sigma=1}^{K} \overline{Z}_\sigma$

From (i)-(iv) it follows that $\overline{Z} = \bigcup_{\sigma=1}^{K} \overline{Z}_\sigma$. Now let

$$\Sigma_\sigma(C) \triangleq \{\zeta : y_o + V^*(z) = C , z \in Z_\sigma \} ,$$

$$\sigma = 1,2,\dots K$$

$\{\Sigma_1(C), \Sigma_2(C), \ldots \Sigma_K(C)\}$ is a *decomposition of* $\Sigma(C)$ associated with the above decomposition of Z.

Assumption 7. There exists a decomposition $\{Z_1, Z_2, \ldots Z_K\}$ of Z such that

(a) φ^* and ψ^* agree on each Z_σ , $\sigma = 1,2,\ldots K$, with some functions φ^σ and ψ^σ , which are of class C^1 on some domain $R_\sigma \supset \overline{Z}_\sigma$;

(b) on each non-empty $M_{k\ell} \triangleq \overline{Z}_k \cap \overline{Z}_\ell$, $k \neq \ell$, φ^* and ψ^* agree with some functions $\varphi^{k\ell}$ and $\psi^{k\ell}$, which are of class C^1 on some domain $R_{k\ell} \supset M_{k\ell}$.

10. Regular Points of a Game Surface

We shall say that $\zeta = (y_0,z) \in \Sigma(C)$ is a *regular point* of $\Sigma(C)$ if there exists an open ball $B(\sigma,z) \subset Z$ of radius σ, $\sigma > 0$, and center z, such that, for all $z' \in B(\sigma,z)$, V^* has a Frechet derivative

$$V^*_z(z' \; ; \; \cdot \;) : \; E \to R$$

Let A and B be two Banach spaces and let $L(A ,B)$ be the set of all linear continuous mappings from A into B. We shall define the *dot product* L.h, where $L \in L(A, B)$ and $h \in A$, by

$$L.h \triangleq L(h) \in B$$

One can easily verify that this dot product has similarities with a scalar product in a vector space, but it is not commutative.

If L and h are time dependent, and if their time derivatives $\dot{L}$ and $\dot{h}$ are defined at time t, the time derivative of L.h at time t is $\dot{L}.h + L.\dot{h}$.

In what follows we shall consider mappings from $R \times \mathcal{E}$ into R, of the form (A,B), where $A \in \mathcal{L}(R,R)$ and $B \in \mathcal{L}(\mathcal{E},R)$.

For $q \triangleq (h_o,h) \in R \times \mathcal{E}$ we shall let

$$(A,B).q \triangleq A(h_o) + B(h) \in R$$

In particular we shall define the *gradient* of $\Phi(\zeta) \triangleq y_o + V^*(z)$, at a regular point ζ of $\Sigma(C)$, by

$$\text{grad } \Phi(\zeta) \triangleq (1, V_z^*(z;\cdot))$$

For $q \triangleq (h_o,h) \in R \times \mathcal{E}$ we shall let

$$\text{grad } \Phi(\zeta)\cdot q \triangleq h_o + V_z^*(z;h)$$

11. Variational Equations and their Adjoints

For given initial point $\zeta^i = (y_o^i, z^i) \in R_\sigma$, $R_\sigma \triangleq R \times \mathcal{R}_\sigma$, let $\tilde{z} : [t_i, t_j] \to \mathcal{R}_\sigma$ be a solution of equation

$$(14) \qquad \dot{z} = \overline{F}(z), \qquad \overline{F}(z) \triangleq F(z,\varphi^\sigma(z),\psi^\sigma(z))$$

such that $\tilde{z}(t_i) = z^i$; and let

$$\tilde{y}_o(t) = y_o^i + \int_{t_i}^{t} f_o(\tilde{z}(s),\varphi^\sigma(\tilde{z}(s)),\psi^\sigma(\tilde{z}(s)))ds$$

that is

$$(15) \qquad \dot{y}_o = \overline{f}_o(z), \qquad \overline{f}_o(z) \triangleq f_o(z,\varphi^\sigma(z),\psi^\sigma(z))$$

We shall associate with equations (14) and (15) the *variational equations*

$$(16) \qquad \dot{h} = \overline{F}_z(z;h)$$

$$(17) \qquad \dot{h}_o = \overline{f}_{oz}(z;h)$$

where $\overline{F}_z(z;\cdot)$ and $\overline{f}_{oz}(z;\cdot)$ are the Frechet derivatives of $\overline{F}$ and $\overline{f}_o$, respectively .

These equations can be rewritten as

$$\dot{q} = M_z\, q \tag{18}$$

where we use for q the new definition

$$q \triangleq \begin{bmatrix} h_o \\ h \end{bmatrix}$$

and where

$$M_z \triangleq \begin{bmatrix} 0 & \overline{f}_{oz}(z;\cdot) \\ 0 & \overline{F}_z\,(z;\cdot) \end{bmatrix}$$

and

$$M_z q \triangleq \begin{bmatrix} \overline{f}_{oz}(z;h) \\ \overline{F}_z\,(z;h) \end{bmatrix}$$

For given initial condition $q = q^i$ at $t = t_i$, the solution $\tilde{q} : t \mapsto q = \tilde{q}(t)$, $t \in [t_i, t_j]$ of equation (18) is unique and continuous. Furthermore $\tilde{q}(t)$, $t \in [t_i, t_j]$, is nonzero provided q^i is nonzero.

The equations adjoint to variational equations (16)(17) are

$$\dot{\lambda}_o = 0 \tag{19}$$

$$\dot{\lambda} = -\,(\lambda_o\, \overline{f}_{oz}(z;\cdot) + \lambda(\overline{F}_z(z;\cdot))) \tag{20}$$

where $\Lambda \triangleq (\lambda_o, \lambda)$ is a mapping from $R \times E$ into R.

These equations can be rewritten as

$$\dot{\Lambda} = -\,\Lambda\, M_z \tag{21}$$

where we use for Λ the new definition

$$\Lambda \triangleq (\lambda_o\ \lambda)$$

and where

$$\Lambda M_z \triangleq (0 \qquad \lambda_o \overline{f}_{oz}(z;\cdot) + \lambda(\overline{F}_z(z;\cdot)))$$

For given initial condition $\Lambda = \Lambda^i$ at $t = t_i$, the solution $\tilde{\Lambda} : t \mapsto \Lambda = \tilde{\Lambda}(t)$, $t \in [t_i, t_j]$, of equation (21) is unique and continuous. Furthermore, $\tilde{\Lambda}(t)$ is nonzero provided Λ^i is nonzero.

It follows from (18) and (21) that

$$\frac{d}{d\,t}(\tilde{\Lambda}(t)\cdot\tilde{q}(t)) = 0$$

so that

$$\tilde{\Lambda}(t)\cdot\tilde{q}(t) = \text{constant} \qquad \forall t \in [t_i, t_j] \tag{22}$$

If the initial vectors $\Lambda^i = \tilde{\Lambda}(t_i)$ and $q^i = \tilde{q}(t_i)$ are such that

$$\Lambda^i \cdot q^i = 0$$

then it follows from (22) that

$$\tilde{\Lambda}(t)\cdot\tilde{q}(t) = 0 \qquad \forall t \in [t_i, t_j] \tag{23}$$

We shall say that $\tilde{\Lambda}(t)$ is *perpendicular to* $\tilde{q}(t)$ for all $t \in [t_i, t_j]$.

Equation (18) defines a linear transformation $T(t_i, t)$ of q^i such that

$$\tilde{q}(t) = T(t_i, t)q^i$$

This transformation is nonsingular ; that is, the inverse transformation $T^{-1}(t_i, t)$ such that

$$q^i = T^{-1}(t_i, t)\,\tilde{q}(t)$$

is defined for all $t \in [t_i, t_j]$. We shall let

$$T(t, t_i) \triangleq T^{-1}(t_i, t)$$

12. Contingent of a Set

Let Q be a non empty subset of $R \times E$. We shall say that q, $q \in R \times E$, is *tangent* to Q at point $\zeta \in \overline{Q}$, if the following condition is fulfilled :

There exists an infinite sequence in $R \times E$ $\{q^\nu : \nu = 1,2,\ldots$ and $\|q^\nu - q\| \to 0$ as $\nu \to \infty\}$ and an infinite sequence of strictly positive numbers $\{\varepsilon^\nu : \nu = 1,2,\ldots$ and $\varepsilon^\nu \to 0$ as $\nu \to \infty\}$ such that

$$\zeta + \varepsilon^\nu q^\nu \in Q \qquad \nu = 1,2,\ldots$$

The *contingent* $C(Q,\zeta)$ of Q at point ζ is defined by $C(Q,\zeta) \triangleq \{\zeta + q : q \text{ is tangent to } Q \text{ at point } \zeta\}$.

Properties of the contigent of Q at a point $\zeta \in \partial Q$ will be called *local properties* of ∂Q.

Lemma 1. *If* $Q' \subseteq Q'' \subseteq R \times E$ *and* $\zeta \in \overline{Q}'$, *then* $C(Q',\zeta) \subseteq C(Q'',\zeta)$.

Lemma 1 follows directly from the definition of a contingent.

13. Contingent of a Game Surface at a Regular Point

Let $\zeta = (y_o, z) \in \Sigma(C)$ be a regular point of $\Sigma(C)$.

Let $q = (h_o, h)$ be tangent to $\Sigma(C)$ at point ζ. Then, there exists an infinite sequence in $R \times E$
$\{q^\nu = (h_o^\nu, h^\nu) : \nu = 1,2,\ldots$ and $\|q^\nu - q\| \to 0$ as $\nu \to \infty\}$
and an infinite sequence of strictly positive numbers
$\{\varepsilon^\nu : \nu = 1,2,\ldots$ and $\varepsilon^\nu \to 0$ as $\nu \to \infty\}$
such that

$$y_o + \varepsilon^\nu h_o^\nu + V^*(z + \varepsilon^\nu h^\nu) = C, \qquad \nu = 1, 2, \ldots$$

It follows that

$$y_o + \varepsilon^\nu h_o^\nu + V^*(z) + V_z^*(z; \varepsilon^\nu h^\nu) + o(\varepsilon^\nu) = C, \qquad \nu = 1, 2,\ldots$$

where $\lim_{\nu \to \infty} \dfrac{|o(\varepsilon^\nu)|}{\varepsilon^\nu} = 0$

Since $y_o + V^*(z) = C$, we have

$$\varepsilon^\nu h_o^\nu + V_z^*(z ; \varepsilon^\nu h^\nu) + o(\varepsilon^\nu) = 0, \qquad \nu = 1, 2, \ldots$$

Dividing by ε^{ν}, and taking account of the linearity of $V_z^*(z;\cdot)$, we obtain

$$h_o^{\nu} + V_z^*(z\ ;\ h^{\nu}) + \frac{o(\varepsilon^{\nu})}{\varepsilon^{\nu}} = 0, \qquad \nu = 1, 2, \ldots$$

Then, letting $\nu \to \infty$, and taking account of the continuity of $V_z^*(z;\cdot)$, we obtain

$$h_o + V_z^*(z;\ h) = 0$$

This condition can be rewritten

$$\text{(24)} \qquad \text{grad}\ \Phi(\zeta)\cdot q = 0$$

Conversely, let $q = (h_o, h)$ be such that $h_o + V_z^*(z;h) = 0$, and let $\{\ \varepsilon^{\nu} : \nu = 1, 2, \ldots\ \}$ be some infinite sequence of strictly positive numbers such that $\varepsilon^{\nu} \to 0$ as $\nu \to \infty$.

Since $V_z^*(z;\cdot)$ is linear, we have

$$\varepsilon^{\nu} h_o + V_z^*(z;\ \varepsilon^{\nu} h) = 0 \qquad \nu = 1, 2, \ldots$$

and since $y_o + V^*(z) = C$, we have

$$y_o + \varepsilon^{\nu} h_o + V^*(z) + V_z^*(z;\ \varepsilon^{\nu} h) = C\ , \qquad \nu = 1, 2, \ldots$$

It follows that

$$y_o + \varepsilon^{\nu} h_o + V^*(z + \varepsilon^{\nu} h) + o(\varepsilon^{\nu}) = C\ , \qquad \nu = 1, 2, \ldots$$

where $\lim\limits_{\nu \to \infty} \dfrac{|o(\varepsilon^{\nu})|}{\varepsilon^{\nu}} = 0$

This condition can be rewritten

$$y_o + \varepsilon^{\nu}\left(h_o + \frac{o(\varepsilon^{\nu})}{\varepsilon^{\nu}}\right) + V^*(z + \varepsilon^{\nu} h) = C\ , \qquad \nu = 1, 2, \ldots$$

Accordingly, there exists an infinite sequence in $R \times E$, namely

$\{ q^\nu = (h_o + \frac{o(\varepsilon^\nu)}{\varepsilon^\nu}, h) : \nu = 1, 2, \ldots\}$ such that $\| q^\nu - q \| \to 0$ as $\nu \to \infty$, and

$$\zeta + \varepsilon^\nu q^\nu \in \Sigma(C)$$

It follows that q is tangent to $\Sigma(C)$ at point ζ. Hence

Lemma 2. *The contingent of* $\Sigma(C)$ *at a regular point* $\zeta \in \Sigma(C)$ *is the set* $P(\zeta) \triangleq \{\zeta + q : \text{grad}\, \Phi(\zeta) \cdot q = 0\}$

14. Transformation of a Contingent

Let $\tilde{\zeta}^\sigma : [t_i, t_j] \to R_\sigma$ be a solution of the system (14)(15). Let Δ^i be a subset of R_σ such that $\tilde{\zeta}^\sigma(t_i) \in \bar{\Delta}^i$, and $\| \zeta - \tilde{\zeta}^\sigma(t_i) \| < \varepsilon$ for all $\zeta \in \Delta^i$.

It follows from (a) of Assumption 7 that, to each initial point $\zeta' \in \Delta^i$, there corresponds a unique solution $\tilde{\zeta} : t \mapsto \tilde{\zeta}(t)$, $t \in [t', t'']$ of the system (14)(15), such that $\tilde{\zeta}(t') = \tilde{\zeta}'$ where t' is the time coordinate of ζ', and $t'' - t' = t_j - t_i$, provided that ε is sufficiently small.

Suppose that ε is sufficiently small, and let $\Delta^j \subset R_\sigma$ be the set of all transforms $\tilde{\zeta}(t'')$, for all initial points $\zeta' \in \Delta^i$.

We shall denote by $T(t_i, t_j)$ the transformation thus defined, and we shall let

$$\Delta^j = T(t_i, t_j)\Delta^i$$

Let

$$\tilde{\zeta}^\sigma(t_i) + q^i \in C(\Delta^i, \tilde{\zeta}^\sigma(t_i))$$

Then, there exists q^ν and $\varepsilon^\nu (\varepsilon^\nu > 0)$, $\nu = 1,2,\ldots$ such that $\tilde{\zeta}^\sigma(t_i) + \varepsilon^\nu q^\nu \in \Delta^i$ for $\nu = 1,2,\ldots$, and $q^\nu \to q^i$ and $\varepsilon^\nu \to 0$ as $\nu \to \infty$.

Let $\zeta' = \tilde{\zeta}^\sigma(t_i) + \varepsilon^\nu q^\nu$. Then, from the dependence on initial conditions of the solution $\tilde{\zeta}$, it follows that

$$\tilde{\zeta}(t'') = \tilde{\zeta}^{\sigma}(t_j) + \varepsilon^{\nu} T(t_i, t_j) q^{\nu} + o(\varepsilon^{\nu})$$

where $\lim\limits_{\nu \to \infty} \dfrac{|o(\varepsilon^{\nu})|}{\varepsilon^{\nu}} = 0$; and furthermore

$$\tilde{\zeta}^{\sigma}(t_j) + \varepsilon^{\nu}\left(T(t_i, t_j) q^{\nu} + \frac{o(\varepsilon^{\nu})}{\varepsilon^{\nu}}\right) \in \Delta^j$$

It follows that $\tilde{\zeta}^{\sigma}(t_j) \in \overline{\Delta}^j$ and since

$$T(t_i, t_j) q^{\nu} + \frac{o(\varepsilon^{\nu})}{\varepsilon^{\nu}} \to q^j = T(t_i, t_j) q^i \quad \text{as } \nu \to \infty$$

we conclude that

$$\tilde{\zeta}^{\sigma}(t_j) + q^j \in C(\Delta^j, \tilde{\zeta}^{\sigma}(t_j))$$

By similar arguments, based on the fact that $T(t_i, t_j)$ is non singular, one can prove easily that

$$\tilde{\zeta}^{\sigma}(t_j) + q^j \in C(\Delta^j, \tilde{\zeta}^{\sigma}(t_j)) \Rightarrow$$

$$\tilde{\zeta}^{\sigma}(t_i) + q^i \in C(\Delta^i, \tilde{\zeta}^{\sigma}(t_i))$$

These results are embedded in

Lemma 3. *If* $\tilde{\zeta}^{\sigma} : [t_i, t_j] \to R_{\sigma}$ *is a solution of the system* (14) (15) ; *and if* Δ^i *is a subset of* R_{σ} *such that* $\tilde{\zeta}^{\sigma}(t_i) \in \overline{\Delta}^i$, *and* $\|\zeta - \tilde{\zeta}^{\sigma}(t_i)\| < \varepsilon$ *for all* $\zeta \in \Delta^i$; *and if* ε *is sufficiently small so that transform* $\Delta^j = T(t_i, t_j)\Delta^i$ *is defined, then*

$$\tilde{\zeta}^{\sigma}(t_i) + q^i \in C(\Delta^i, \tilde{\zeta}^{\sigma}(t_i)) \Leftrightarrow \tilde{\zeta}^{\sigma}(t_i) + q^j \in C(\Delta^j, \tilde{\zeta}^{\sigma}(t_j))$$

where $q^j = T(t_i, t_j) q^i$.

Corollary 1. *If* $\tilde{\zeta}^{\sigma}(t_i)$ *is a regular point of* $\Sigma_{\sigma}(C)$, *and* $\Delta^i \triangleq \{ \zeta : \zeta \in \Sigma_{\sigma}(C), \|\zeta - \tilde{\zeta}^{\sigma}(t_i)\| < \varepsilon \}$, *then for* ε *sufficiently small* $\Delta^j = T(t_i, t_j)\Delta^i$ *is defined and*

$$C(\Delta^j, \tilde{\zeta}^{\sigma}(t_j)) = \{\tilde{\zeta}^{\sigma}(t_j) + q : \tilde{\Lambda}(t_j) \cdot q = 0 \}$$

where $\tilde{\Lambda} : t \mapsto \tilde{\Lambda}(t)$ *is the solution of adjoint equation* (21) *defined on* $[t_i, t_j]$, *for initial condition* $\tilde{\Lambda}(t_i) = \text{grad } \Phi(\tilde{\zeta}^{\sigma}(t_i))$.

Corollary 1 is a direct consequence of the definition of a contingent, Lemma 2, relation (22) and Lemma 3 .

15. A Relation Between Adjoint and Gradient Vectors

Now suppose that there exists a restriction† of function $\tilde{\zeta}^{\sigma}$, namely $\zeta^* : [t_\alpha, t_\beta] \to Z_\sigma$, where $[t_\alpha, t_\beta] \subseteq [t_i, t_j]$ and $Z_\sigma \triangleq R \times Z_\sigma$.

From (a) of Assumption 7, together with the definition of $\tilde{\zeta}^{\sigma}$, it follows that the set of points $\{ \zeta : \zeta = \zeta^*(t), t \in [t_\alpha, t_\beta] \}$ is a portion of an optimal path $\Pi^*(C) \subset \Sigma(C)$. We shall say that it is a *regular portion* of $\Pi^*(C)$ if all of its points are regular points of $\Sigma(C)$.

Now suppose that this set is a regular portion of $\Pi^*(C)$ .Let us denote by $\pi^{\alpha\beta}$ its projection on Z_σ , and let

$$\Delta^\alpha \triangleq \{ \zeta : \zeta \in \Sigma_\sigma(C), \ \| \zeta - \zeta^*(t_\alpha) \| < \varepsilon \} .$$

Suppose that ε is sufficiently small so that $\Delta^\beta = T(t_\alpha, t_\beta)\Delta^\alpha$ is defined .

From the definition of Δ^β, and from the fact that $\overline{F}$ and $\overline{f}_0$ are C^1 functions of z on some neighborhood of $\pi^{\alpha\beta}$ (namely,on R_σ), and from (c) of Theorem 1, it follows that there exists $\varepsilon' > 0$ such that

$$B(\zeta^*(t_\beta), \varepsilon') \cap \Sigma_\sigma(C) = B(\zeta^*(t_\beta), \varepsilon') \cap \Delta^\beta$$

where

$$B(\zeta^*(t_\beta), \varepsilon') \triangleq \{ \zeta : \| \zeta - \zeta^*(t_\beta) \| < \varepsilon' \}$$

Then, from the definition of a contingent we have

$$C(\Delta^\beta, \zeta^*(t_\beta)) = C(\Sigma(C), \zeta^*(t_\beta)) \tag{25}$$

---

† that is, $\zeta^*(t) = \tilde{\zeta}^{\sigma}(t), \ \forall t \in [t_\alpha, t_\beta]$

From (25) and Corollary 1 we have

$$(26) \qquad C(\Sigma(C), \zeta^*(t_\beta)) = \{ \zeta^*(t_\beta) + q : \tilde{\Lambda}(t_\beta) \cdot q = 0 \}$$

where $\tilde{\Lambda} : t \mapsto \tilde{\Lambda}(t)$ is the solution of adjoint equation (21) defined on $[t_\alpha, t_\beta]$, for initial condition $\tilde{\Lambda}(t_\alpha) = \text{grad } \Phi(\zeta^*(t_\alpha))$.

From Lemma 2, and from the fact that $\zeta^*(t_\beta)$ is a regular point of $\Sigma(C)$, we have

$$(27) \qquad C(\Sigma(C), \zeta^*(t_\beta)) = \{\zeta^*(t_\beta)+q : \text{grad } \Phi(\zeta^*(t_\beta))\cdot q = 0 \}$$

At last, from (26) and (27) we deduce (Appendix B)

$$(28) \qquad \tilde{\Lambda}(t_\beta) = k \text{ grad } \Phi(\zeta^*(t_\beta))$$

where $k \in R$.

Let $\tilde{\Lambda}(t) = (\tilde{\lambda}_o(t), \tilde{\lambda}(t))$, $t \in [t_\alpha, t_\beta]$.

From the initial condition at time $t = t_\alpha$ we have $\tilde{\lambda}_o(t_\alpha) = 1$, and from condition (28) we have $\tilde{\lambda}_o(t_\beta) = k$.

Since, from equation (19), $\tilde{\lambda}_o(t)$ is constant for all $t \in [t_\alpha, t_\beta]$, we have $k = 1$, and hence

$$\tilde{\Lambda}(t_\beta) = \text{grad } \Phi(\zeta^*(t_\beta))$$

Since the above arguments hold for any subinterval $[t_\alpha, t_\beta] \subseteq [t_\alpha, t_\beta]$, we see that

$$(29) \qquad \tilde{\Lambda}(t) = \text{grad } \Phi(\zeta^*(t)), \quad \forall t \in [t_\alpha, t_\beta]$$

In other words, the transfer of grad $\Phi(\zeta^*(t))$ along a regular portion of an optimal path is governed by the adjoint equation (21).

16. Regular Portion of an Optimal Path

Again consider a regular portion of an optimal path, given by $\zeta^* : [t_\alpha, t_\beta] \to Z_\sigma$, and point $\zeta = (y_o, z) = \zeta^*(t)$, for some

$t \in [t_\alpha, t_\beta]$. Consider also P- and E- optimal paths generated by (φ^*, ψ) and (φ, ψ^*), respectively, through ζ.

Let $\zeta + \delta\zeta = (y_o + \delta y_o, z + \delta z) \in Z_\sigma$. We have

$$\Phi(\zeta + \delta\zeta) = y_o + \delta y_o + V^*(z + \delta z)$$

$$= \Phi(\zeta) + \delta y_o + V^*_z(z; \delta z) + \varepsilon \, \|\delta z\|$$

where $\varepsilon \to 0$ as $\|\delta z\| \to 0$.

It follows directly from Theorem 1 and from the linearity of $V^*_z(z\, ;\, \cdot)$ that

$$(30) \qquad f_o(z,\varphi^*(z),\psi(z)) + V^*_z(z;\, F(z,\varphi^*(z),\psi(z))) \leqslant 0$$

$$(31) \qquad f_o(z,\varphi(z),\psi^*(z)) + V^*_z(z;\, F(z,\varphi(z),\psi^*(z))) \geqslant 0$$

$$(32) \qquad f_o(z,\varphi^*(z),\psi^*(z)) + V^*_z(z;\, F(z,\varphi^*(z),\psi^*(z))) = 0$$

Clearly (30) and (31) hold for all P- and E- optimal paths through ζ.

These relations may be rewritten

$$(30)' \qquad \text{grad } \Phi(\zeta)\cdot(f_o(z,\varphi^*(z),\psi(z)) \,,\, F(z,\varphi^*(z),\psi(z))) \leqslant 0$$

$$(31)' \qquad \text{grad } \Phi(\zeta)\cdot(f_o(z,\varphi(z),\psi^*(z)) \,,\, F(z,\varphi(z),\psi^*(z))) \geqslant 0$$

$$(32)' \qquad \text{grad } \Phi(\zeta)\cdot(f_o(z,\varphi^*(z),\psi^*(z)),\, F(z,\varphi^*(z),\psi^*(z))) = 0$$

We shall add to these relations the transfer equation of grad $\Phi(\zeta)$ along the portion of optimal path, obtained in the earlier paragraphs.

Let

$$H(\lambda,z,\mu,\nu) \triangleq f_o(z,\mu,\nu) + \lambda(F(z,\mu,\nu))$$

Then relations (30)-(32) become :

(30)" $H(\tilde{\lambda}(t), z^*(t), \varphi^*(z^*(t)), \psi(z^*(t))) \leqslant 0$

(31)" $H(\tilde{\lambda}(t), z^*(t), \varphi(z^*(t)), \psi^*(z^*(t))) \geqslant 0$

(32)" $H(\tilde{\lambda}(t), z^*(t), \varphi^*(z^*(t)), \psi^*(z^*(t))) = 0$

for all strategy pairs (φ^*, ψ) and (φ, ψ^*), where $z^*(t)$ is the projection of $\zeta^*(t)$ on Z_σ.

Relations (30) - (32) can also be written

(30)"' $\underset{\nu \in K_E}{\text{Max}} \; H(\tilde{\lambda}(t), z^*(t), \varphi^*(z^*(t)), \nu) = 0$

(31)"' $\underset{\mu \in K_P}{\text{Min}} \; H(\tilde{\lambda}(t), z^*(t), \mu, \psi^*(z^*(t))) = 0$

(32)"' $H(\tilde{\lambda}(t), z^*(t), \varphi^*(z^*(t)), \psi^*(z^*(t))) = 0$

Since $z \triangleq z^*(t)$ belongs to Z_σ, $\varphi^*_z(z;\cdot)$ and $\psi^*_z(z;\cdot)$ are defined, and we have

$$\varphi^*(z+\varepsilon h) = \varphi^*(z) + \varepsilon\varphi^*_z(z;h) + o_1(\varepsilon) \in K_P$$

$$\psi^*(z+\varepsilon h) = \psi^*(z) + \varepsilon\psi^*_z(z;h) + o_2(\varepsilon) \in K_E$$

where $h \in E$, and $|\varepsilon|$ ($\varepsilon > 0$ or $\varepsilon < 0$) is sufficiently small . Indeed

$$\lim_{|\varepsilon| \to 0} \frac{|o_1(\varepsilon)|}{|\varepsilon|} = \lim_{|\varepsilon| \to 0} \frac{|o_2(\varepsilon)|}{|\varepsilon|} = 0$$

Then, from relation (31), since K_P and K_E do not depend on the state, we have

$$\begin{aligned} &f_o(z, \varphi^*(z) + \varepsilon\varphi^*_z(z;h) + o_1(\varepsilon), \psi^*(z)) \\ &+ V^*_z(z; F(z, \varphi^*(z) + \varepsilon\varphi^*_z(z;h) + o_1(\varepsilon), \psi^*(z))) \geqslant 0 \end{aligned} \tag{33}$$

Since f_o and F are C^1 on $G \times U \times V$, relation (33) may be rewritten :

$$f_o(z,\varphi^*(z),\psi^*(z)) + V_z^*(z;F(z,\varphi^*(z),\psi^*(z)))$$

$$+ f_{o\mu}(z,\varphi^*(z),\psi^*(z);\ \varepsilon\varphi_z^*(z;h) + o_1(\varepsilon))$$

$$+ V_z^*(z;F_\mu(z,\varphi^*(z),\psi^*(z)\ ;\ \varepsilon\varphi_z^*(z;h) + o_1(\varepsilon)))$$

$$+ o(\varepsilon) \geqslant 0$$

where $\lim\limits_{|\varepsilon| \to 0} \dfrac{|o(\varepsilon)|}{|\varepsilon|} = 0$

Taking account of (32), then dividing by ε, for $\varepsilon > 0$ and for $\varepsilon < 0$, and letting $|\varepsilon|$ tend to zero, we obtain

$$f_{o\mu}(z,\varphi^*(z),\psi^*(z)\ ;\ \varphi_z^*(z;h))$$

$$+ V_z^*(z;F_\mu(z,\varphi^*(z),\psi^*(z)\ ;\ \varphi_z^*(z;h))) = 0$$

and since this relation holds for all $h \in E$, we have

$$f_{o\mu}(z,\varphi^*(z),\psi^*(z)\ ;\ \varphi_z^*(z;\cdot))$$

$$+ V_z^*(z;F_\mu(z,\varphi^*(z),\psi^*(z)\ ;\ \varphi_z^*(z;\cdot))) = 0$$

By similar arguments, one can deduce from (30) and (32) that

$$f_{o\nu}(z,\varphi^*(z),\psi^*(z)\ ;\ \psi_z^*(z;\cdot))$$

$$+ V_z^*(z;F_\nu(z,\varphi^*(z),\psi^*(z)\ ;\ \psi_z^*(z;\cdot))) = 0$$

It follows that

$$\lambda_o\bar{f}_{oz}(z;\cdot) + \lambda(\bar{F}_z(z;\cdot)) = \lambda_o f_{oz}(z;\cdot) + \lambda(F_z(z;\cdot))$$

where $\lambda_o = 1$ and $\lambda = V_z^*(z;\cdot)$.

Accordingly, the adjoint equations (19) and (20) become

(19)' $\quad \dot{\lambda}_o = 0$

(20)' $\quad \dot{\lambda} = -(\lambda_o f_{oz}(z;\cdot) + \lambda(F_z(z;\cdot)))$

17. Transversality Condition

We shall now investigate conditions which must hold at the point where an optimal path $\Pi^*(C)$ reaches Θ .

The intersection of the game surface $\Sigma(C)$ with Θ is the set

$$\Theta \cap \Sigma(C) = \{ \zeta : \Phi(\zeta) = C , \quad z \in \theta \} \tag{34}$$

Since $z \in \theta$ implies that $V^*(z) = 0$ which in turn implies that

$$\Phi(\zeta) \triangleq y_o + V^*(z) = y_o$$

We have

$$\Theta \cap \Sigma(C) = \{ \zeta : \; y_o = C , \quad z \in \theta \} \tag{35}$$

That is, the intersection of $\Sigma(C)$ with Θ may be deduced from θ by translation parallel to the y_o-axis .

We shall restrict the discussion to the case of an optimal path π^* that reaches target θ from Z_σ ; in other words, we shall assume that the terminal point $z^*(t_f)$ of π^* belongs to θ and that there exists $\alpha > 0$ such that $z^*(t) \notin \theta$ for all $t \in [t_f-\alpha, t_f)$.

The terminal point $\zeta^*(t_f)$ of $\Pi^*(C)$ belongs to $\Sigma(C) \cap \Theta$. Let us suppose that the portion of $\Pi^*(C)$ defined by

$\zeta^* : t \mapsto \zeta^*(t), t \in [t_\alpha , t_f)$, is a regular portion in Z_σ .

We can readily verify by direct computation of $V^*(z)$ that this property is a consequence of (a) of Assumption 7, and Assumption 3, and

Assumption 8 : $z^*(t_f)$ is an interior point of θ in the topology induced by E on the set $\{z : \; z \in G, m(z) = 0\}$;

and

Assumption 9 : $m_z(z ; \overline{F}(z)) \neq 0$ for all z in a neighborhood of θ in G ; that is, no optimal path in $Z \cup \theta$ can reach θ tangentially.

Let

$$\Delta^{\alpha} \triangleq \{ \zeta \; : \; \zeta \in \Sigma_{\sigma}(C) \; \|\zeta - \zeta^{*}(t_{\alpha})\| < \varepsilon \}$$

Suppose that ε is sufficiently small so that $\Delta^{f} = T(t_{\alpha},t_{f})\Delta^{\alpha}$ is defined.

From the definition of Δ^{f}, from the fact that $\overline{F}$ and $\overline{f}_{o}$ are C^{1} functions of z on R_{σ} , and from Assumptions 3, 8 and 9, it follows that there exists $\varepsilon' > 0$ such that

$$B(\zeta^{*}(t_{f}),\varepsilon') \cap \Sigma(C) \cap \Theta \subset \Delta^{f}$$

where

$$B(\zeta^{*}(t_{f}),\varepsilon')) \triangleq \{ \zeta \; : \; \|\zeta - \zeta^{*}(t_{f})\| < \varepsilon' \}$$

From Lemma 1 it follows that

$$C(\Sigma(C) \cap \Theta \; , \; \zeta^{*}(t_{f})) \subseteq C(\Delta^{f}, \; \zeta^{*}(t_{f})) \tag{36}$$

From Corollary 1 we have

$$C(\Delta^{f},\zeta^{*}(t_{f})) = \{ \zeta^{*}(t_{f}) + q \; : \; \widetilde{\Lambda}(t_{f})\cdot q = 0 \} \tag{37}$$

where $\widetilde{\Lambda} \; : t \mapsto \widetilde{\Lambda}(t) \triangleq (1, \widetilde{\lambda}(t))$, is the solution of adjoint equation (21) defined on $[t_{\alpha}, t_{f}]$, for initial condition $\widetilde{\Lambda}(t_{\alpha}) = \text{grad } \Phi(\zeta^{*}(t_{\alpha}))$.

From (36) and (37) we deduce the following *transversality condition*

$$\widetilde{\Lambda}(t_{f}) \cdot q = 0 \tag{38}$$

for all q tangent to $\Sigma(C) \cap \Theta$ at point $\zeta^{*}(t_{f})$.

Since $\Sigma(C) \cap \Theta$ may be deduced from θ by translation parallel to the y_{o}-axis, we have $q = (0,h)$, where h is tangent to θ at point $z^{*}(t_{f})$.

Accordingly, the transversality condition may be rewritten

$$\lambda^{f}(h) = 0 \qquad \forall \; z^{*}(t_{f}) + h \in C(\theta, \; z^{*}(t_{f})) \tag{38'}$$

where $\lambda^{f} \triangleq \widetilde{\lambda}(t_{f})$

18. Discontinuity Surfaces

We shall suppose that every non-empty set $M_{k\ell}$ is defined by a single equation $m^{k\ell}(z) = 0$; that is

$$z \in M_{k\ell} \Rightarrow m^{k\ell}(z) = 0$$

where function $m^{k\ell} : \Delta_{k\ell} \to R$ is defined on a neighborhood $\Delta_{k\ell}$ of $M_{k\ell}$ in G .

Assumption 10. For each non-empty $M_{k\ell}$, $k \neq \ell$, $m^{k\ell}$ is of class C^1, and $m_z^{k\ell}(z;\cdot) \neq 0$ on $\Delta_{k\ell}$, where $m_z^{k\ell}(z;\cdot)$ is the Frechet derivative of $m^{k\ell}$ with respect to z, at point $z \in \Delta_{k\ell}$.

A surface $M_{k\ell}$ is termed a *discontinuity surface* .

The different types of discontinuity surfaces are

Transition surface M_T	(transition surface)
Repulsive surface M_R	(dispersal surface)
Attractive surface M_A	(universal surface)
Semi-attractive surface M_{SA}	(semi-universal surface)
Semi-repulsive surface M_{SR}	
Neutral surface M_N	

They differ by the signs of the two expressions

$$S_k(z) \triangleq m_z^{k\ell}(z;\overline{F}^k(z)) \quad z \in \Delta_{k\ell} \cap R_k$$

$$S_\ell(z) \triangleq m_z^{k\ell}(z;\overline{F}^\ell(z)) \quad z \in \Delta_{k\ell} \cap R_\ell$$

where

$$\overline{F}^k(z) \triangleq F(z,\varphi^k(z),\psi^k(z))$$

$$\overline{F}^\ell(z) \triangleq F(z,\varphi^\ell(z),\psi^\ell(z))$$

A table of the above classification is given below ; the ineqalities correspond to grad $m_z^{k\ell}(z;h) \geqslant 0 \quad \forall\, z+h \in C(Z_\ell,z)$ for $z \in M_{k\ell}$.

$M_T \begin{cases} S_k(z) > 0 \quad < 0 \\ \quad\quad\text{or} \\ S_\ell(z) > 0 \quad < 0 \end{cases}$	$M_N \begin{cases} S_k(z) = 0 \\ S_\ell(z) = 0 \end{cases}$
$M_R \begin{cases} S_k(z) < 0 \\ S_\ell(z) > 0 \end{cases}$	$M_{SR} \begin{cases} S_k(z) < 0 \\ S_\ell(z) = 0 \end{cases}$
$M_A \begin{cases} S_k(z) > 0 \\ S_\ell(z) < 0 \end{cases}$	$M_{SA} \begin{cases} S_k(z) > 0 \\ S_\ell(z) = 0 \end{cases}$

From the definitions of M_T, M_R and M_A it follows that no optimal path in $Z \cup \theta$ can reach or leave such a discontinuity surface tangentially.

19. Jump Condition

We shall now investigate conditions which must hold at a point $\zeta' = \zeta^*(t')$ where the optimal path $\Pi^*(C)$ crosses

$$M_{k\ell} \triangleq R \times M_{k\ell}$$

where $M_{k\ell} = M_T$ is a transition surface .

We shall assume that π^*, the projection of $\Pi^*(C)$ on $Z \cup \theta$, crosses $M_{k\ell}$ from Z_k to Z_ℓ ; in the other words $z' = z^*(t')$ belongs to $M_{k\ell}$ and there exists $\alpha > 0$ such that $z^*(t) \in Z_k$ for all $t \in [t'-\alpha , t')$ and $z^*(t) \in Z_\ell$ for all $t \in (t' , t'+\alpha]$.

Assumption 11. $z^*(t')$ is an interior point of $M_{k\ell}$ in the topology induced by E on the set $\{ z : z \in G, m^{k\ell}(z) = 0 \}$.

Assumption 12. There exists a function V^{ℓ} of class C^1 on a neighborhood of $z^*(t')$ in G', such that $V^*(z) = V^{\ell}(z)$ for all $z \in \overline{Z}_{\ell}$ in that neighborhood.

Let $\Phi^{\ell}(\zeta) \triangleq y_o + V^{\ell}(z)$, and $\Delta^{\ell} \triangleq \{\zeta : \Phi^{\ell}(\zeta) = C\}$

For ε, $\varepsilon > 0$, sufficiently small we have

$$B(\zeta',\varepsilon) \cap \Sigma(C) \cap M_{k\ell} = B(\zeta',\varepsilon) \cap \Delta^{\ell} \cap M_{k\ell} \tag{39}$$

where $B(\zeta',\varepsilon) \triangleq \{\zeta : \|\zeta - \zeta'\| < \varepsilon\}$; and furthermore (Appendix D)

$$C(\Delta^{\ell} \cap M_{k\ell},\zeta') = C(\Delta^{\ell},\zeta') \cap C(M_{k\ell},\zeta') \tag{40}$$

One can verify by direct computation of $V^*(z)$ that all points of $\Pi^*(C)$ in $Z_k \triangleq R \times Z_k$ are regular points of $\Sigma(C)$. This property is a consequence of (b) of Assumption 7, and Assumptions 10-12. Let us consider the regular portion of $\Pi^*(C)$ defined by ζ^* : $t \mapsto \zeta^*(t)$, $t \in [t_{\alpha}, t')$, and let

$$\Delta^{\alpha} \triangleq \{\zeta : \zeta \in \Sigma_k(C), \|\zeta - \zeta^*(t_{\alpha})\| < \varepsilon\}$$

Suppose that ε is sufficiently small so that $\Delta^k = T(t_{\alpha},t')\Delta^{\alpha}$ is defined.

From the definition of Δ^k, from the fact that $\overline{F}^k \triangleq F(\cdot, \varphi^k(\cdot), \psi^k(\cdot))$ and $\overline{f}^k_o \triangleq f_o(\cdot, \varphi^k(\cdot), \psi^k(\cdot))$ are C^1 functions of z on R_k, and from Assumptions 10 and 11, together with the fact that no optimal path in $Z \cup \theta$ can reach $M_{k\ell}$ tangentially, it follows that there exists $\varepsilon > 0$ such that

$$B(\zeta',\varepsilon) \cap \Sigma(C) \cap M_{k\ell} \subset \Delta^k \tag{41}$$

In other words, all points of $\Sigma(C) \cap M_{k\ell}$, in a sufficiently small neighborhood of ζ', can be reached by optimal paths emanating from points in Δ^{α}.

From the definition of a contingent, we have

$$C(B(\zeta',\varepsilon) \cap \Delta^{\ell} \cap M_{k\ell},\zeta') = C(\Delta^{\ell} \cap M_{k\ell},\zeta')$$

Then, from (39), (40) and (41) we deduce

(42) $$C(\Delta^{\ell},\zeta') \cap C(M_{k\ell},\zeta') \subset C(\Delta^{k},\zeta')$$

From Corollary 1, with σ replaced by k, we have

(43) $$C(\Delta^{k},\zeta') = \{ \zeta' + q : \quad \tilde{\Lambda}^{k}(t') \cdot q = 0 \}$$

where $\tilde{\Lambda}^{k} : t \mapsto \tilde{\Lambda}^{k}(t)$, is the solution of adjoint equation $\dot{\Lambda} = - \Lambda M_z^k$ on $[t_\alpha, t']$ for initial condition $\tilde{\Lambda}^{k}(t_\alpha)=\text{grad } \Phi(\zeta^*(t_\alpha))$, where

$$M_z^k \triangleq \begin{bmatrix} 0 & \bar{f}_{oz}^{k}(z;\cdot) \\ 0 & \bar{F}_z^{k}(z;\cdot) \end{bmatrix}$$

Let

(44) $$\tilde{\Lambda}^{k}(t) = (1, \tilde{\lambda}^{k}(t))$$

From the definition of Δ^{ℓ}, and Assumption 12, we have

(45) $$C(\Delta^{\ell},\zeta') = \{ \zeta' + q : \quad \text{grad } V^{\ell}(\zeta') \cdot q = 0 \}$$

where

(46) $$\text{grad } V^{\ell}(\zeta') \triangleq (1, V_z^{\ell}(z' ; \cdot))$$

Let

(47) $$\text{grad } m^{k\ell}(\zeta') \triangleq (0, \text{grad } m^{k\ell}(z'))$$

(48) $$\text{grad } m^{k\ell}(z') \triangleq m_z^{k\ell}(z' ; \cdot)$$

From Assumption 10 we deduce

(49) $$C(M_{k\ell},\zeta') = \{ \zeta' + q : \quad \text{grad } m^{k\ell}(\zeta') \cdot q = 0 \}$$

At last, from (42), (43), (45) and (49) it follows (Appendix C) that there exist real constants C_1, C_2, C_3, not all of which are zero, such that

(50) $$C_1\tilde{\Lambda}^{k}(t')+C_2 \text{ grad } V^{\ell}(\zeta')+C_3 \text{ grad } m^{k\ell}(\zeta') = 0$$

In view of (44), (46) and (47) we have $C_1 = - C_2$; and since

$\text{grad}\ m^{k\ell}(\zeta') \neq 0$

$$C_1 = - C_2 \neq 0$$

Thus, it follows from (50) that

$$\lambda^- - \lambda^+ + \frac{C_3}{C_1} m_z^{k\ell}(z' \ ; \ \cdot) = 0 \tag{51}$$

where $\lambda^- \triangleq \tilde{\lambda}^k(t')$ and $\lambda^+ \triangleq V_z^\ell(z' \ ; \ \cdot)$. Constant $\frac{C_3}{C_1}$ can be easily computed.

Letting

$$f_o^- \triangleq f_o(z', \varphi^k(z'), \psi^k(z'))$$

$$F^- \triangleq F(z', \varphi^k(z'), \psi^k(z'))$$

$$f_o^+ \triangleq f_o(z', \varphi^\ell(z'), \psi^\ell(z'))$$

$$F^+ \triangleq F(z', \varphi^\ell(z'), \psi^\ell(z'))$$

we deduce from (32)" and (51) that

$$\frac{C_3}{C_1} = - \frac{f_o^+ + \lambda^-(F^+)}{m_z^{k\ell}(z' \ ; \ F^+)} = \frac{f_o^- + \lambda^+(F^-)}{m_z^{k\ell}(z' \ ; \ F^-)} \tag{52}$$

Equation (51) is the *jump condition* at a point where optimal path $\Pi^*(C)$ crosses $M_{k\ell}$.

20. A Min-Max Theorem

The necessary conditions for an optimal strategy pair are summarized in

Theorem 2. *If there exists a strategy pair* (φ^*, ψ^*) *that is optimal on* Z *and that generates piece-wise regular optimal path* $\Pi^*(C)$, *represented by* $\zeta^* : t \mapsto \zeta = \zeta^*(t)$, $t \in [t_i, t_f]$, $\zeta = (y_o, z)$, *and if* $z^*(t) \in Z_\sigma$ *for all* $t \in (t_\beta, t_\delta)$, *then there exists a non-zero continuous solution*

$$\tilde{\Lambda} : t \mapsto \Lambda = \tilde{\Lambda}(t), \quad t \in [t_\beta, t_\delta], \quad \tilde{\Lambda}(t) = (1, \tilde{\lambda}(t))$$

of equation (21), *such that*

(i) $$\underset{\mu \in K_P}{\text{Min}} \; H(\tilde{\lambda}(t), z^*(t), \mu, \psi^*(z^*(t)))$$

$$= \underset{\nu \in K_E}{\text{Max}} \; H(\tilde{\lambda}(t), z^*(t), \varphi^*(z^*(t)), \nu)$$

$$= H(\tilde{\lambda}(t), z^*(t), \varphi^*(z^*(t)), \psi^*(z^*(t))) = 0$$

(ii) $$\tilde{\lambda}_o(t) = 1$$

for all $t \in (t_\beta, t_\delta)$; *and if there exists* $\alpha > 0$ *such that* $z^*(t) \in Z_\sigma$ *for all* $t \in [t_f - \alpha, t_f)$, *and* $z^*(t_f) \in \theta$, *then*

(iii) $$\lambda^f(h) = 0 \qquad \forall z^*(t_f) + h \in C(\theta, z^*(t_f))$$

where $\lambda^f \triangleq \tilde{\lambda}(t_f)$

21. A Linear Case

Let

$$F(z,\mu,\nu) = F_1(z,\mu) + F_2(z,\nu)$$

where, for all $z \in G$, $F_1(z,\cdot)$ and $F_2(z,\cdot)$ are linear mappings from U and V, respectively, into E; and let

$$f_o : G \to R$$

Suppose that constraint sets K_P and K_E contain the null vectors of U and V, respectively.

We have

$$H(\lambda,z,\mu,\nu) = f_o(z) + \lambda(F_1(z,\mu) + F_2(z,\nu))$$

Suppose that there exists a strategy pair (φ^*, ψ^*) that is optimal on Z and that generates piece-wise regular optimal path $\Pi^*(C)$, represented by $\zeta^* : t \mapsto \zeta = \zeta^*(t)$, $t \in [t_i, t_f]$; $\zeta = (y_o, z)$.

Let $z = z^*(t) \in Z_\sigma$, $\mu = \varphi^*(z^*(t))$, $\nu = \psi^*(z^*(t))$

From condition (i) of Theorem 2 we have

$$\lambda(F_1(z,\mu+\delta\mu)) \geqslant \lambda(F_1(z,\mu)) \qquad \forall \mu+\delta\mu \in K_P \tag{53}$$

$$\lambda(F_2(z,\nu+\delta\nu)) \leqslant \lambda(F_2(z,\nu)) \qquad \forall \nu+\delta\nu \in K_E \tag{54}$$

From the linearity of λ, $F_1(z,\cdot)$ and $F_2(z,\cdot)$, it follows that

$$\lambda(F_1(z,\delta\mu)) \geqslant 0 \qquad \forall \mu+\delta\mu \in K_P \tag{55}$$

$$\lambda(F_2(z,\delta\nu)) \leqslant 0 \qquad \forall \nu+\delta\nu \in K_E \tag{56}$$

Consider planes H_1 and H_2 through points μ and ν, respectively, namely

$$H_1 = \{ \mu+\delta\mu : \delta\mu \in U, \quad \lambda(F_1(z,\delta\mu)) = 0 \}$$

$$H_2 = \{ \nu+\delta\nu : \delta\nu \in V, \quad \lambda(F_2(z,\delta\nu)) = 0 \}$$

From (55) and (56) it appears that K_P belongs to one of the closed half-spaces determined by H_1, and K_E to one of the closed half-spaces determined by H_2 .

Furthermore, letting $\mu+\delta\mu = 0$ and $\nu+\delta\nu = 0$ in (53) and (54), respectively, we have

$$\lambda(F_1(z,\mu)) \leqslant 0 \text{ ; and}$$

$$\lambda(F_2(z,\nu)) \geqslant 0$$

22. Example 1

Here we shall consider a game described by $\dot{z} = \mu + \nu$ where $z \triangleq (t,x)$, $\mu + \nu \triangleq (1, u + v)$

Let $G = U = V = R \times C^o [0, d]$, $d > 0$, where $C^o[0, d]$ is the set of continuous mappings $[0, d] \to R$; that is, x, u and v belongs to $C^o [0, d]$.

Constraints are defined by

(57) $$\sup_{s \in [0,d]} |u(s)| \leqslant 1 , \qquad \sup_{s \in [0,d]} |v(s)| \leqslant 1$$

Target θ is defined by

$$m(z) \triangleq t + \int_0^d x(s)ds = 0$$

so that $m_z(z;\cdot) = m$.

We shall consider the time-optimal problem ; namely, J_P seeks to minimize while J_E seeks to maximize the time of transfer from an initial point to the target.

To apply the necessary conditions we have need of the H-function ; here

$$H(\lambda,z,\mu,\nu) = \lambda_o + \lambda(\mu+\nu)$$

The adjoint equations are

$$\dot{\lambda}_o = 0 \qquad \dot{\lambda} = 0$$

so that

(58) $$\lambda = \tilde{\lambda}(t) = \text{constant}$$

and from condition (ii) of Theorem 2 we have $\lambda_o = 1$.

From the transversality condition we obtain

(59) $$\lambda^f = \tilde{\lambda}(t_f) = c\, m \quad , \quad c \in R$$

By (58) and (59) we have

$$\lambda = \lambda^f = c\, m$$

so that

(60) $$\lambda(\mu+\nu) = c(1 + \int_0^d u(s)ds + \int_0^d v(s)ds)$$

From (57) and (60), and from condition (i) of Theorem 2 we obtain

(61) $$1 + c(1 + \int_0^d u^*(s)ds + \int_0^d v^*(s)ds) = 0$$

and

$$\left.\begin{array}{lll}(62) & u^*(s) = -\operatorname{sgn} c \\ (63) & v^*(s) = \operatorname{sgn} c\end{array}\right\} \quad \forall\, s \in [0,d]$$

where $u^* \triangleq \varphi^*(z)$ and $v^* \triangleq \psi^*(z)$ at point z of a supposed optimal path π^* .

From (61), (62) and (63) we deduce $c = -1$

$$\left.\begin{array}{l}u^*(s) = 1 \\ v^*(s) = -1\end{array}\right\} \quad \forall\, s \in [0,d]$$

Consequently, the strategy pair that satisfies the necessary conditions for optimality is given by

$$\varphi^*(z) = \tilde{1}$$

$$\psi^*(z) = -\tilde{1}$$

where $\tilde{1}$ denotes the mapping from $[0,d]$ into R such that $\tilde{1}(s) = 1$ for all $s \in [0,d]$.

A path π^* that emanates from $z^i \triangleq (t_i , x^i)$, and that reaches θ at time t_f , is given by $\tilde{z} : t \mapsto \tilde{z}(t)$, $t \in [t_i , t_f]$, $\tilde{z}(t) = (t, x^i)$.

Since $t_f = - \int_0^d x^i(s)ds$, the value of the game at z^i is

$$V^*(z^i) = -(t_i + \int_0^d x^i(s)ds) = -m(z^i)$$

Since $V^*(z^i)$ is a transfer time, it is non negative, and hence z^i must be such that

$$m(z^i) \leqslant 0$$

23. Example 2

Next let us consider a game described by

$$(64) \qquad \dot{z} = \mu+\nu \;, \quad z \triangleq (t, x) \;, \qquad \mu+\nu \triangleq (1, u+v)$$

Again, let $G = U = V = R \times C^o[0,d]$.

Constraints are defined by

$$0 \leq u(s) \leq 1 \;, \quad 0 \leq v(s) \leq 1 \;, \quad \forall\, s \in [0,d]$$

Target θ is defined by

$$(65) \qquad m(z) \triangleq \int_0^d x(s)ds + C = 0, \qquad C < 0$$

The performance index associated with a path π^{ij}, represented by $\tilde{z} : t \to z = \tilde{z}(t)$, $t \in [t_i, t_j]$, is

$$\int_{t_i}^{t_j} f_o(\tilde{z}(t))dt$$

where

$$(66) \qquad f_o(z) \triangleq \int_0^d x(s)ds$$

The H-function is

$$(67) \qquad H(\lambda, z, \mu, \nu) = f_o(z) + \lambda(\mu+\nu)$$

The adjoint equations are

$$\dot{\lambda}_o = 0 \qquad\qquad \dot{\lambda} = -\lambda_o f_{oz}(z\,;\,\cdot)$$

and we have $f_{oz}(z\,;\,\cdot) = f_o$, so that

$$(68) \qquad \lambda = \tilde{\lambda}(t) = \lambda_o(t_f - t)f_o + \tilde{\lambda}(t_f)$$

From the transversality condition we obtain

$$(69) \qquad \lambda^f = \tilde{\lambda}(t_f) = c\, m_z(z^f; \cdot), \quad z^f \triangleq \tilde{z}(t_f) \in \theta, \quad c \in R$$

In view of (65), (66), (68) and (69), (67) rewrites

(70) $$H(\lambda,z,\mu,\nu) = \int_0^d x(s)ds + (c+\lambda_o(t_f-t)) \int_0^d (u(s)+v(s))ds$$

From condition (ii) of Theorem 2 we have $\lambda_o = 1$, and from condition (i) of that theorem we obtain

$$\left.\begin{array}{l} u^*(s) = 0 \\ v^*(s) = 1 \end{array}\right\} \forall\ s \in [0,d] \quad \text{if} \quad c + t_f - t > 0$$

$$\left.\begin{array}{l} u^*(s) = 1 \\ v^*(s) = 0 \end{array}\right\} \forall\ s \in [0,d] \quad \text{if} \quad c + t_f - t < 0$$

where $u^* \triangleq \varphi^*(z)$ and $v^* \triangleq \psi^*(z)$ at point z of a supposed optimal path π^*.

From (65), (70) and (71) we have, at the terminal point of π^*

$$-C + cd = 0$$

from which we deduce

(72) $$c = \frac{C}{d}$$

Since, from (65), c is negative, there exists a time t_c such that $c + t_f - t = 0$, namely

(73) $$t_c = t_f + c$$

Upon integration of (64) we find

(74) $$x = \tilde{x}(t) = (t-t_f)(u^* + v^*) + \tilde{x}(t_f)$$

By (72), (73) and (74), the sign of $c + t_f - t$ switches at point x^c

$$x^c = \frac{C}{d}\ (u^* + v^*) + \tilde{x}(t_f)$$

and from (65) and (71) we deduce

(75) $$\int_0^d x^c(s)ds = 0$$

Consequently, the strategy pair that satisfies the necessary conditions for optimality is given by

$$\left.\begin{aligned} \varphi^*(z) &= \tilde{1} \\ \psi^*(z) &= \tilde{0} \end{aligned}\right\} \quad \text{for} \quad 0 < \int_0^d x(s)ds < -C$$

$$\left.\begin{aligned} \varphi^*(z) &= \tilde{0} \\ \psi^*(z) &= \tilde{1} \end{aligned}\right\} \quad \text{for} \quad \int_0^d x(s)ds < 0$$

where $\tilde{1}$ and $\tilde{0}$ denote the mappings from $[0,d]$ into R such that $\tilde{1}(s) = 1$ and $\tilde{0}(s) = 0$ for all $s \in [0,d]$.

Thus

$$M \triangleq \{ z \triangleq (t, x) : \int_0^d x(s)ds = 0 \}$$

is a discontinuity surface. Paths π^* emanating from points $z \triangleq (t,x)$ such that $\int_0^d x(s)ds < 0$ cross M .

Since $f_o^+ = f_o^-$ and $F^+ = F^-$, the jump condition (51) becomes

$$\lambda^- = \lambda^+$$

Hence λ is continuous at the point where a path π^* crosses M .

24. Sufficiency Conditions

In this section we shall deduce a sufficiency theorem for optimal strategies.

Theorem 3. *If* $(\varphi^\circ,\psi^\circ)$ *is a playable strategy pair on* Z†, *and if there exist*

(i) *a function* $V^\circ : Z \cup \theta \to R$ *defined and continuous on* $Z \cup \theta$, *such that* $V^\circ(z) = 0$ *for all* $z \in \theta$;

† that is, $(\varphi^\circ, \psi^\circ) \in I(z) \quad \forall z \in Z$

(ii) *a decomposition of* Z, *namely* $\{Z_1, Z_2, \dots Z_K\}$, *such that* $M_{ij} \cap M_{k\ell} = \phi$ *for* $ij \neq k\ell \neq ji$; *and for all* Z_σ, $\sigma = 1, 2, .. K$, *a function* V^σ *which is of class* C^1 *on some domain* $R_\sigma \supset \overline{Z}_\sigma$; *and on each non-empty* $M_{k\ell}$, $k \neq \ell$, $k,\ell = 1, 2,.. K$, *a function* $V^{k\ell}$ *which is of class* C^1 *on some domain* $R_{k\ell} \supset M_{k\ell}$; *such that*

(a)

$V^\circ(z) = V^\sigma(z)$ *for all* $z \in Z_\sigma$, $\sigma = 1, 2, \dots K$, *and*
$V^\circ(z) = V^{k\ell}(z)$ *for all* $z \in M_{k\ell}$, $k \neq \ell$, $k,\ell = 1, 2,..K$;

(b)

$$f_o(z,\varphi^\circ(z),\psi^\circ(z)) + V^\circ_z(z;F(z,\varphi^\circ(z),\psi^\circ(z))) = 0 \tag{1}$$

$$f_o(z,\varphi^\circ(z), \psi(z)) + V^\circ_z(z;F(z,\varphi^\circ(z), \psi(z))) \leqslant 0 \tag{2}$$

$$f_o(z, \varphi(z),\psi^\circ(z)) + V^\circ_z(z;F(z, \varphi(z),\psi^\circ(z))) \geqslant 0 \tag{3}$$

for all $z \in Z$, *and for all strategy pairs* (φ°,ψ) *and* (φ, ψ°) *playable on* Z, *where* $V^\circ_z(z;\cdot) \triangleq V^\sigma_z(z;\cdot)$, *if* $z \in Z_\sigma$ $(\sigma = 1, 2, \dots K)$

and

$V^\circ_z(z;\cdot) \triangleq V^{k\ell}_z(z;\cdot)$ if $z \in M_{k\ell}$, $k \neq \ell$ $(k,\ell = 1, 2, ..K)$;

then $(\varphi^\circ, \psi^\circ)$ *is optimal on* Z .

Consider a terminating path in $Z \cup \Theta$, generated by strategy pair (φ,ψ), represented by $\tilde{\zeta}: t \mapsto \zeta = \tilde{\zeta}(t)$, $t \in [t_i, t_f]$.

First, let us prove that *if* $d\tilde{\zeta}(t)/dt$ *is defined at time* $t_c \in (t_i, t_f)$, *then* $d\Phi(\tilde{\zeta}(t))/dt$ *is defined at that time* , $\Phi(\zeta) \triangleq y_o + V^\circ(z)$.

This property is trivial when $\tilde{\zeta}(t_c)$ belongs to some domain Z_σ .

Let us suppose that $\tilde{\zeta}(t_c)$ belongs to some set $M_{k\ell}$. Two cases need be considered, namely.

1. There exists $\alpha > 0$ such that $\tilde{\zeta}(t_c+\varepsilon) \notin M_{k\ell}$ for all $\varepsilon \in (0,\alpha)$. Then, in view of continuity of $\tilde{\zeta}$, $\tilde{\zeta}(t_c+\varepsilon)$ lies in Z_k, or in Z_ℓ, for all $\varepsilon \in (0,\alpha)$

2. No matter how small α, $\alpha > 0$, there exists $\varepsilon \in (0,\alpha)$ such that $\tilde{\zeta}(t_c+\varepsilon) \in M_{k\ell}$

Let

$$\Phi^k(\zeta) \triangleq y_o + V^k(z)$$

$$\Phi^\ell(\zeta) \triangleq y_o + V^\ell(z)$$

$$\Phi^{k\ell}(\zeta) \triangleq y_o + V^{k\ell}(z)$$

In case 1, $d\Phi(\tilde{\zeta}(t))/dt = d\Phi^\sigma(\tilde{\zeta}(t))/dt$, at $t = t_c$, where $\sigma = k$, if $\tilde{\zeta}(t_c+\varepsilon) \in Z_k$ for all $\varepsilon \in (0,\alpha)$, and $\sigma = \ell$ if $\tilde{\zeta}(t_c+\varepsilon) \in Z_\ell$ for all $\varepsilon \in (0,\alpha)$. Hence, $d\Phi(\tilde{\zeta}(t))/dt$ is defined at time t_c.

In case 2, since $d\tilde{\zeta}(t)/dt$ is defined at $t = t_c$, $F(z^c, \varphi(z^c), \psi(z^c))$ is tangent to $M_{k\ell}$ at point z^c, the projection of $\tilde{\zeta}(t_c)$ on Z.

Let us consider the sets

$$\bar{\Sigma}_k(C) \triangleq \{\zeta : y_o + V^k(z) = C, \quad z \in \bar{Z}_k\}$$

$$\bar{\Sigma}_\ell(C) \triangleq \{\zeta : y_o + V^\ell(z) = C, \quad z \in \bar{Z}_\ell\}$$

$$\Sigma_{k\ell}(C) \triangleq \{\zeta : y_o + V^{k\ell}(z) = C, \quad z \in M_{k\ell}\}$$

In view of condition (a) of Theorem 3 together with the continuity of V^o, we have $V^k(z) = V^\ell(z) = V^{k\ell}(z) = V^o(z)$ for all $z \in M_{k\ell}$, and since $M_{k\ell} \triangleq \bar{Z}_k \cap \bar{Z}_\ell$, we have

$$\Sigma_{k\ell}(C) = \bar{\Sigma}_k(C) \cap \bar{\Sigma}_\ell(C)$$

Let C be such that $\tilde{\zeta}(t_c) \in \overline{\Sigma}_k(C) \cap \overline{\Sigma}_\ell(C)$, and let $q \triangleq (h_o,h) \in R \times E$ be a vector tangent to $\overline{\Sigma}_k(C) \cap \overline{\Sigma}_\ell(C)$ at point $\tilde{\zeta}(t_c)$.

We have

$$h_o + V_z^k(z^c\ ;\ h) = 0\ , \text{ and} \tag{76}$$

$$h_o + V_z^\ell(z^c\ ;\ h) = 0 \tag{77}$$

and since q is tangent to $\Sigma_{k\ell}(C)$, we have also

$$h_o + V_z^{k\ell}(z^c\ ;\ h) = 0 \tag{78}$$

Now, since $F(z^c, \varphi(z^c), \psi(z^c))$ is tangent to $M_{k\ell}$ at point z^c, we can let $h \triangleq F(z^c, \varphi(z^c), \psi(z^c))$. Then, it follows from (76), (77) and (78) that

$$\begin{aligned} V_z^k(z^c;F(z^c, \varphi(z^c), \psi(z^c))) &= V_z^\ell(z^c;F(z^c, \varphi(z^c), \psi(z^c))) \\ &= V_z^{k\ell}(z^c;F(z^c, \varphi(z^c), \psi(z^c))) \end{aligned} \tag{79}$$

Three sub-cases need be considered, namely

2.1 There exists $\beta > 0$ such that $\tilde{\zeta}(t_c+\varepsilon) \in M_{k\ell}$ for all $\varepsilon \in (0,\beta)$.

2.2 No matter how small β, $\beta > 0$, there exists $\varepsilon_k \in (0,\beta)$ such that

$$\tilde{\zeta}(t_c+\varepsilon_k) \in Z_k$$

and, for β sufficiently small, $\tilde{\zeta}(t_c+\varepsilon) \in M_{k\ell} \cup Z_k$ for all $\varepsilon \in (0,\beta)$.

2.3 No matter how small β, $\beta > 0$, there exist ε_k and ε_ℓ in $(0,\beta)$ such that

$$\tilde{\zeta}(t_c+\varepsilon_k) \in Z_k\ ;\ \text{and}$$

$$\tilde{\zeta}(t_c+\varepsilon_\ell) \in Z_\ell$$

In case 2.1, $d\Phi(\tilde{\zeta}(t))/dt$ is defined at time t_c, since we have $d\Phi(\tilde{\zeta}(t))/dt = J^{k\ell}$ at that time, where

$$J^{k\ell} \triangleq f_o(z^c,\varphi(z^c),\psi(z^c)) + V_z^{k\ell}(z^c;F(z^c,\varphi(z^c),\psi(z^c)))$$

Cases 2.2 and 2.3 can be treated by invoking similar arguments. Consider for instance case 2.3.

From condition (ii) of Theorem 3 we have

$$J^k = f_o(z^c,\varphi(z^c),\psi(z^c)) + V_z^k(z^c;F(z^c,\varphi(z^c),\psi(z^c)))$$

$$J^\ell = f_o(z^c,\varphi(z^c),\psi(z^c)) + V_z^\ell(z^c;F(z^c,\varphi(z^c),\psi(z^c)))$$

where $J^k \triangleq d\Phi^k(\tilde{\zeta}(t))/dt$ and $J^\ell \triangleq d\Phi^\ell(\tilde{\zeta}(t))/dt$, at $t = t_c$.

From (79) we deduce $J^k = J^\ell = J^{k\ell}$, and hence $d\Phi(\tilde{\zeta}(t))/dt$ is defined at $t = t_c$.

Since $d\tilde{\zeta}(t)/dt$ is defined for all $t \in [t_i, t_f]$, except on a subset of $[t_i, t_f]$ at most denumerable, the same holds for $d\Phi(\tilde{\zeta}(t))/dt$.

Now, for a terminating path in $Z \cup \Theta$ generated by (φ^o,ψ), we deduce from relation (2) of Theorem 3 that $d\Phi(\tilde{\zeta}(t))/dt$ is non positive almost everywhere. It follows that $\Phi(\tilde{\zeta}(t))$ is non increasing on $[t_i, t_f]$.

Likewise, for a terminating path in $Z \cup \Theta$ generated by (φ,ψ^o) we deduce from relation (3) of Theorem 3 that $\Phi(\tilde{\zeta}(t))$ is non decreasing on $[t_i, t_f]$.

At last, for a terminating path in $Z \cup \Theta$ generated by (φ^o,ψ^o) we deduce from relation (1) of Theorem 3 that $\Phi(\tilde{\zeta}(t)) = \text{constant}$, for all $t \in [t_i, t_f]$.

The proof of Theorem 3 follows at once .

It should be noted that Isaacs' "Verification Theorem" (Ref.1) constitutes a special case of Theorem 3 .

We shall leave it to the reader to verify that the strategies deduced in Examples 1 and 2 satisfy the conditions of Theorem 3 and hence are optimal in the sense of the theorem.

25. Appendix A

Let $\mathcal{A}$ and $\mathcal{B}$ be two Banach spaces, and let f be a mapping from $A \subseteq \mathcal{A}$ into $\mathcal{B}$, continuous at point $a \in \overset{\circ}{A}$†. f is said to be Frechet differentiable at point a if there exists a continuous linear mapping L from $\mathcal{A}$ into $\mathcal{B}$, which is tangent to f at point a ; that is, for all $\eta > 0$ there exists an open ball $B(\varepsilon, a)$ of radius ε and center a such that

$$\begin{matrix} x \in B(\varepsilon, a) \\ x \neq a \end{matrix} \quad \Rightarrow \quad \frac{\| f(x) - f(a) - L(x-a) \|}{\| x-a \|} < \eta \tag{80}$$

In other words

$$\lim_{\substack{x \to a \\ x \neq a}} \frac{\| f(x) - f(a) - L(x-a) \|}{\| x-a \|} = 0$$

L is unique. It is termed the Frechet derivative of f at point a . From (80) it follows that

$$f(x) = f(a) + L(x-a) + o(x-a)$$

where

$$\lim_{\substack{x \to a \\ x \neq a}} \frac{\| o(x-a) \|}{\| x-a \|} = 0$$

---

† $\overset{\circ}{A}$ denotes the interior of A

26. Appendix B

Let A be a Banach space ; let $L_2 \in L(A, R)$, $L_2 \neq 0$; and

$$H \triangleq \{ h : \quad h \in A, \quad L_2.h = 0 \}$$

Let L_1, $L_1 \in L(A, R)$, be such that

$$L_1.h = 0 \qquad \forall\, h \in H$$

Since $L_2 \neq 0$, there exists h_o, $h_o \in A$, such that

$$L_2.h_o = 1$$

Since $L_2.h - (L_2.h)(L_2.h_o) = 0$, we have

$$h - (L_2.h)h_o \in H \qquad \forall\, h \in A$$

which implies that

$$L_1.h - (L_2.h)(L_1.h_o) = 0 \qquad \forall\, h \in A$$

Let $L_1.h_o = \alpha$, we have

$$(L_1 - \alpha L_2).h = 0 \qquad \forall\, h \in A$$

and accordingly

$$L_1 - \alpha L_2 = 0$$

27. Appendix C

Let A be a Banach space. Let $L_2 \in L(A, R)$ and $L_3 \in L(A, R)$, $L_2 \neq 0$, $L_3 \neq 0$. Let

$$H_2 \triangleq \{ h : \quad h \in A, \quad L_2.h = 0 \}$$

$$H_3 \triangleq \{ h : \quad h \in A \quad L_3.h = 0 \}$$

Let L_1, $L_1 \in L(A, R)$, be such that

$$L_1.h = 0 \qquad \forall\, h \in H_2 \cap H_3$$

If $L_2.h = 0 \quad \forall\, h \in H_3$, then from Appendix B there exists $\alpha \in R$ such that†

$$L_2 = \alpha\, L_3$$

Suppose that this is not the case, then there exists $h_o \in H_3$ such that $L_2.h_o = 1$; that is, $L_2.h_o = 1$ *and* $L_3.h_o = 0$.

Likewise, there exists h_1 such that $L_3.h_1 = 1$ *and* $L_2.h_1 = 0$. Clearly

$$h - (L_2.h)h_o - (L_3.h)h_1 \in H_2 \cap H_3 \qquad \forall\, h \in A$$

and hence

$$(L_1.h) - (L_2.h)(L_1.h_o) - (L_3.h)(L_1.h_1) = 0$$

Let $L_1.h_o = \beta$ and $L_1.h_1 = \gamma$, we have

$$(L_1 - \beta L_2 - \gamma L_3).h = 0 \qquad \forall\, h \in A$$

and accordingly

$$L_1 - \beta L_2 - \gamma L_3 = 0$$

28. Appendix D

Let us prove that

$$C(\Delta^{\ell},\zeta') \cap C(M_{k\ell},\zeta') \subseteq C(\Delta^{\ell} \cap M_{k\ell},\zeta') \tag{81}$$

Let

$$\zeta' + (h_o,h) \in C(\Delta^{\ell},\zeta') \cap C(M_{k\ell},\zeta'), \qquad \zeta' \triangleq (y'_o,z')$$

Since $\zeta' + (h_o,h) \in C(\Delta^{\ell},\zeta')$ we have

$$h_o + V^{\ell}_z(z';h) = 0 \tag{82}$$

Since $\zeta' + (h_o,h) \in C(M_{k\ell},\zeta')$, there exist $\varepsilon^{\nu}(\varepsilon^{\nu} > 0)$ and h^{ν} , $\nu = 1, 2, \ldots$, $\varepsilon^{\nu} \to 0$ and $h^{\nu} \to h$ as $\nu \to \infty$, such that

$$\zeta' + \varepsilon^{\nu}(h^{\nu}_o, h^{\nu}) \in M_{k\ell} \qquad \nu = 1, 2, \ldots \tag{83}$$

---

† Here, $\alpha \neq 0$ since $L_2 \neq 0$

From

$$V^{\ell}(z'+\varepsilon^{\nu}h^{\nu}) = V^{\ell}(z') + V_z^{\ell}(z';\varepsilon^{\nu}h^{\nu}) + o(\varepsilon^{\nu})$$

where $\lim\limits_{\nu \to \infty} \dfrac{|o(\varepsilon^{\nu})|}{\varepsilon^{\nu}} = 0$, and $y_o' + V(z') = C$, we deduce

$$(84) \qquad y_o' + \varepsilon^{\nu}h_o^{\nu} + V^{\ell}(z'+\varepsilon^{\nu}h^{\nu}) = C \qquad \nu = 1, 2, \ldots$$

where

$$(85) \qquad h_o^{\nu} = -\frac{1}{\varepsilon^{\nu}}(V_z^{\ell}(z';\varepsilon^{\nu}h^{\nu}) + o(\varepsilon^{\nu})) = -(V_z^{\ell}(z';h^{\nu}) + \frac{o(\varepsilon^{\nu})}{\varepsilon^{\nu}})$$

From relation (84) it follows that

$$(86) \qquad \zeta' + \varepsilon^{\nu}(h_o^{\nu}, h^{\nu}) \in \Delta^{\ell} \qquad \nu = 1, 2, \ldots$$

and from (83) and (86) we have

$$(87) \qquad \zeta' + \varepsilon^{\nu}(h_o^{\nu}, h^{\nu}) \in \Delta^{\ell} \cap M_{k\ell} \qquad \nu = 1, 2, \ldots$$

From (82) and (85) it follows that $h_o^{\nu} \to h_o$ as $\nu \to \infty$, and hence

$$(88) \qquad (h_o^{\nu}, h^{\nu}) \to (h_o, h) \quad \text{as} \quad \nu \to \infty$$

At last, from (87) and (88) we deduce that

$$\zeta' + (h_o, h) \in C(\Delta^{\ell} \cap M_{k\ell}, \zeta')$$

which proves relation (81) .

Conversely, from the definition of a contingent it follows directly that

$$(89) \qquad C(\Delta^{\ell} \cap M_{k\ell}, \zeta') \subseteq C(\Delta^{\ell}, \zeta') \cap C(M_{k\ell}, \zeta')$$

From (81) and (89) it follows that

$$C(\Delta^{\ell} \cap M_{k\ell}, \zeta') = C(\Delta^{\ell}, \zeta') \cap C(M_{k\ell}, \zeta')$$

References

(1) R. ISAACS, Differential Games (Wiley, 1965)

(2) A. BLAQUIERE, F. GERARD and G. LEITMANN, Quantitative and Qualitative Games (Academic Press Inc., 1969)

(3) A. BLAQUIERE and G. LEITMANN, Jeux Quantitatifs, Mémorial des Sciences Mathématiques, Fasc. 168 (Gauthier-Villars, 1969)

(4) G. LEITMANN and G. MON, On a Class of Differential Games, Proc. Colloquium on Adv. Problems and Methods of Space Flight Optimization, Liège (Pergamon, 1968) pp.25-46

(5) A. BLAQUIERE, An Introduction to Differential Games, in "Differential Games and Related Topics" Edited by H.W. KUHN and G.P.SZEGÖ (North-Holland, 1971) .

DIFFERENTIAL GAMES WITH TIME LAG

A. BLAQUIERE and P. CAUSSIN*

1. Point of the Game, State of the Game, Strategies

We shall be concerned with a system whose dynamical behavior is governed by

$$(1) \qquad \dot{y} = f(y, u, v)$$

where $y \triangleq (y_1, \ldots, y_n) \in D \subseteq E^n$,

$u \triangleq (u_1, \ldots, u_r) \in U \subseteq E^r$, $v \triangleq (v_1, \ldots, v_q) \in V \subseteq E^q$, and

$f \triangleq (f_1, \ldots, f_n) : D \times U \times V \rightarrow E^n$.

y will be termed the *point* of the system or, as we shall say the point of the game. u and v are *control variables*. We assume that D, U, V are open sets in E^n, E^r and E^q , respectively, and that f is of class C^1 on $D \times U \times V$. We shall take $y_n \equiv t$ so that $f_n(y, u, v) \equiv 1$.

We shall assume that the *state of the game*, at any instant of time, belongs to the set $\mathcal{D}$,

$$\mathcal{D} \triangleq \{ \; x \; : \; [0,d] \rightarrow D, \quad x \in C^o \; \}$$

where d, d > 0, is the *time lag*. We shall say that x, $x \in \mathcal{D}$, is *at time* t if the n-th component of x(d) is t. For any s, $s \in [0,d]$, point x(s) in D will be called a *point of the state* x .

* Laboratoire d'Automatique Théorique, Université de Paris 7, France .

Players J_P and J_E make their decisions through choosing the values of control variables u and v, respectively, at each instant of time. These choices are governed by strategies which J_P and J_E select from two prescribed sets of strategies S_P and S_E, respectively .

We shall consider strategies for players J_P and J_E to be functions of y, $p : y \mapsto p(y)$ and $e : y \mapsto e(y)$, respectively, defined on a domain $Y \subset D$, belonging to prescribed classes of functions, and such that $p(y) \in K_u \subseteq U$ and $e(y) \in K_v \subseteq V$, for all $y \in Y$. K_u and K_v are *constraint sets*. We shall assume that they do not depend on y .

We shall call members (p, e) of $S_P \times S_E$ *strategy pairs.*

Assumption 1. For any p', p'' $\in S_P$ and e', e'' $\in S_E$, and for any $y^j \in Y$, the functions $p''' : y \mapsto p'''(y)$ and $e''' : y \mapsto e'''(y)$, $y \in Y$, defined by

$$\left.\begin{aligned} p'''(y) &= p'(y) \\ e'''(y) &= e'(y) \end{aligned}\right\} \quad \text{for} \quad y_n \leqslant y_n^j$$

$$\left.\begin{aligned} p'''(y) &= p''(y) \\ e'''(y) &= e''(y) \end{aligned}\right\} \quad \text{for} \quad y_n > y_n^j$$

are strategies ; that is $p''' \in S_P$ and $e''' \in S_E$.

2. Trajectories and Paths

For given initial point y^i in D, at time t_i , and control functions

$$\tilde{u} \; : \; [t_i \, , t_j] \to K_u$$

$$\tilde{v} \quad [t_i \, , t_j] \to K_v$$

where $t_i < t_j$, we shall denote by $C(y^i, \tilde{u}, \tilde{v}, t_j)$ the set of functions $\tilde{y} : [t_i, t_j] \to D$, continuous on $[t_i, t_j]$, such that $\tilde{y}(t_i) = y^i$ and

$$(2) \qquad d\tilde{y}(t)/dt = f(\tilde{y}(t), \tilde{u}(t), \tilde{v}(t))$$

for all $t \in [t_i, t_j]$, except on a subset of $[t_i, t_j]$ at most denumerable. Note that $C(y^i, \tilde{u}, \tilde{v}, t_j)$ may be an empty set .

When $C(y^i, \tilde{u}, \tilde{v}, t_j)$ is non empty, a *trajectory* ρ^{ij} is defined by

$$\rho^{ij} \triangleq \tilde{\rho}^{ij}(\tilde{y}) \triangleq \{\tilde{y}(t) : t \in [t_i, t_j]\} \quad \tilde{y} \in C(y^i, \tilde{u}, \tilde{v}, t_j)$$

Let

$$X \triangleq \{ x : \; x \in \mathcal{D},\ x(s) \in Y, \quad s \in [0,d] \}$$

For given initial state x^i in X, at time t_i , and strategy pair (p,e), we shall denote by $S(x^i,p,e,t_j)$, where $t_i < t_j$, the set of functions $\tilde{y} : [t_i - d, t_j] \to D$, continuous on $[t_i - d, t_j]$, such that

(i) $\tilde{y}(t) \in Y$ for all $t \in [t_i - d, t_j - d)$; and

(ii) $\tilde{y}(t) = x^i(t + d - t_i)$ for all $t \in [t_i - d, t_i]$; and

(iii)

$$(3) \qquad d\tilde{y}(t)/dt = f(\tilde{y}(t), p(\tilde{y}(t-d), e(\tilde{y}(t-d)))$$

for all $t \in [t_i, t_j]$, except on a subset of $[t_i, t_j]$ at most denumerable.

Indeed $S(x^i,p,e,t_j)$ may be an empty set. When $S(x^i,p,e,t_j)$ is non empty, a *trajectory* γ^{ij} is defined by

$$\gamma^{ij} \triangleq \tilde{\gamma}^{ij}(\tilde{y}) \triangleq \{\tilde{y}(t) : t \in [t_i, t_j]\} \quad \tilde{y} \in S(x^i,p,e,t_j)$$

A *trajectory* δ^{ij} is defined by

$$\delta^{ij} \triangleq \tilde{\delta}^{ij}(\tilde{y}) \triangleq \{\tilde{y}(t) : t \in [t_i - d, t_j]\}, \tilde{y} \in S(x^i,p,e,t_j)$$

Note that trajectories ρ^{ij}, γ^{ij} and δ^{ij} are ordered sets of points in D . Trajectories ρ^{ij} will be termed *open-loop trajectories* since they are generated by controls (i.e., open-loop strategies), while trajectories γ^{ij} and δ^{ij} will be termed *closed-loop trajectories* .

Now let us associate with a function $\tilde{y} : [t_i - d, t_j] \to D$, $t_i < t_j$, the function $q(\cdot, \cdot, \tilde{y}) : [t_i, t_j] \times [0, d] \to D$, defined by $q(t,s,\tilde{y}) = \tilde{y}(t+s-d)$ for all $s \in [0,d]$ and for all $t \in [t_i, t_j]$.

When $S(x^i,p,e,t_j)$ is non empty, a *path* π^{ij} is an ordered set of states in $\mathcal{D}$ defined by

$$\pi^{ij} \triangleq \tilde{\pi}^{ij}(\tilde{y}) \triangleq \{ x : \; x = q(t,\cdot,\tilde{y}), \; t \in [t_i, t_j] \}$$

$$\tilde{y} \in S(x^i,p,e,t_j)$$

In other words, for $\tilde{y} \in S(x^i,p,e,t_j)$, $\tilde{\pi}^{ij}(\tilde{y})$ is the set of states x all of whose points belong to $\tilde{\delta}^{ij}(\tilde{y})$.

3. Terminating Trajectories and Paths, Playable Strategy Pair

Target θ is a given set of points in D. We shall assume that θ is an (n-1)-dimensional surface belonging to the boundary ∂Y of Y, defined by a single equation $m(y) = 0$; that is,

$$y \in \theta \;\Rightarrow\; m(y) = 0$$

where function m is of class C^1 and grad $m(y) \neq 0$ on a neighborhood of θ .

A *terminating trajectory* is a trajectory $\gamma^{ij} \triangleq \tilde{\gamma}^{ij}(\tilde{y})$ (or $\delta^{ij} \triangleq \tilde{\delta}^{ij}(\tilde{y})$), $\tilde{y} \in S(x^i,p,e,t_j)$ such that $\tilde{y}(t) \notin \theta$ for all $t \in [t_i - d, t_j)$ and $\tilde{y}(t_j) \in \theta$.

Let h be the mapping that associates with each path π^{ij} the trajectory $\gamma^{ij} = h(\pi^{ij})$

$$h(\pi^{ij}) \triangleq \{ x(d) : \quad x \in \pi^{ij} \}$$

We shall say that $h(\pi^{ij})$ is the *projection of* π^{ij} *on* E^n .

π^{ij} is a *terminating path* if $h(\pi^{ij})$ is a terminating trajectory.

A strategy pair (p,e) is *playable at* $x^i \in X$, x^i at time t_i, if there exists a time t_j, $t_j > t_i$, and a function $\tilde{y}$, $\tilde{y} \in S(x^i,p,e,t_j)$, such that $\tilde{y}(t_j) \in \theta$. We shall let $I(x^i)$ denote the set of all strategy pairs playable at x^i .

Likewise we shall say that $(\tilde{u},\tilde{v})$, $\tilde{u} : [t_i, t_j] \to K_u$, $\tilde{v} : [t_i, t_j] \to K_v$, $t_i < t_j$, is playable at point $y^i \in D$, at time $y^i_n = t_i$, if there exists a time t_k, $t_k \in [t_i, t_j]$, and a function $\tilde{y}$, $\tilde{y} \in C(y^i,\tilde{u},\tilde{v},t_j)$, such that $\tilde{y}(t_k) \in \theta$. We shall let $J(y^i)$ denote the set of all such pairs .

4. Cost, Optimality of a Strategy Pair

From now on we shall restrict the classes of strategies as follows. S_P and S_E will be the sets of functions $p : Y \to K_u$ and $e : Y \to K_v$, respectively, that generate piece-wise continuous controls, for all x^i in X ; that is, for all x^i in X, functions $\tilde{u} : [t_i, t_i+d] \to K_u$ and $\tilde{v} : [t_i, t_i+d] \to K_v$, such that

(4) $$\tilde{u}(t) = p(x^i(t-t_i)), \text{ and}$$

(5) $$\tilde{v}(t) = e(x^i(t-t_i))$$

for all $t \in [t_i, t_i+d]$, are piece-wise continuous .

For given (x^i,p,e,t_j), x^i in X at time t_i, $t_j = t_i+d$, we shall denote by $g_P(x^i,p)$ and $g_E(x^i,e)$ the functions $\tilde{u}$ and $\tilde{v}$, respectively, defined by (4) and (5) .

For given (x^i,p,e,γ^{ij}), x^i in X at time t_i, $\gamma^{ij} \triangleq \tilde{\gamma}^{ij}(\tilde{y})$, $\tilde{y} \in S(x^i,p,e,t_j)$, we shall denote by $g(x^i,p,e,\gamma^{ij})$ the pair $(\tilde{u},\tilde{v})$ where $\tilde{u} : [t_i, t_j] \to K_u$ and $\tilde{v} : [t_i, t_j] \to K_v$ are such that

$$\tilde{u}(t) = p(\tilde{y}(t-d)), \text{ and}$$

$$\tilde{v}(t) = e(\tilde{y}(t-d))$$

for all $t \in [t_i, t_j)$.

Let us denote by $C_P[t, t+d]$ and $C_E[t, t+d]$ the sets of piecewise continuous functions of time defined on $[t, t+d]$, with ranges in K_u and K_v, respectively. We have

Lemma 1. *For all* $x \in X$, x *at time* t, *for all* $\tilde{u} \in C_P[t, t+d]$ *and for all* $\tilde{v} \in C_E[t, t+d]$, *there exists a strategy pair* (p,e) *such that* $g_P(x,p) = \tilde{u}$ *and* $g_E(x,e) = \tilde{v}$.

This lemma is a direct consequence of the definitions of S_P and S_E .

The cost of $\rho^{ij} \triangleq \tilde{\rho}^{ij}(\tilde{y})$, $\tilde{y} \in C(y^i,\tilde{u},\tilde{v},t_j)$ is

$$V_\rho(y^i,y^j,\tilde{u},\tilde{v},\rho^{ij}) \triangleq \int_{t_i}^{t_j} f_o(\tilde{y}(t), \tilde{u}(t), \tilde{v}(t))dt + w(y^j)$$

where $f_o : D \times U \times V \to R$, and $w : D \to R$ such that $w(y^j) = 0$ for $y^j \notin \theta$, $y^j \triangleq \tilde{y}(t_j)$. We assume that f_o is of class C^1 on $D\times U\times V$, and that w agrees on θ with some function which is of class C^1 on some neighborhood of θ.

The cost of $\gamma^{ij} \triangleq \tilde{\gamma}^{ij}(\tilde{y})$, $\tilde{y} \in S(x^i,p,e,t_j)$ is

$$V_\gamma(x^i,y^j,p,e,\gamma^{ij}) \triangleq V_\rho(y^i,y^j,\tilde{u},\tilde{v},\rho^{ij})$$

where $y^i \triangleq x^i(d)$, $y^j \triangleq \tilde{y}(t_j)$, $(\tilde{u},\tilde{v}) = g(x^i,p,e,\gamma^{ij})$, $\rho^{ij} = \gamma^{ij}$.

The cost of $\pi^{ij} \triangleq \tilde{\pi}^{ij}(\tilde{y})$, $\tilde{y} \in S(x^i,p,e,t_j)$ is

$$V_\pi(x^i,x^j,p,e,\pi^{ij}) \triangleq V_\gamma(x^i,y^j,p,e,\gamma^{ij})$$

where $y^j \triangleq \tilde{y}(t_j)$, $\gamma^{ij} = h(\pi^{ij})$.

If π^{ij} is a terminating path, we shall let

$$V_\pi(x^i,\theta;p,e) \triangleq V_\pi(x^i,x^j,p,e,\pi^{ij})$$

Since, for given x^i and $(p,e) \in I(x^i)$, terminating path π^{ij} need not be unique, $V_\pi(x^i,\theta;p,e)$ need not be unique .

For all x^i such that $y^i \triangleq x^i(d) \in \theta$, we shall let

$$V_\pi(x^i,\theta;p,e) \triangleq w(y^i)$$

Definition 1. (p^*,e^*) is *optimal at* $x \in X$ if

(i) $(p^*,e^*) \in I(x)$

(ii) $V_\pi(x,\theta;p^*,e^*)$ is defined ; it will be termed the *value of the game* at state x, and denoted by

$$V^*(x) \triangleq V_\pi(x,\theta;p^*,e^*)$$

(iii) the *saddle-point condition*

$$V_\pi(x,\theta;p^*,e) \leqslant V_\pi(x,\theta;p^*,e^*) \leqslant V_\pi(x,\theta;p,e^*)$$

is satisfied for all $(p^*,e) \in I(x)$, for all $(p,e^*) \in I(x)$, and for all terminating paths generated by (p^*,e) and (p,e^*) from state x .

Let $I^*(x)$ denote the set of all strategy pairs optimal at x .

A terminating path generated from x^i, $x^i \in X$, by a strategy pair (p^*,e^*), $(p^*,e^*) \in I^*(x^i)$, will be called an *optimal path.*

We shall say that (p^*,e^*) is *optimal on* X if

$$(p^*,e^*) \in I^*(x) \qquad \forall\, x \in X$$

Assumption 2. There exists a strategy pair (p^*,e^*) optimal on X .

5. Trajectories and Paths in Augmented Spaces

Let us introduce another variable y_o, and consider the spaces E^{n+1} and $R \times N$, $N \triangleq \{ x : [0, d] \to E^n , x \in C^o \}$, of points η and ξ, respectively, where $\eta \triangleq (y_o , y) \in E^{n+1}$ and $\xi \triangleq (y_o, x) \in R \times N$.

Let us define a *path* $\Pi(\pi^{ij}, C)$ *in* $R \times \mathcal{D}$, where $\pi^{ij} \triangleq \tilde{\pi}^{ij}(\tilde{y})$, $\tilde{y} \in S(x^i,p,e,t_j)$, $C \in R$, by

$$\Pi(\pi^{ij},C) \triangleq \{ \xi^k \triangleq (y_o^k, x^k) : y_o^k + V_\pi(x^k,x^j,p,e,\pi^{kj}) = C,$$

$$\pi^{kj} \subseteq \pi^{ij} \}$$

Likewise, let us define a P-*path in* $R \times \mathcal{D}$, by

$$\Pi(\pi_P^{ij}, C) \triangleq \{\xi^k : y_o^k + V_\pi(x^k,x^j,p^*,e,\pi^{kj}) = C$$

$$\pi^{kj} \subseteq \pi_P^{ij} \}$$

where $\pi_P^{ij} \triangleq \tilde{\pi}^{ij}(\tilde{y}_P)$, $\tilde{y}_P \in S(x^i,p^*,e,t_j)$; an E-*path in* $R \times \mathcal{D}$, by

$$\Pi(\pi_E^{ij}, C) \triangleq \{\xi^k : y_o^k + V_\pi(x^k,x^j,p,e^*,\pi^{kj}) = C,$$

$$\pi^{kj} \subseteq \pi_E^{ij} \}$$

where $\pi_E^{ij} \triangleq \tilde{\pi}^{ij}(\tilde{y}_E)$, $\tilde{y}_E \in S(x^i,p,e^*,t_j)$; a PE-*path in* $R \times \mathcal{D}$, by

$$\Pi(\pi_{PE}^{ij}, C) \triangleq \{\xi^k : y_o^k + V_\pi(x^k,x^j,p^*,e^*,\pi^{kj}) = C,$$

$$\pi^{kj} \subseteq \pi_{PE}^{ij} \}$$

where $\pi_{PE}^{ij} \triangleq \tilde{\pi}^{ij}(\tilde{y}_{PE})$, $\tilde{y}_{PE} \in S(x^i,p^*,e^*,t_j)$.

We shall define also a *closed-loop trajectory* $\Gamma(\pi^{ij}, C)$ in $R \times D$ by

$$\Gamma(\pi^{ij},C) \triangleq \{\eta^k \triangleq (y_o^k,x^k(d)) : (y_o^k,x^k) \in \Pi(\pi^{ij},C)\}$$

$\eta^k = (y_o^k \, , \, x^k(d))$ is the *projection of* $\xi^k = (y_o^k, x^k)$ *on* E^{n+1}. It will be denoted by $H(\xi^k)$. Likewise $\Gamma(\pi^{ij}, C)$ is the *projection of* $\Pi(\pi^{ij}, C)$ *on* E^{n+1}.

We shall say that $\Gamma(\pi^{ij}, C)$ is *generated from* $\xi^i = (y_o^i, x^i)$ by strategy pair (p,e), and that it *emanates from* $\eta^i = H(\xi^i)$.

Let us define a P-*trajectory* $\Gamma(\pi_P^{ij}, C)$ *in* $R \times D$, by

$$\Gamma(\pi_P^{ij}, C) \triangleq \{\eta^k : \eta^k = H(\xi^k), \; \xi^k \in \Pi(\pi_P^{ij}, C) \}$$

an E-*trajectory* $\Gamma(\pi_E^{ij}, C)$ *in* $R \times D$, by

$$\Gamma(\pi_E^{ij}, C) \triangleq \{\eta^k : \eta^k = H(\xi^k), \; \xi^k \in \Pi(\pi_E^{ij}, C) \}$$

and a PE-*trajectory* $\Gamma(\pi_{PE}^{ij}, C)$ *in* $R \times D$ by

$$\Gamma(\pi_{PE}^{ij}, C) \triangleq \{\eta^k : \eta^k = H(\xi^k), \; \xi^k \in \Pi(\pi_{PE}^{ij}, C) \}$$

PE-paths in $R \times \mathcal{D}$ and PE-trajectories in $R \times D$ associated with a terminating path in $\mathcal{D}$ are termed *optimal paths in* $R \times \mathcal{D}$ and *optimal trajectories in* $R \times D$, respectively .

At last we define an *open-loop trajectory* $P(\rho^{ij}, C)$ *in* $R \times D$, where $\rho^{ij} \triangleq \tilde{\rho}^{ij}(\tilde{y})$, $\tilde{y} \in C(y^i, \tilde{u}, \tilde{v}, t_j)$, by

$$P(\rho^{ij}, C) \triangleq \{\eta^k \triangleq (y_o^k, y^k) : y_o^k + V_\rho(y^k, y^j, \tilde{u}, \tilde{v}, \rho^{kj}) = C$$
$$\rho^{kj} \subseteq \rho^{ij} \}$$

6. Neighborhoods of Trajectories and Paths

For any two functions m_1 and m_2 defined on $[a,b]$ and with ranges in E^n, continuous on $[a,b]$,we shall let

$$d(m_1, m_2) \triangleq \sup \; \|m_1(s) - m_2(s)\| \qquad s \in [a,b]$$

Indeed, for functions in D we have $[a,b] = [0,d]$. Let

$$\Theta \triangleq \{ x : \; x(d) \in \theta \}$$

and let π^{ij} be a path in $X \cup \Theta$ generated from $x^i \in X$ at time t_i by strategy pair (p,e), namely $\pi^{ij} \triangleq \tilde{\pi}^{ij}(\tilde{y})$, $\tilde{y} \in S(x^i,p,e,t_j)$. Let $\rho^{ij} \triangleq h(\pi^{ij})$.

We define a *neighborhood* $\Delta(\pi^{ij})$ *of* π^{ij} *in* $\mathcal{D}$, and a *neighborhood* $B(\rho^{ij})$ *of* ρ^{ij} *in* D, by

$$\Delta(\pi^{ij}) \triangleq \{ x : \exists x^k \in \pi^{ij} \quad d(x,x^k) < \alpha \}$$

$$B(\rho^{ij}) \triangleq \{ y : \exists y^k \in \rho^{ij} \; \| y-y^k \| < \beta \}$$

where α and β are any strictly positive numbers.

For $\alpha = \beta$, it follows directly from these definitions that

(i) $B(\rho^{ij}) = \{ x(d) : \quad x \in \Delta(\pi^{ij}) \}$

and

(ii) if $x(s) \in B(\rho^{ij})$ for all $s \in [0,d]$, then $x \in \Delta(\pi^{ij})$.

7. A Basic Theorem

Let π^* be an optimal path generated from $x^i \in X$ at time t_i by (p^*,e^*), namely $\pi^* \triangleq \tilde{\pi}^{if}(y^*)$, $y^* \in S(x^i,p^*,e^*,t_f)$. Let $\rho^* \triangleq h(\pi^*)$.

Assumption 3. $y^*(t) \in Y$ for all $t \in [t_i-d,\ t_f)$.

Assumption 4. p^* and e^* are of class C^1 on Y.

Assumption 5. $y^*(t_f)$ is an interior point of θ in the topology induced by D on the set $\{ y : \ y \in D,\ m(y) = 0 \}$

Assumption 6. ρ^* does not reach θ tangentially .

Theorem 1. *If Assumptions* 1-6 *are satisfied, then there exists a neighborhood* $B(\rho^*)$ *of* ρ^* *in* D, *a neighborhood* $\Delta(\pi^*)$ *of* π^* *in* $\mathcal{D}$, *and a function* W^* :

$$W^* : Y^* \to R \qquad\qquad Y^* \triangleq Y \cap B(\rho^*)$$

such that

$$V_\pi(x,\theta;p^*,e^*) = W^*(x(d))$$

for all $x \in \Delta^*$, $\Delta^* \triangleq X \cap \Delta(\pi^*)$.

Let x', $x'' \in X$ be such that $x'(d) = x''(d) \triangleq y^k$. From Assumption 2 we have $(p^*,e^*) \in I^*(x') \cap I^*(x'')$. Let $\tilde{u}' \triangleq g_P(x',p^*)$, $\tilde{v}' \triangleq g_E(x',e^*)$, $\tilde{u}'' \triangleq g_P(x'',p^*)$, $\tilde{v}'' \triangleq g_E(x'',e^*)$. From Lemma 1 and Assumption 1 there exist strategies p', e', p'', e'' such that $p'(y) = p''(y) = p^*(y)$ and $e'(y) = e''(y) = e^*(y)$ for all $y \in Y$ such that $y_n > y_n^k$, and $\tilde{u}' = g_P(x'',p')$, $\tilde{v}' = g_E(x'',e')$, $\tilde{u}'' = g_P(x',p'')$, $\tilde{v}'' = g_E(x',e'')$.

Then from Assumptions 3-6 and from the fact that f is C^1 on $D \times U \times V$, it follows that there exists a neighborhood $\Delta(\pi^*)$ of π^* in $\mathcal{D}$, such that

$$\left.\begin{array}{l} x',\ x'' \in X \cap \Delta(\pi^*) \\ \\ x'(d) = x''(d) \end{array}\right\} \Rightarrow \left\{\begin{array}{l} (p^*,e''),(p'',e^*) \in I(x') \\ \\ (p^*,e'),(p',e^*) \in I(x'') \end{array}\right.$$

Let π_P' and π_E' be the terminating paths generated from x'' by (p^*,e') and (p',e^*), respectively, and π_P'' and π_E'' the terminating paths generated from x' by (p^*,e'') and (p'',e^*), respectively . We have $h(\pi_P'') = h(\pi_E')$ and $h(\pi_E'') = h(\pi_P')$.

It follows that

$$V_\pi(x',\theta;p^*,e'') = V_\pi(x'',\theta;p',e^*) \ ; \text{ and}$$

$$V_\pi(x',\theta;p'',e^*) = V_\pi(x'',\theta;p^*,e')$$

At last since

$$V_\pi(x',\theta;p^*,e'') \leqslant V_\pi(x',\theta;p^*,e^*) \leqslant V_\pi(x',\theta;p'',e^*)$$

$$V_\pi(x'',\theta;p^*,e') \leqslant V_\pi(x'',\theta;p^*,e^*) \leqslant V_\pi(x'',\theta;p',e^*)$$

we have

$$V_\pi(x',\theta;p^*,e^*) = V_\pi(x'',\theta;p^*,e^*)$$

Since $V_\pi(x,\theta;p^*,e^*)$ depends on $x(d)$ only, provided that $x \in \Delta^*$, we let

$$V_\pi(x,\theta;p^*,e^*) = W^*(x(d))$$

where function W^* is defined on the set Y^*, $Y^* \triangleq Y \cap B(\rho^*)$, $B(\rho^*) \triangleq \{ x(d) : x \in \Delta(\pi^*) \}$. Clearly we have also $Y^* = \{ x(d) : x \in \Delta^* \}$. In other words, Y^* is the *projection of* Δ^* *on* E^n .

Note that Y^* and Δ^* are associated with the optimal path π^* defined above. From now on the definitions of π^*, Y^* and Δ^* will remain unchanged .

Also, note that if x' and x" in X are such that $x'(d) = x''(d) \triangleq y^k$ with $y^k \in \theta$, we have

$$V_\pi(x',\theta;p^*,e^*) = V_\pi(x'',\theta;p^*,e^*) = w(y^k)$$

Accordingly we can extend the definition of function W^* by letting $W^*(y) \triangleq w(y)$ for $y \in \theta$

8. Game Surfaces

Since $V^*(x) \triangleq V_\pi(x,\theta;p^*,e^*)$ is defined for all $x \in X$, and for all x such that $x(d) \in \theta$, we can use an approach similar to the one developed in our previous chapter, devoted to further geometric aspects of differential games ; that is, we can define a game surface in $R \times N$.

We shall leave it to the reader to develop this method for the problem stated above. Here it will be convenient to rely on Theorem 1 and to use another approach .

Since $W^*(y)$ is defined for all $y \in Y^* \cup \theta$, we shall define a *game surface* $\Sigma(C)$ *in* E^{n+1} by

$$(6) \qquad \Sigma(C) \triangleq \{ \eta : y_o + W^*(y) = C, \qquad y \in Y^* \cup \theta \}$$

where $C \in R$ is a constant parameter.

$\Sigma(C)$ is a set of points which are in one-to-one correspondance with the points of $Y^* \cup \theta$.

As the value of parameter C is varied, equation (6) defines a one-parameter family of surfaces, namely $\{ \Sigma(C) \}$, whose members may be deduced from one another by translation parallel to the y_o-axis .

For a game surface $\Sigma(C)$ corresponding to parameter value C, we define

$$A/\Sigma(C) \triangleq \{ \eta : y_o > C - W^*(y), \quad y \in Y^* \cup \theta \} \tag{7}$$

$$B/\Sigma(C) \triangleq \{ \eta : y_o < C - W^*(y), \quad y \in Y^* \cup \theta \} \tag{8}$$

A point $\eta \in A/\Sigma(C)$ will be called an A-*point relative to* $\Sigma(C)$, and a point $\eta \in B/\Sigma(C)$ a B-*point relative to* $\Sigma(C)$.

Let us also define the sets $\Xi(C)$, $A/\Xi(C)$ and $B/\Xi(C)$, namely

$$\Xi(C) \triangleq \{ \xi : \; y_o + V^*(x) = C \, , \quad x \in \Delta^* \cup \Theta \} \tag{9}$$

$$A/\Xi(C) \triangleq \{ \xi : \; y_o > C - V^*(x) \, , \quad x \in \Delta^* \cup \Theta \} \tag{10}$$

$$B/\Xi(C) \triangleq \{ \xi : \; y_o < C - V^*(x) \, , \quad x \in \Delta^* \cup \Theta \} \tag{11}$$

A point $\xi \in A/\Xi(C)$ will be called an A-*point relative to* $\Xi(C)$, and a point $\xi \in B/\Xi(C)$ a B-*point relative to* $\Xi(C)$.

Since $Y^* \cup \theta$ is the projection of $\Delta^* \cup \Theta$ on E^n, and since $V^*(x) = W^*(x(d))$ for all $x \in \Delta^* \cup \Theta$, it follows from the definitions (6), (7), (8), (9), (10) and (11) that

Lemma 2. (a) *If* $\xi \in A/\Xi(C)$, *then* $H(\xi) \in A/\Sigma(C)$.
(b) *If* $\xi \in B/\Xi(C)$, *then* $H(\xi) \in B/\Sigma(C)$.
(c) *If* $\xi \in \Xi(C)$, *then* $H(\xi) \in \Sigma(C)$.

In other words, $\Sigma(C)$, $A/\Sigma(C)$ and $B/\Sigma(C)$ are the projections on E^{n+1} of $\Xi(C)$, $A/\Xi(C)$ and $B/\Xi(C)$, respectively .

9. A Property of Game Surfaces $\Xi(C)$

By the same arguments as in [1] for games without retardation, one can easily prove that[†]

Lemma 3. (a) *No point of a* P-*path in* $R \times (\Delta^* \cup \Theta)$, *emanating from* $\xi^\circ \in \Xi(C)$, *is a* A-*point relative to* $\Xi(C)$.

(b) *No point of an* E-*path in* $R \times (\Delta^* \cup \Theta)$, *emanating from* $\xi^\circ \in \Xi(C)$, *is a* B-*point relative to* $\Xi(C)$

(c) *All points of a* PE-*path in* $R \times (\Delta^* \cup \Theta)$, *emanating from* $\xi^\circ \in \Xi(C)$, *belong to* $\Xi(C)$.

10. Some Properties of Game Surfaces $\Sigma(C)$

From Lemmas 2 and 3, and from the definitions of Δ^* and Y^*, it follows directly that

Lemma 4. (a) *No point of a* P-*trajectory in* $R \times (Y^* \cup \theta)$, *emanating from* $H(\xi^\circ) \in \Sigma(C)$, *where* $\xi^\circ \in \Xi(C)$, *is an* A-*point relative to* $\Sigma(C)$.

(b) *No point of an* E-*trajectory in* $R \times (Y^* \cup \theta)$, *emanating from* $H(\xi^\circ) \in \Sigma(C)$, *where* $\xi^\circ \in \Xi(C)$, *is a* B-*point relative to* $\Sigma(C)$.

(c) *All points of a* PE-*trajectory in* $R \times (Y^* \cup \theta)$, *emanating from* $H(\xi^\circ) \in \Sigma(C)$, *where* $\xi^\circ \in \Xi(C)$, *belong to* $\Sigma(C)$.

From Assumptions 3 and 4 it follows that there exist $u^* : [t_i, t_f+d] \to K_u$ and $v^* : [t_i, t_f+d] \to K_v$, such that

$$\begin{aligned} u^*(t) &\triangleq p^*(y^*(t-d)) \\ v^*(t) &\triangleq e^*(y^*(t-d)) \quad , \text{ for all } t \in [t_i, t_f+d) \end{aligned} \tag{12}$$

† See also the chapter "Further Geometric Aspects of Differential Games" .

From Assumptions 5 and 6 and from the fact that f is C^1 on $D \times U \times V$, it follows that there exists α, $\alpha > 0$, such that

$$y^\circ \in T(y^*(t_i),\alpha) \Rightarrow (u^*, v^*) \in J(y^\circ)$$

where

$$T(y^*(t_i),\alpha) \triangleq \{ y : y_n = t_i, \; \|y-y^*(t_i)\| < \alpha \}$$

Suppose that $y^\circ \in T(y^*(t_i),\alpha)$ and let ρ^{os} be the open-loop trajectory in D, generated by (u^*, v^*) emanating from y° at time $t_o = t_i$, that reaches θ at point y^s at time t_s. Furthermore, suppose that α is sufficiently small so that $\rho^{os} \subset Y^* \cup \theta$.

Let $P(\rho^{os},C')$ be the corresponding open-loop trajectory in $R \times (Y^* \cup \theta)$, emanating from $\eta^\circ \in \Sigma(C)$, namely

$$P(\rho^{os},C') \triangleq \{ \eta^k \triangleq (y_o^k,y^k) : y_o^k+V_\rho(y^k,y^s,u^*,v^*,\rho^{ks}) = C' \quad \rho^{ks} \subseteq \rho^{os} \}$$

We shall prove

Lemma 5. *For* α *sufficiently small, all points of* $P(\rho^{os}, C')$ *belong to* $\Sigma(C)$.

Let $\eta^j \triangleq (y_o^j, y^j)$ and $\eta^k \triangleq (y_o^k, y^k)$ be any two points of $P(\rho^{os}, C')$ at times t_j and t_k, respectively, such that $t_k-t_j=d-\varepsilon$, $0 < \varepsilon < d$, and $t_j < t_f$. Note that $|t_s-t_f| < d$ for α sufficiently small, so that there exists ε, $0 < \varepsilon < d$, and $t_j < t_f$ such that $t_s-t_j = d-\varepsilon$.

Consider state x^j, $x^j \in \pi^*$ such that $x^j(d) = y^*(t_j)$. Since $\rho^{os} \subset Y^* \cup \theta$ and $t_j < t_f$, for α sufficiently small, there exists a state $\hat{x}^j$ such that $\hat{x}^j(s) \in Y^*$ for all $s \in [0,d]$, and

(i) $\hat{x}^j(d) = y^j$; and

(ii) $\hat{x}^j(s) = x^j(s)$ for all $s \in [0, d-\varepsilon]$

From the definitions of Y^* and Δ^* it follows that $\hat{x}^j \in \Delta^*$, and from Theorem 1 and condition (i) above we have

$$V^*(\hat{x}^j) = W^*(y^j)$$

Now, if $\eta^j \in \Sigma(C)$, we have

$$y_o^j + W^*(y^j) = y_o^j + V^*(\hat{x}^j) = C$$

that is, $\xi^j \triangleq (y_o^j , \hat{x}^j) \in \Xi(C)$.

From condition (ii) above we have

$$u^*(t) \triangleq p^*(y^*(t-d)) = p^*(\hat{x}^j(t-t_j))$$

$$v^*(t) \triangleq e^*(y^*(t-d)) = e^*(\hat{x}^j(t-t_j))$$

for all $t \in [t_j, t_k]$; so that the portion of $P(\rho^{os}, C')$ begining at η^j and ending at η^k is a PE-trajectory in $R \times (Y^* \cup \theta)$, generated from ξ^j . It follows from Lemma 4 that all of its points belong to $\Sigma(C)$.

Since there exists a finite number ν of time intervals of duration $d-\varepsilon$ such that $\nu(d-\varepsilon) \geqslant t_s-t_o$, since $\eta^o \in \Sigma(C)$, and since $t_s-t_f < d$ for α sufficiently small, by degrees, the above property of a portion of trajectory $P(\rho^{os}, C')$ is extended to the whole trajectory $P(\rho^{os}, C')$ which concludes the proof .

From Lemma 5 it follows that $C' = C$ and, for all $\eta^k \in P(\rho^{os}, C)$ we have

$$V_\rho(y^k, y^s, u^*, v^*, \rho^{ks}) = W^*(y^k), \qquad \rho^{ks} \subseteq \rho^{os} \tag{13}$$

12. A Min-Max Principle

Assumption 7. x^i is of class C^1 .

From Assumptions 3, 4 and 7, it follows that u^* and v^* are of class C^o on $[t_i, t_f+d]$, and C^1 on $[t_i, t_i+d]$ and $[t_i+d, t_f+d]$. Then for all t, $t \in (t_i , t_i+d)$, and $t \in (t_i+d, t_f)$ if $t_i+d < t_f$, there exists a neighborhood $B(y^*(t))$ of $y^*(t)$ in Y^* on which W^* is twice differentiable.

This can be shown by direct computation, based on relation (13) and on the fact that, from each point of $T(y^*(t_i),\alpha)$, for α sufficiently small, there emanates an open-loop trajectory in $Y^* \cup \theta$, generated by (u^*, v^*), that reaches θ.

Let

$$\Phi(\eta) \triangleq y_o + W^*(y)$$

$$\text{grad } \Phi \triangleq (1, \text{grad } W^*)$$

$$\text{grad } W^* \triangleq (\partial W^*/\partial y_1, \ldots \partial W^*/\partial y_n)$$

grad Φ is defined and continuous on $B(y^*(t))$ for all t, $t \in (t_i, t_i+d)$, and $t \in (t_i+d, t_f)$ if $t_i+d < t_f$.

At a point $y^*(t)$ where grad W^* is defined, we shall let

$$\tilde{\Lambda}(t) \triangleq (1,\tilde{\lambda}(t)), \quad \tilde{\lambda}(t) \triangleq \text{grad } W^*(y^*(t))$$

Consider state $x^k \in \pi^*$ such that $x^k(d) = y^*(t_k)$, $t_k \in (t_i, t_i+d)$, or $t_k \in (t_i+d, t_f)$ if $t_i+d < t_f$, and let y_o^k be such that $\eta^k \triangleq (y_o^k, y^*(t_k)) \in \Sigma(C)$. Since $V^*(x^k) = W^*(x^k(d))$, we have $\xi^k \triangleq (y_o^k , x^k) \in \Xi(C)$.

By considering a P-, an E-, and a PE-trajectory in $R\times(Y^* \cup \theta)$, emanating from η^k, generated from ξ^k by strategy pairs (p^*, e), (p, e^*) and (p^*, e^*), respectively, and by invoking similar arguments as in (1) for games without retardation, based on Lemma 4, we obtain†

$$(14) \quad \begin{aligned} &\min_{u \in K_u} H(\tilde{\lambda}(t),y^*(t), u, e^*(y^*(t-d))) = \\ &\max_{v \in K_v} H(\tilde{\lambda}(t),y^*(t), p^*(y^*(t-d)), v) = \\ &H(\tilde{\lambda}(t),y^*(t),p^*(y^*(t-d),e^*(y^*(t-d))) \end{aligned}$$

† See also the chapter "Further Geometric Aspects of Differential Games".

(15) $$H(\widetilde{\lambda}(t),\ y^*(t),\ p^*(y^*(t-d),\ e^*(y^*(t-d)) = 0$$

for $t = t_k$; where

$$H(\lambda,y,u,v) \triangleq f_o(y,u,v) + \lambda \cdot f(y,u,v)$$

From Lemma 5 one can easily deduce that, on $(t_i,\ t_i+d)$ and on $(t_i+d,\ t_f)$ if $t_i+d < t_f$, $\widetilde{\lambda}$ is a solution of *adjoint equation.*

(16) $$d\lambda/dt = -\ (\partial f/\partial y)^T\ \lambda$$

where

$$\frac{\partial f}{\partial y} \triangleq \left|\frac{\partial f_\nu(y,u,v)}{\partial y_\alpha}\right| \qquad \begin{matrix} \nu = 1,\dots\ n \\ \alpha = 1,\dots\ n \end{matrix}$$

evaluated for $y = y^*(t)$, $u = p^*(y^*(t-d))$, $v = e^*(y^*(t-d))$.

The *transversality condition* is obtained as follows .

Let (ζ_o,ζ), where $\zeta_o \in R$ and $\zeta \in E^n$, be a vector tangent to the set $(R \times \theta) \cap \Sigma(C)$ at point $y^f \triangleq y^*(t_f)$. Since θ is defined by $m(y) = 0$ where function m is of class C^1 on a neighborhood of θ , we have

(17) $$\zeta \cdot \text{grad } m(y^f) = 0$$

where $\text{grad } m \triangleq (\partial m/\partial y_1 \dots\ \partial m/\partial y_n)$.

Since $W^*(y) = w(y)$ for all $y \in \theta$, and w agrees on θ with some function which is of class C^1 on some neighborhood of θ , we have $y_o+w(y) = C$ on $(R \times \theta) \cap \Sigma(C)$, and accordingly

(18) $$\zeta_o + \zeta \cdot \text{grad } w(y^f) = 0$$

where $\text{grad } w \triangleq (\partial w/\partial y_1 \dots \partial w/\partial y_n)$.

Since $y^*(t)$, $p^*(y^*(t-d))$ and $e^*(y^*(t-d))$, in adjoint equation (16), are defined on $[t_i,\ t_f]$, $\widetilde{\lambda}(t)$ tends to a limit $\widetilde{\lambda}(t_f)$ as $t \to t_f$. Since $\widetilde{\Lambda}(t)$ is normal to $\Sigma(C)$ for all t on some interval $(t_f-\varepsilon,\ t_f)$, $\varepsilon > 0$, one can prove easily that $\widetilde{\Lambda}(t_f) \triangleq (1,\widetilde{\lambda}(t_f))$

is normal to $(R \times \theta) \cap \Sigma(C)$ at point y^f, and hence

$$(19) \qquad \zeta_o + \zeta \cdot \tilde{\lambda}(t_f) = 0$$

From (17), (18) and (19) it follows that there exist constants c_1, c_2, c_3, not all of which are zero, such that

$$(20) \qquad c_1(1,\tilde{\lambda}(t_f))+c_2(1,\text{grad } w(y^f))+c_3(0, \text{grad } m(y^f)) = 0$$

Since the zeroth component of the last term, in the left hand side of (20), is zero, we have $c_1 = -c_2$; and since grad $m(y^f) \neq 0$

$$c_1 = - c_2 \neq 0$$

Thus (20) can be rewritten as

$$(21) \qquad \tilde{\lambda}(t_f) = \text{grad } w(y^f) + c \text{ grad } m(y^f)$$

where $c = - c_3/c_1$.

Letting

$$\bar{f}_o \triangleq f_o(y^*(t_f), p^*(y^*(t_f-d)), e^*(y^*(t_f-d)))$$

$$\bar{f} \triangleq f(y^*(t_f), p^*(y^*(t_f-d)), e^*(y^*(t_f-d)))$$

we deduce from (21), and from relation (15) where we let $t_k \to t_f$

$$(22) \qquad c = - \frac{\bar{f}_o + \bar{f} \cdot \text{grad } w(y^f)}{\bar{f} \cdot \text{grad } m(y^f)}$$

When $t_i+d < t_f$, grad W^* may not be defined at point $y^*(t_i+d)$. The jump condition through the plane $y_n = t_i+d$ can be obtained by invoking the same arguments as in [1] .

Since $f_o(y^*(t), p^*(y^*(t-d)), e^*(y^*(t-d)))$, and $f(y^*(t), p^*(y^*(t-d)), e^*(y^*(t-d)))$ are continuous at $t = t_i+d$, the jump condition becomes

$$\tilde{\lambda}(t_i+d-0) = \tilde{\lambda}(t_i+d+0)$$

Hence, $\tilde{\lambda}$ is continuous at $t = t_i+d$.

At last we arrive at the following

Theorem 2. *If there exists a strategy pair* (p^*, e^*) *optimal on* X *that generates from* $x^i \in X$, x^i *at time* t_i, *optimal path* $\pi^* \triangleq \tilde{\pi}^{if}(y^*)$, $y^* \in S(x^i, p^*, e^*, t_f)$, *and if Assumptions* 1 *and* 3-7 *are satisfied, then for the problem stated above, there exists a nonzero vector function* $\tilde{\lambda} : [t_i, t_f] \to E^n$, *continuous on* $[t_i, t_f]$ *which is a solution of Equation* (16), *such that*

(i)
$$\min_{u \in K_u} H(\tilde{\lambda}(t), y^*(t), u, e^*(y^*(t-d))) =$$

$$\max_{v \in K_v} H(\tilde{\lambda}(t), y^*(t), p^*(y^*(t-d)), v) =$$

$$H(\tilde{\lambda}(t), y^*(t), p^*(y^*(t-d)), e^*(y^*(t-d)))$$

(ii) $H(\tilde{\lambda}(t), y^*(t), p^*(y^*(t-d)), e^*(y^*(t-d))) = 0$

for all $t \in (t_i, t_f)$.

(iii) $\tilde{\lambda}(t_f) = \text{grad}\, w(y^*(t_f)) + c\ \text{grad}\, m(y^*(t_f))$

where c *is given by* (22)

12. Closed vs Open Loop Control

If we let $d = 0$ in the statement of the above problem, termed Problem 1, we obtain a differential game without retardation,termed Problem 2 ; and the statement reduces to the one given in (2) and (3). Then, the state of the game is a point in n-dimensional Euclidean space E^n .

Assumptions 1 and 2, and 4-6, reduce to similar assumptions previously introduced in (2)(3), Assumption 3 is of course satisfied, and Assumption 7 is useless. Most of the arguments introduced in the present chapter are no longer valid, nevertheless, as shown in (2)(3), the following theorem which is similar to Theorem 2 holds. In its statement, we shall use the same notation as in Theorem 2, as far as is allowed by consistency, keeping in mind the fact that $X = Y$, and that the concepts of paths and trajectories coalesce .

Theorem 3. *If* $d = 0$, *and if there exists a strategy pair* (p^*, e^*) *optimal on* Y *that generates from* $y^i \in Y$, y^i *at time* t_i, *optimal path* π^*, *represented by* $y^* : t \mapsto y = y^*(t)$, $t \in [t_i, t_f]$, *and if Assumptions* 1 *and* 4-6 *are satisfied, then there exists a nonzero vector function* $\tilde{\lambda} : [t_i, t_f] \to E^n$, *continuous on* $[t_i, t_f]$, *which is a solution of adjoint equation, such that*

$$\text{(i)} \quad \operatorname*{Min}_{u \in K_u} H(\tilde{\lambda}(t), y^*(t), u, e^*(y^*(t))) =$$

$$\operatorname*{Max}_{v \in K_v} H(\tilde{\lambda}(t), y^*(t), p^*(y^*(t)), v) =$$

$$H(\tilde{\lambda}(t), y^*(t), p^*(y^*(t)), e^*(y^*(t)))$$

$$\text{(ii)} \quad H(\tilde{\lambda}(t), y^*(t), p^*(y^*(t)), e^*(y^*(t))) = 0$$

for all $t \in (t_i, t_f)$.

$$\text{(iii)} \quad \tilde{\lambda}(t_f) = \text{grad } w(y^*(t_f)) + c \text{ grad } m(y^*(t_f)) .$$

Another step is taken when Problem 2 is replaced by Problem 3 where strategies are functions of time only†, that is when one considers open-loop controls $\tilde{u} : R \to K_u$ and $\tilde{v} : R \to K_v$. If strategies are piece-wise continuous functions of time, Assumptions 1 and 4 can be released. Theorem 3 is replaced by

Theorem 4. *If* $d=0$, *and if there exists an open-loop control pair* (u^*, v^*) *optimal on* Y *that generates from* $y^i \in Y$, y^i *at time* t_i , *optimal path* π^*, *represented by* $y^* : t \mapsto y = y^*(t)$, $t \in [t_i, t_f]$, *and if Assumptions* 5 *and* 6 *are satisfied, then there exists a*

---

† This is a special case of strategies depending on the state.

nonzero vector function $\tilde{\lambda} : [t_i, t_f] \to E^n$, *continuous on* $[t_i, t_f]$, *which is a solution of adjoint equation, such that*

$$\text{(i)} \quad \underset{u \in K_u}{\text{Min}} \; H(\tilde{\lambda}(t), y^*(t), u, v^*(t)) =$$

$$\underset{v \in K_v}{\text{Max}} \; H(\tilde{\lambda}(t), y^*(t), u^*(t), v) =$$

$$H(\tilde{\lambda}(t), y^*(t), u^*(t), v^*(t))$$

$$\text{(ii)} \quad H(\tilde{\lambda}(t), y^*(t), u^*(t), v^*(t)) = 0$$

for all $t \in (t_i, t_f)$

$$\text{(iii)} \quad \tilde{\lambda}(t_f) \text{ grad } w(y^*(t_f)) + c \text{ grad } m(y^*(t_f))$$

When one applies conditions (i)-(iii) of Theorem 4, for given initial point $y^i \in Y$, one may obtain a control pair that satisfies the necessary conditions for optimality in Problem 3,and that depends on y^i. We shall denote such a control pair by $(\bar{u}(y^i; \cdot), \bar{v}(y^i; \cdot))$

$$\bar{u}(y^i; \cdot) : R \to K_u$$

$$\bar{v}(y^i; \cdot) : R \to K_v$$

Now, in Problem 1, if we define u^* and v^* by relations (12), and if we replace in Theorem 2 $p^*(y^*(t-d))$ and $e^*(y^*(t-d))$ by $u^*(t)$ and $v^*(t)$, respectively, we see that trajectory $\rho^* \triangleq h(\pi^*)$ is a candidate to be an optimal path in Problem 3 .

Likewise, in Problem 2, if we let $u^*(t) = p^*(y^*(t))$ and $v^*(t) = e^*(y^*(t))$, $t \in [t_i, t_f]$, we see that optimal path π^* is a candidate to be an optimal path in Problem 3.

Now let us consider the case where, in Problem 3, there emanates from each point of Y one and only one path, that reaches target θ and that satisfies the necessary conditions of Theorem 4.

For initial point $y^i \in Y$, let us denote by $\bar{u}(y^i;\cdot)$ and $\bar{v}(y^i;\cdot)$ the controls that generate this path.

Then $\rho^* \triangleq h(\pi^*)$, where π^* is an optimal path in Problem 1, is one of these paths, namely the one that emanates from $x^i(d)$, and according to relations (12), for $t_k \in [t_i, t_f]$, we have

$$p^*(y^*(t_k-d)) = u^*(t_k)$$
$$e^*(y^*(t_k-d)) = v^*(t_k)$$

Let β be a strictly positive number and, for $t_k \in [t_i-d, t_f]$, let

$$T(y^*(t_k),\beta) \triangleq \{ y : y_n = t_k, \ \|y-y^*(t_k)\| < \beta \}$$

For β sufficiently small and $y \in T(y^*(t_k-d),\beta)$, $t_k \in [t_i, t_f]$, there exists a state $x \in X$ such that $x(0) = y$ and $x(d) = y^*(t_k)$.

From Assumption 2, (p^*, e^*) is optimal at that state x and, accordingly

$$p^*(y) = u^*(t_k)$$
$$e^*(y) = v^*(t_k)$$

Clearly, for β sufficiently small, this property holds for all $y \in T(y^*(t_k-d),\beta)$ and for all $t_k \in [t_i, t_f]$. In other words, if we let

$$T^*(t_i-d, t_f-d, \beta) \triangleq \bigcup_{t_k \in [t_i, t_f]} T(y^*(t_k-d),\beta)$$

for β sufficiently small, p^* and e^* agree on $T^*(t_i-d, t_f-d, \beta)$ with functions of time $R \to K_u$ and $R \to K_v$, respectively.

Now let $y \in Y \cap T(y^*(t_k), \beta)$ for some $t_k \in [t_i, t_f]$. For β sufficiently small, there exists a state $x \in X$ such that $x(0) = y^*(t_k-d)$ and $x(d) = y$. y is the initial point of a path that reaches target θ and that satisfies the necessary conditions of Theorem 4.

Since (p^*, e^*) is optimal at state x, we have

$$\bar{u}\,(y\ ;\ t_k) = p^*\,(y^*(t_k - d))$$

$$\bar{v}\,(y\ ;\ t_k) = e^*\,(y^*(t_k - d))$$

and, for β sufficiently small, this property holds for all $y \in Y \cap T(y^*(t_k), \beta)$ and for all $t_k \in [t_i, t_f]$. In other words, if we let

$$T^*(t_i, t_f, \beta) \triangleq \bigcup_{t_k \in [t_i, t_f]} T(y^*(t_k), \beta)$$

for β sufficiently small, there exists a control pair $(\tilde{u}, \tilde{v})$, $\tilde{u} : R \to K_u$ and $\tilde{v} : R \to K_v$, that satisfies the necessary conditions of Theorem 4, such that

$$\tilde{u} = \bar{u}(y\ ;\ \cdot)$$

$$\tilde{v} = \bar{v}(y\ ;\ \cdot)$$

for all $y \in Y \cap T^*(t_i, t_f, \beta)$.

Indeed, it may well be that $(\tilde{u}, \tilde{v})$ is not an optimal pair in the sense of Theorem 4.

13. Example

Next let us consider a game described by

$$\frac{dy_1}{dt} = 0, \qquad \frac{dy_2}{dt} = u + v, \qquad \frac{dy_3}{dt} = 1 \tag{23}$$

with constraints

$$K_u\ :\ -1 \leqslant u \leqslant 0$$

$$K_v\ :\ -1 \leqslant v \leqslant 0$$

Target θ is defined by : $\theta \triangleq \{\ y = (y_1, y_2, y_3)\ :\ \ y_2 \leqslant 0\ \}$

The cost functions $f_o(y,u,v)$ and $w(y)$ are

$$f_o(y,u,v) \triangleq 1-y_1y_2$$

$$w(y) \triangleq 0$$

The time lag is d, $d > 0$.

It is understood that y_2 is non-negative.

The H-function is

$$H(\lambda,y,u,v) = 1-y_1y_2 + \lambda_2(u+v) + \lambda_3$$

The adjoint equations are

$$\frac{d\lambda_1}{dt} = y_2, \qquad \frac{d\lambda_2}{dt} = y_1, \qquad \frac{d\lambda_3}{dt} = 0$$

so that

(24) $\qquad \lambda_3(t) \equiv \text{constant}$

From the transversality condition, we obtain

$$\lambda_1(t_f) = \lambda_3(t_f) = 0$$

so that, in view of (24)

$$\lambda_3(t) \equiv 0$$

From condition (i) of Theorem 2 we obtain

$$u^*(t) = -1 \quad \text{if} \quad \lambda_2(t) > 0$$

$$u^*(t) = 0 \quad \text{if} \quad \lambda_2(t) < 0$$

$$v^*(t) = 0 \quad \text{if} \quad \lambda_2(t) > 0$$

$$v^*(t) = -1 \quad \text{if} \quad \lambda_2(t) < 0$$

and so, by condition (ii)

(25) $\qquad 1 - y_1^*(t)\, y_2^*(t) - \tilde{\lambda}_2(t) = 0$

For $u = u^*(t)$ and $v = v^*(t)$, Eqs. (23) may be rewritten as

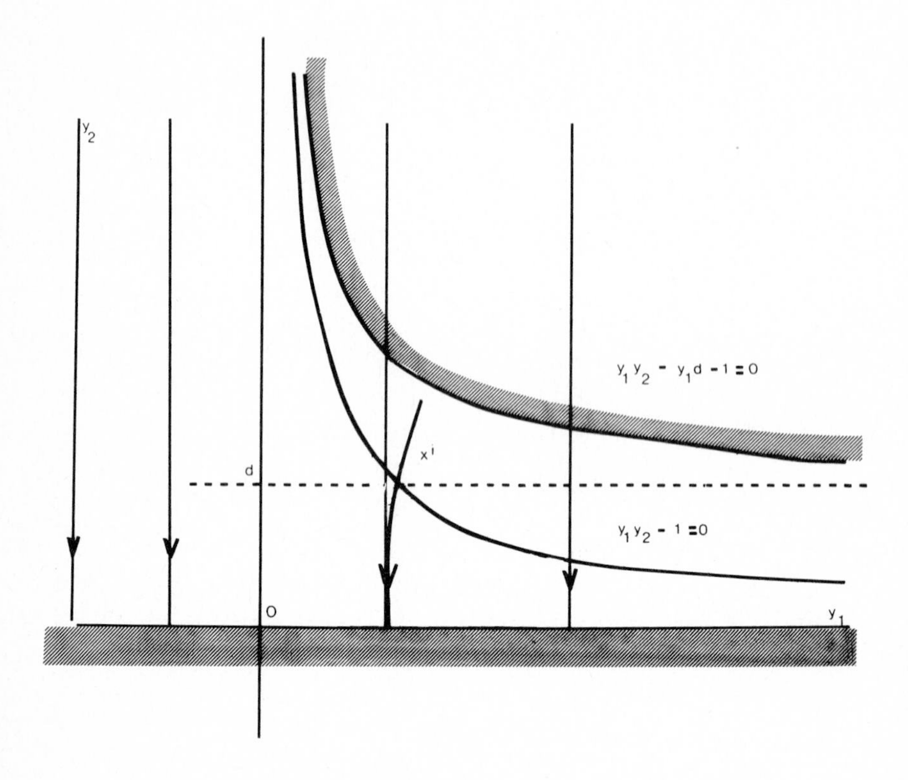

Fig.1. Example. Projection on the y_1-y_2 plane of target θ, paths and domain Y .

(26) $$\frac{dy_1}{dt} = 0, \quad \frac{dy_2}{dt} = -1, \quad \frac{dy_3}{dt} = 1$$

Upon integration of (26) we obtain

$$y_1^*(t) = y_1^\circ$$

$$y_2^*(t) = y_2^\circ - t$$

By (25) the sign of $\tilde{\lambda}_2(t)$ switches on the surface

$$M \triangleq \{ y = (y_1, y_2, y_3) : \quad 1 - y_1 y_2 = 0 \}$$

Consequently, the strategy pair that satisfies conditions (i) - (iii) of Theorem 2 is given by

$$p^*(y) = 0$$

$$e^*(y) = -1$$

for $y_1 > 0$ together with $y_1 y_2 - y_1 d - 1 > 0$; and

$$p^*(y) = -1$$

$$e^*(y) = 0$$

for $y_1 \leqslant 0$, or $y_1 > 0$ together with $y_1 y_2 - y_1 d - 1 < 0$.

In order that Assumption 4 be satisfied, we shall let

$$Y = \{ y : y_1 \leqslant 0 \} \cup \{ y : \ y_1 > 0, \ y_1 y_2 - y_1 d - 1 < 0 \}$$

and accordingly

$$X = \{ x : \ x(s) \in Y \quad s \in [0, d] \}$$

At last, from the discussion of the paragraph before, it follows that there is no optimal path emanating from an initial state $x^i \in X$ if the end point of x^i, namely $y^i \triangleq x^i(d)$, is such that $1 - y_1^i y_2^i < 0$.

References

(1) A. BLAQUIERE, F. GERARD and G. LEITMANN, Quantitative and Qualitative Games, Academic Press Inc., 1969

(2) A. BLAQUIERE and G. LEITMANN, Jeux Quantitatifs, Mémorial des Sciences Mathématiques, Fasc. 168, Gauthier-Villars, 1969

(3) A. BLAQUIERE, An Introduction to Differential Games, in "Differential Games and Related Topics" Edited by H.W. KUHN and G.P. SZEGÖ, North-Holland, 1971

(4) A. BLAQUIERE, Differential Games with Time Lag, in "Techniques of Optimization", Academic Press Inc., 1972

(5) A. BLAQUIERE and P. CAUSSIN, Jeux différentiels avec retard, CRAS, Paris, t.272, p.1607-1609, 14 Juin 1971, Série A

(6) A. BLAQUIERE and P. CAUSSIN, Jeux différentiels avec retard, propriétés globales d'une surface du jeu, CRAS, Paris, t.273, p.326-328, 2 Août 1971, Série A

(7) A. BLAQUIERE and P. CAUSSIN, Jeux différentiels avec retard ; démonstration du théorème de base, CRAS, Paris, t.273, p.936-938, 15 Novembre 1971, Série A

(8) A. BLAQUIERE and P. CAUSSIN, Jeux différentiels avec retard, théorème de min-max, CRAS, Paris, t.273, p.1178-1180, 8 Décembre 1971, Série A

(9) A. BLAQUIERE and P. CAUSSIN, Jeux différentiels avec retard, extension du théorème de base, CRAS, Paris, t.274, p.1008-1010, 20 Mars 1972, Série A .

DIFFERENTIAL GAMES WITH INFORMATION TIME LAG

Michael D. CILETTI†

1. Introduction

In discussing some possible extensions of his work on differential games (DG), Isaacs mentioned the need for a relaxation of the assumption that each player in a DG has perfect information (Ref.1). In many practical situations of dynamic conflict, the assumption cannot be justified. A particular problem which Isaacs cited is that in which a player in a DG has a time lag associated with the availability of his measurements of the opponent's state vector. In practice, such lags can occur in a DG when data processing and/or transmission takes place prior to the generation of control signals. The importance of differential games with information time lag (DGWITL) stems from the practical need to know precisely how much information delay leads to intolerable performance degradation and the need to synthesize a strategy to use time-delayed data. At the same time, knowledge of the effects of time delay might permit relaxation of system design specifications to achieve an overall economy of design while maintaining acceptable performance.

† Systems Control, Inc., 260 Sheridan Ave., Palo Alto, California, 94306, USA.

The historical basis for the problem appears to be the classical "bomber-battleship duel", in which a bomber must take into account ordnance delivery time in deciding when to strike an evasive battleship. Since 1950, a number of papers have sucessfully treated bomber-battleship duel as a multistage game with probabilistic strategies (Refs.2-9). However, little was known about the effects of information delay for more general dynamic games. Subsequently, finite games, multistage games and linear differential games in which one player has an information time lag were treated (Refs. 10-13), and a class of linear N player differential games with information time lag was solved by function space methods in (Ref. 14). Recently, an extension of the well-known Hamilton-Jacobi theory for optimal control and the "main equation" analysis of Isaacs was developed to treat DGWITL (Ref. 15). In an independent work, Sokolov and Chernous'ko (Refs. 16-17) showed that, under certain conditions of separability, a class of DGWITL can be considered equivalent to a related DG without information time lag. Petrosjan (Ref. 18) treated a discretization of a DGWITL within the context of mixed strategies.

The presentation of this chapter will include a development of the main theoretical results for DGWITL and the solution of two examples. We show that the notion of "value" in a DG with perfect information leads to a notion of "potential value" in differential games with information time lag. Under appropriate smoothness assumptions, the potential value will be shown to satisfy a generalized Hamilton Jacobi equation (Main Equation). This and other results will be related to the well known theory for DG with perfect information ; our examples will be used to clarify the theory and demonstrate the practical importance of information delay in dynamic conflict. In particular, information delay will be considered in relation to the problem of strategy synthesis.

2. Differential Games with Perfect Information

As a point of reference for our discussion of DGWITL we begin with a brief summary of differential games with perfect information. The dynamic systems controlled, respectively, by the pursuer (P) and the evader (E) are described by :

$$\dot{x} = f(x,\tilde{u}(t),t) \tag{1a}$$

$$\dot{y} = g(y,\tilde{v}(t),t) \tag{1b}$$

where $x,\ y \in R^n$, $\tilde{u}(t) \in R^q$, $\tilde{v}(t) \in R^m$. At each time t the control inputs $\tilde{u}(t)$ and $\tilde{v}(t)$ are to be determined by strategies p and e which are members of sets S_P and S_E, where

$$S_P = \{p : R^n \times R^n \times R^1 \to K_u \subset R^q\},$$

$$S_E = \{e : R^n \times R^n \times R^1 \to K_v \subset R^m\},$$

K_u and K_v are locally compact, and members of S_P and S_E are continuous in t and Lipschitzian in (x,y). Given regions G_P and G_E to which motions of (1a) and (1b) are restricted and a closed target set $\theta \subset G = G_P \times G_E$, we form $S_a \subset S_P \times S_E$ to include only those strategy pairs which produce for any initial phase in $G-\theta$ a solution which reaches θ without leaving G . Accordingly, the terminal time, t_f (depending on x_o, y_o, t_o, p, e), is the first instant at which a solution to (1) penetrates θ. With $(p,e) \in S_a$ and $(x_o,\ t_o,\ y_o,\ t_o) \in G$ we associate a scalar payoff made by P to E and defined by :

$$\begin{aligned} J(x_o,y_o,t_o,p,e) = & \int_{t_o}^{t_f} L_1(x(\alpha),\tilde{u}(\alpha),\alpha)d\alpha \\ & + \int_{t_o}^{t_f} L_2(y(\alpha),\tilde{v}(\alpha),\alpha)d\alpha \\ & + \tilde{W}(x(t_f),y(t_f),t_f) \end{aligned} \tag{2}$$

where $x(\cdot)$ and $y(\cdot)$ are understood to be the solution to (2) associated with the strategy pair (p,e), and $\tilde{u}$ and $\tilde{v}$ at each α are specified by p and e respectively.

The methodology of game theory then centers about the task of finding a saddlepoint of $J(\cdot)$ w.r.t. p and e. The well known results are that under appropriate smoothness assumptions the saddlepoint† or value, $V^o(\cdot)$, satisfies a Hamilton-Jacobi type partial differential equation. The analysis also produces an implicit definition of the optimal strategies in terms of the states and the components of the gradient of $V^o(\cdot)$. Solution of the HJE leads to closed form explicit definitions of the strategies p^o and e^o which induce the saddlepoint. Thus, implementation of the strategies requires that $x(t)$ and $y(t)$ be available to each player at every t .

3. Differential Games with Information Time Lag (DGWITL)

3.1 - *Problem introduction*

The class of differential games with information time lag which are discussed in this chapter are those in which the evader, E, has access to $x(t-\sigma)$ instead of $x(t)$ at each time t, where σ is a fixed non-negative parameter. First, the problem will be introduced and discussed in a heuristic manner. Then we reformulate the problem with attention to detail and develop for DGWITL an extension of the Hamilton Jacobi theory for optimal control and the Main Equation analysis of Isaacs. Although the analysis is presented for games in which E has information time lag, it is easy to extend it to apply to the case where P has delay or the case where both P and E have delay.

---

† Under an existence assumption

We begin with a discussion of some basic notions in DGWITL. Consider the situation in which two dynamic systems are governed by (1) and two players (P and E) choose $\tilde{u}(t)$ and $\tilde{v}(t)$ at each time t in competition as represented by a real-valued payoff (2), made by P to E at the end of the game. It is inherent to game theory that a player must base his choice of strategy on a consideration of his opponent's possible present and future actions as well as his own. This leads naturally to a "max min" strategy for E, which enables E to establish a lower bound on the payoff, and a "min max" strategy for P, which establishes an upper bound on the payoff. For systems with dynamics prescribed by (1) (i.e., no memory) and perfect information†, it's clear that P and E's choice of control values at time t should not be conditioned on past values of x and y. That is, each player need only consider

$$J(x(t),y(t),t,p,e) = \int_t^{t_f} L_1(x(\alpha),\tilde{u}(\alpha),\alpha)d\alpha + \int_t^{t_f} L_2(y(\alpha),\tilde{v}(\alpha),\alpha)d\alpha + \tilde{W}(x(t_f),y(t_f),t_f) \tag{3}$$

and its dependence on admissible p and e. The quantity given by (3) describes, for a given pair (p,e), the payoff that remains to be determined by present and future behavior of P and E, given x(t) and y(t). In a similar manner, we observe that in our differential game with information time lag the quantity given by

† Information patterns for P and E are such that x(t) and y(t) are available at each time t during play.

$$J(x(t-\sigma),y(t-\sigma),t-\sigma,p,e) = \int_{t-\sigma}^{t_f} L_1(x(\alpha),\tilde{u}(\alpha),\alpha)d\alpha + \int_{t-\sigma}^{t_f} L_2(y(\alpha),\tilde{v}(\alpha),\alpha)d\alpha + \tilde{W}(x(t_f),y(t_f),t_f) \tag{4}$$

represents the payoff that remains to be determined at time $t-\sigma$ as seen by P. (We will, for the moment, not be concerned with relating $\tilde{u}$ and p, $\tilde{v}$ and e).

At time t in a differential game with information time lag, the payoff that remains to be determined is not the same for both players. From P's point of view, with perfect information, (3) defines the payoff that remains to be determined. On the other hand, from E's point of view, the payoff that remains to be determined at time t is given by

$$\begin{aligned} J_r(x(t-\sigma),y(t),t,p,e) &\triangleq J(x(t-\sigma),y(t-\sigma),t-\sigma,p,e) - \int_{t-\sigma}^{t} L_2(y(\alpha),\tilde{v}(\alpha),\alpha)d\alpha \\ &= \int_{t-\sigma}^{t_f} L_1(x(\alpha),\tilde{u}(\alpha),\alpha)d\alpha + \int_{t}^{t_f} L_2(y(\alpha),\tilde{v}(\alpha),\alpha)d\alpha + \tilde{W}(x(t_f),y(t_f),t_f), \end{aligned} \tag{5}$$

since $y(t-\sigma)$ and $\tilde{v}(\alpha)$, $t-\sigma \leqslant \alpha < t$ are known to E. Inspection of (5) reveals that the remaining payoff from E's point of view depends on E's present and future control values and P's past, present and future control values.

Figure 3.1 illustrates the situation in which at time t, E has access to $y(\alpha)$, $t-\sigma \leqslant \alpha \leqslant t$, and $x(t-\sigma)$. At time t, E has no means of knowing the P-system trajectory over $[t-\sigma, t]$; possible trajectories are indicated by the dashed lines, with the actual trajectory being the solid line. It is important to observe that E can (1) examine all possible E moves emanating from $y(t)$, (2) examine all possible P moves emanating from $x(t-\sigma)$ and occuring over $t-\sigma$ to t, and (3) examine all possible P moves emanating from the possible $x(t)$ which are conditioned on $x(t-\sigma)$ and possible P moves over $t-\sigma$ to t.

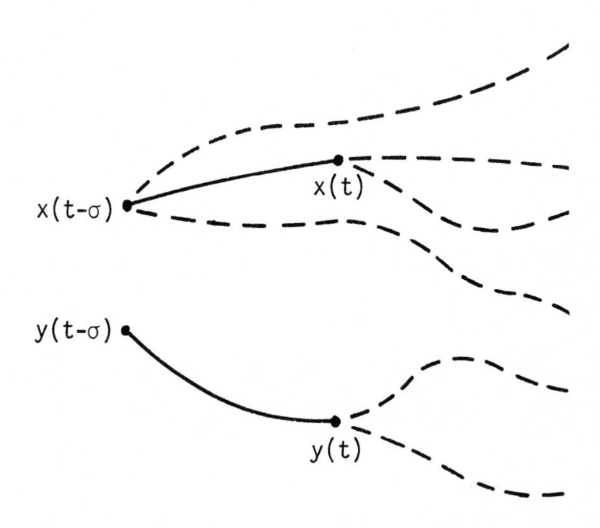

Fig. 3.1 Pursuer Trajectory Uncertainty

In this manner, E can examine the possible outcomes of the game, even though $x(t)$ is unknown, and choose a strategy which bounds-from below-the possible payoffs. Because the information is continuously available, E should choose $\tilde{v}(t)$ in consideration of P's possible choices for $\tilde{u}(t)$ and in anticipation of the evolution of data that results in the future availability of $x(t+\Delta t-\sigma)$ and $y(t+\Delta t)$. Hence, we seek for E a payoff bound that is conditioned on the data and a feedback strategy that employs the data in an optimal manner.

Observe that

$$J_r(x(t-\sigma),y(t),t,p,e) \cong \tilde{W}(x(t_f),y(t_f),t_f)$$

$$+ \int_{t-\sigma+\Delta t}^{t_f} L_1(x(\alpha),\tilde{u}(\alpha),\alpha)d\alpha + \int_{t+\Delta t}^{t_f} L_2(y(\alpha),\tilde{v}(\alpha),\alpha)d\alpha$$

$$+ L_1(x(t-\sigma),\tilde{u}(t-\sigma),t-\sigma)\Delta t + L_2(y(t),\tilde{v}(t),t)\Delta t$$

$$\cong J_r(x(t+\Delta t-\sigma),y(t+\Delta t),t+\Delta t,p,e) + L_1(x(t-\sigma),\tilde{u}(t-\sigma),t-\sigma)\Delta t$$

$$+ L_2(y(t),\tilde{v}(t),t)\Delta t$$

$$\cong J_r(x(t-\sigma),y(t),t,p,e) + \nabla_x J_r'(x(t-\sigma),y(t),t,p,e)\Delta x$$

$$+ \nabla_y J_r'(x(t-\sigma),y(t),t,p,e)\Delta y$$

$$+ (\partial/\partial t)J_r(x(t-\sigma),y(t),t,p,e)\Delta t + L_1(x(t-\sigma),\tilde{u}(t-\sigma),t-\sigma)\Delta t$$

$$+ L_2(y(t),\tilde{v}(t),t)\Delta t,$$

and note that†

$$\Delta x \cong f(x(t-\sigma),\tilde{u}(t-\sigma),t-\sigma)\Delta t \qquad \Delta y \cong g(y(t),t,\tilde{v}(t),t)\Delta t$$

$$0 \cong (\partial J_r/\partial t)\Delta t+\nabla_x J_r'(x(t-\sigma),y(t),t,p,e)f(x(t-\sigma),\tilde{u}(t-\sigma),t-\sigma)\Delta t$$

$$+ \nabla_y J_r'(x(t-\sigma),y(t),t,p,e)g(y(t),\tilde{v}(t),t)\Delta t$$

$$+ L_1(x(t-\sigma),\tilde{u}(t-\sigma),t-\sigma)\Delta t + L_2(y(t),\tilde{v}(t),t)\Delta t$$

---

† Prime denotes vector transpose.

At this point, it becomes more apparent that the following generalized Hamilton-Jacobi equation can be obtained[†]

$$\max_{\tilde{v}(t)} \min_{\tilde{u}(t-\sigma)} \{ < V_{x(t-\sigma)}, f(x(t-\sigma),\tilde{u}(t-\sigma),t-\sigma) >$$

$$+ < V_{y(t)}, g(y(t),\tilde{v}(t),t) >$$

$$+ L_1(x(t-\sigma),\tilde{u}(t-\sigma),t-\sigma) + L_2(y(t),\tilde{v}(t),t)\}$$

$$= - \partial V(x(t-\sigma),y(t),t)/\partial t$$

provided we can define a lower value as

$$V(x(t-\sigma),y(t),t) \triangleq \max_{e} \min_{p} J_r(x(t-\sigma),y(t),t,p,e).$$

The boundary condition for the equation is obtained from the following optimal control problem :

$$V(x(t_f-\sigma),y(t_f),t_f) = \min_{\tilde{u}(\cdot)} \int_{t_f-\sigma}^{t_f} L_1(x(\alpha),\tilde{u}(\alpha),\alpha)d\alpha + G(x(t_f),y(t_f)t_f)$$

The boundary condition for the lower value corresponds to the game beginning and ending at the same instant. To determine the payoff bound, E must examine all possible evader control functions over $t_f-\sigma$ to t_f.

It should be noted here that $V_{x(t-\sigma)}$ is the gradient of V with respect to $x(t-\sigma)$ and that the max min operation is performed with respect to control choices at two time instants, $t-\sigma$ and t. In relation to the principle of optimality, we have tacitly assumed that $V(\cdot)$ is known at time $t+\Delta t$ and over the states emanating from $x(t-\sigma)$ and $y(t)$. Therefore, we optimize with respect to $\tilde{u}(\cdot)$ over $t-\sigma$ to $t-\sigma+\Delta t$ and $\tilde{v}(\cdot)$ over t to $t+\Delta t$ and shrink Δt to zero to obtain the generalized Hamilton-Jacobi equation (HJE). Note too that, as $\sigma \to 0$, the HJE and the boundary condition agree with the results for perfect information DG.

---

† $<,>$ denotes the usual inner product for R^n

3.2 - *Problem formulation - DGWITL*

In this section we present a detailed development of the generalized Hamilton Jacobi equation for differential games with information time lag.

Consider two dynamical systems described by

$$\dot{x} = f(x,\tilde{u}(t),t) \tag{6}$$

$$\dot{y} = g(y,\tilde{v}(t),t) \tag{7}$$

where x and y are "*plant states*", $f : (x,\mu,t) \mapsto f(x,\mu,t)$ and $g : (y,\beta,t) \mapsto g(y,\beta,t)$ are n-vector functions continuous in t and Lipschitzian in the pairs (x,μ) and (y,β) respectively for $x,y \in R^n$, $\mu \in R^q$ and $\beta \in R^m$. Also, $\tilde{u}(t)$ and $\tilde{v}(t)$ are q and m vectors determined by functions $\tilde{u}(\cdot)$ and $\tilde{v}(\cdot)$ defined on $R^n \times R^1 \times R^n \times R^1$.

Let S_P denote the class of functions p which are defined on $R^n \times R^1 \times R^n \times R^1$, continuous in τ and t, and Lipschitzian in the pair (x,y), and the values of which belong to the locally compact set $K_u \subset R^q$. Members of S_P will be called admissible ideal exploitation strategies for P, the pursuer. Likewise, let S_E denote the class of functions e which are defined on $R^n \times R^1 \times R^n \times R^1$, continuous in τ and t, and Lipschitzian in the pair (x,y) and the values of which belong to the locally compact set $K_v \subset R^m$. Members of S_E will be called admissible strategies for E, the evader.

Let t_o and $\sigma \geqslant 0$ be given, and let $\tau_o = t_o - \sigma$. For a given $p \in S_P$ and $e \in S_E$ we let φ_{pe} and ψ_{pe} denote the unique solutions to :

$$(d/dt)x(t-\sigma) = f(x(t-\sigma),p(x(t-\sigma),t-\sigma,y(t),t),t-\sigma) \tag{8}$$

$$(d/dt)y(t) = g(y(t),e(x(t-\sigma),t-\sigma,y(t),t),t) \qquad t \geqslant t_o \tag{9}$$

with

$$x(\tau_o) = x_{o\sigma}, \qquad y(t_o) = y_o$$

Formally :

$$\varphi_{pe} = \varphi_{pe}(t;x_{o\sigma},\tau_o,y_o,t_o), \quad \psi_{pe} = \psi_{pe}(t;x_{o\sigma},\tau_o,y_o,t_o), \quad t \geqslant t_o$$

and

$$(d/dt)\,\varphi_{pe}(t;x_{o\sigma},\tau_o,y_o,t_o) = f(\varphi_{pe}(t;x_{o\sigma},\tau_o,y_o,t_o),$$

$$p(\varphi_{pe}(t;x_{o\sigma},\tau_o,y_o,t_o),\ t-\sigma,\psi_{pe}(t;x_{o\sigma},\tau_o,y_o,t_o),t),t-\sigma)$$

$$(d/dt)\,\psi_{pe}(t;x_{o\sigma},\tau_o,y_o,t_o) = g(\psi_{pe}(t;x_{o\sigma},\tau_o,y_o,t_o),$$

$$e(\varphi_{pe}(t;x_{o\sigma},\tau_o,y_o,t_o),\ t-\sigma,\psi_{pe}(t;x_{o\sigma},\tau_o,y_o,t_o),t),t)$$

and

$$\varphi_{pe}(t_o;x_{o\sigma},\tau_o,y_o,t_o) = x_{o\sigma} = x(t_o-\sigma)\,,$$

$$\psi_{pe}(t_o;x_{o\sigma},\tau_o,y_o,t_o) = y_o$$

Note that $\varphi_{pe}(t)$ is the observed output of (8), that is, $\varphi_{pe}(t) \equiv x(t-\sigma)$. Where confusion is unlikely, $x(t-\sigma)$, $y(t)$, $\varphi_{pe}(t)$, $\psi_{pe}(t)$ will be used in lieu of the more cumbersome and rigorous notation.

To (8) and (9), we append the equation : $\dot{\tau} = 1$, $\tau(t_o) = \tau_o = t_o-\sigma$, that is, $\tau(t) = t-\sigma$.

Definition. Corresponding to solutions of (8) and (9), define

$$\hat{M}_{pe}(t) = \{(\varphi_{pe}(\alpha;x_{o\sigma},\tau_o,y_o,t_o),\tau(\alpha)) : t_o \leqslant \alpha \leqslant t\ \}$$

$$\check{M}_{pe}(t) = \{(\psi_{pe}(\alpha;x_{o\sigma},\tau_o,y_o,t_o),\alpha) : t_o \leqslant \alpha \leqslant t\ \}$$

$\hat{M}_{pe}(t)$ and $\check{M}_{pe}(t)$ are motions of P and E up to time t which can be observed by E. Points in $\hat{M}_{pe}(t)$ are points along the path of (8) from $t_o-\sigma$ to $t-\sigma$.

Definition. The set of observation quadruples M_{pe} is defined by :

$$M_{pe} = \{(\varphi_{pe}(\alpha;x_{o\sigma},\tau_o,y_o,t_o),\tau(\alpha),\psi_{pe}(\alpha;x_{o\sigma},\tau_o,y_o,t_o),\alpha) : t_o \leqslant \alpha\}$$

It consists of points along the E-viewable motions of P and E, and will be called an *E-viewable-motion*. For a given $t \geqslant t_o$, E's knowledge of the system is updated by the past phase of P, $(x(t-\sigma),t-\sigma) = (\varphi_{pe}(t),\tau(t))$, and E's present phase, $(y(t),t) = (\psi_{pe}(t),t)$.

Let G_P and G_E be subsets of $R^n \times R^1$ to which all motions of (6) and (7) are restricted with $G = G_P \times G_E$. Let the target set θ be a closed subset of G. Let $\overline{C}_u$ denote the set of all q-vector piecewise continuous functions $\overline{u}(\cdot)$ defined on $[0,\sigma]$. For a given t, $\overline{u} \in \overline{C}_u$ and $\varphi_{pe}(t;x_{o\sigma},\tau_o,y_o,t_o)$, let $\overline{\varphi}_u = \overline{\varphi}_u(\cdot;\varphi_{pe}(t;x_{o\sigma},\tau_o,y_o,t_o),\tau(t))$ denote the unique solution to :

$$(d/d\alpha)\,\overline{\varphi}_u(\alpha) = f(\overline{\varphi}_u(\alpha),\overline{u}(\alpha),\alpha) \tag{10 a}$$

$$\overline{\varphi}_u(t-\sigma) = \varphi_{pe}(t;x_{o\sigma},\tau_o,y_o,t_o) = x(t-\sigma) \tag{10 b}$$

Definition. A piecewise-continuous function $\overline{v}(\cdot) : [0,\sigma] \to K_v$ will be called an *admissible conditioning control* for E. $\overline{C}_v$ will denote the set of admissible conditioning controls for E.

For a given $\overline{v} \in \overline{C}_v$ we let $\overline{\psi}_v$ denote the unique solution to

$$(d/dt)y(t) = g(y,\overline{v}(t),t), \quad t_o-\sigma \leqslant t \leqslant t_o \tag{11}$$

$$y(t_o-\sigma) = y_o$$

In the work that follows we will consider y_o to be obtained from $\overline{v}$ through (11), that is, $y_o = y(t_o) = \overline{\psi}(t_o)$

Definition. The observed quadruple $(\varphi_{pe}(t),\tau(t),\psi_{pe}(t),t) = (x(t-\sigma),t-\sigma,y(t),t)$ is a *potentially terminal phase* if there exists a $\bar{u} \in \bar{C}_u$, a corresponding $\bar{\varphi}_u$ which solves (10) subject to $\bar{\varphi}_u(t-\sigma) = \varphi_{pe}(t)$ and $(\bar{\varphi}_u(t),t,y(t),t) \in \theta$. $\bar{u}$ will be called a P-*terminal control*. Note that $\bar{u} = \bar{u}(\cdot;\varphi_{pe}(t),\tau(t),\psi_{pe}(t),t)$.

Definition. The *preliminary target set*, θ_p, $\theta_p \subset R^n \times R^1 \times R^n \times R^1$, is the set of all potentially terminal observed phases.

Definition. For a given $x_{o\sigma},\tau_o,y_o,t_o$, $p \in S_P$, $e \in S_E$, the *potential terminal time* $t_p = t_p(x_{o\sigma},\tau_o,y_o,t_o,p,e)$, is the smallest t such that a member of M_{pe} is a potentially terminal observed phase.

Figure 3.2 illustrates a potentially terminal observed phase with terminating and non-terminating P paths.

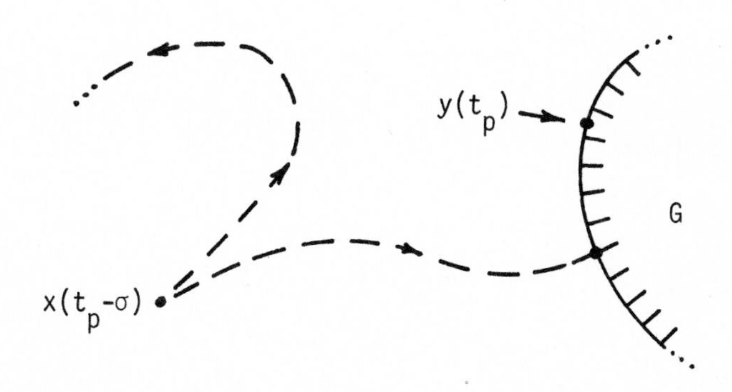

Fig. 3.2 Potential Termination

Definition. The *actual terminal time* t_f is the smallest $t \geq t_o$ such that $(x(t),t,y(t),t) \in \theta$.

Definition. Let $x_{o\sigma}, \tau_o, y_o, t_o$, $p \in S_P$, $e \in S_E$, $\bar{v} \in \bar{C}_v$ be given, with $t_o \leqslant t_p(x_{o\sigma},\tau_o,y_o,t_o,p,e)$. Assume that $t_p(x_{o\sigma},\tau_o,y_o,t_o,p,e)$ is finite and let $\bar{u}$ be a member of $\bar{C}_u$ such that

$$(\bar{\varphi}_u(t_p;\varphi_{pe}(t_p)\tau(t_p),t_p,\psi_{pe}(t_p),t_p) \in \theta$$

$\bar{u}$ need not be unique. The *payoff function* J is defined to be

$$J(x_{o\sigma},\tau_o,y_{o\sigma},y_o,t_o,p,\bar{u},e) = \tilde{W}(\bar{\varphi}_u(t_p),t_p,\psi_{pe}(t_p),t_p)$$

$$+ \int_{\tau_o}^{t_p-\sigma} L_1(\varphi_{pe}(\alpha+\sigma),p(\varphi_{pe}(\alpha+\sigma),\alpha,\psi_{pe}(\alpha+\sigma),\alpha+\sigma)\alpha)d\alpha$$

$$+ \int_{t_p-\sigma}^{t_p} L_1(\bar{\varphi}_u(\alpha),\bar{u}(\alpha),\alpha)d\alpha + \int_{\tau_o}^{t_o} L_2(\bar{\psi}_v(\alpha),\bar{v}(\alpha),\alpha)d\alpha$$

$$+ \int_{t_o}^{t_p} L_2(\psi_{pe}(\alpha),e(\varphi_{pe}(\alpha),\alpha-\sigma,\psi_{pe}(\alpha),\alpha),\alpha)d\alpha$$

and the *remaining payoff function* is defined to be

$$J_r(x_{o\sigma},\tau_o,y_o,t_o,p,\bar{u},e) = J(x_{o\sigma},\tau_o,y_o,t_o,p,\bar{u},e)$$

$$- \int_{\tau_o}^{t_o} L_2(\bar{\psi}(\alpha),\bar{v}(\alpha),\alpha)d\alpha$$

where

$$\varphi_{pe}(\alpha) = \varphi_{pe}(\alpha;x_{o\sigma},\tau_o,y_o,t_o), \qquad \psi_{pe}(\alpha) = \psi_{pe}(\alpha;x_{o\sigma},\tau_o,y_o,t_o)$$

$$\bar{\varphi}_u(\alpha) = \bar{\varphi}_u(\alpha;\varphi_{pe}(t_p),\tau(t_p)), \qquad \bar{\psi}_v(\alpha) = \bar{\psi}_v(\alpha;y_{o\sigma},\tau_o), \text{ and}$$

$$t_p = t_p(x_{o\sigma},\tau_o,y_o,t_o,p,e) .$$

Definition. For a given $p,e,x_{o\sigma},\tau_o,y_o$, and $t_p(x_{o\sigma},\tau_o,y_o,t_o,p,e)$, we define the *potential terminal payoff function on* θ_p as follow :

$$W(\varphi_{pe}(t_p;x_{o\sigma},\tau_o,y_o,t_o),\tau(t_p),\psi_{pe}(t_p;x_{o\sigma},\tau_o,y_o,t_o),t_p)$$

$$= \inf_{\overline{u}\in\overline{C}_u} \tilde{W}(\overline{\varphi}_u(t_p),t_p,\psi_{pe}(t_p;x_{o\sigma},\tau_o,y_o,t_o),t_p) + \int_{t_p-\sigma}^{t_p} L_1(\overline{\varphi}_u(\alpha),\overline{u}(\alpha),\alpha)d\alpha$$

where $\overline{\varphi}_u$ solves (10) subject to $\overline{\varphi}_u(t_p-\sigma) = \varphi_{pe}(t_p;x_{o\sigma},\tau_o,y_o,t_o)$.

Definition. The *potential payoff function* J_p is defined to be

$$J_p(x_{o\sigma},\tau_o,y_o,t_o,p,e) = W(\varphi_{pe}(t_p),\tau(t_p),\psi_{pe}(t_p),t_p) + \int_{\tau_o}^{t_p-\sigma} L_1(\varphi_{pe}(\alpha+\sigma),p(\varphi_{pe}(\alpha+\sigma),\alpha,\psi_{pe}(\alpha+\sigma),\alpha+\sigma),\alpha)d\alpha + \int_{t_o}^{t_p} L_2(\psi_{pe}(\alpha),e(\varphi_{pe}(\alpha),\alpha-\sigma,\psi_{pe}(\alpha),\alpha),\alpha)d\alpha \tag{12}$$

where $t_p = t_p(x_{o\sigma},\tau_o,y_o,t_o,p,e)$.

3.2.1 - Properties of dynamical systems and DGWITL. We make the following observations. The solutions of (8) and (9) have the property that

$$\varphi_{pe}(\alpha;\varphi_{pe}(t;x_{o\sigma},\tau_o,y_o,t_o),\tau(t),\psi_{pe}(t;x_{o\sigma},\tau_o,y_o,t_o),t) = \varphi_{pe}(\alpha;x_{o\sigma},\tau_o,y_o,t_o) \tag{13}$$

and

$$\psi_{pe}(\alpha;\varphi_{pe}(t;x_{o\sigma},\tau_o,y_o,t_o),\tau(t),\psi_{pe}(t;x_{o\sigma},\tau_o,y_o,t_o),t) = \psi_{pe}(\alpha;x_{o\sigma},\tau_o,y_o,t_o) \tag{14}$$

for $t_o \leq t \leq \alpha$. Therefore, if $(x_{o\sigma},\tau_o,y_o,t_o) \in G-\theta_p$ and (p,e) is a potentially terminal admissible pair, the potential terminal time satisfies

$$t_p(\varphi_{pe}(t;x_{o\sigma},\tau_o,y_o,t_o),\tau(t),\psi_{pe}(t;x_{o\sigma},\tau_o,y_o,t_o),t,p,e) = \tag{15}$$
$$t_p(x_{o\sigma},\tau_o,y_o,t_o,p,e)$$

for $t_o \leq t \leq t_p(x_{o\sigma},\tau_o,y_o,t_o)$. It follows from (12)-(15) that

$$J_p(\varphi_{pe}(t;x_{o\sigma},\tau_o,y_o,t_o),\tau(t),\psi_{pe}(t;x_{o\sigma},\tau_o,y_o,t_o),t,p,e)$$
$$= W(\varphi_{pe}(t_p;x_{o\sigma},\tau_o,y_o,t_o),\tau(t),\psi_{pe}(t_p;x_{o\sigma},\tau_o,y_o,t_o,t_p)$$
$$+ \int_{\tau(t)}^{t_p-\sigma} L_1(\varphi_{pe}(\alpha+\sigma),p(\varphi_{pe}(\alpha+\sigma),\alpha,\psi_{pe}(\alpha+\sigma),\alpha+\sigma),\alpha)d\alpha \tag{16}$$
$$+ \int_t^{t_p} L_2(\psi_{pe}(\alpha),e(\varphi_{pe}(\alpha),\alpha-\sigma,\psi_{pe}(\alpha),\alpha),\alpha)d\alpha$$

where $t_p = t_p(x_{o\sigma},\tau_o,y_o,t_o,p,e)$, $\varphi_{pe}(\alpha) = \varphi_{pe}(\alpha;x_{o\sigma},\tau_o,y_o,t_o)$, and $\psi_{pe}(\alpha) = \psi_{pe}(\alpha;x_{o\sigma},\tau_o,y_o,t_o)$. Consequently, the Eulerian derivative of J_p along E-viewable motions of (8) and (9) is given by

$$(d/dt)J_p(\varphi_{pe}(t;x_{o\sigma},\tau_o,y_o,t_o),\tau(t),\psi_{pe}(t;x_{o\sigma},\tau_o,y_o,t_o),t,p,e)$$
$$= - L_1(\varphi_{pe}(t),p(\varphi_{pe}(t),t-\sigma,\psi_{pe}(t),t\ ),t-\sigma)$$
$$- L_2(\psi_{pe}(t),e(\varphi_{pe}(t),t-\sigma,\psi_{pe}(t),t),t) \tag{17}$$
$$= - L_1(x(t-\sigma),p(x(t-\sigma),t-\sigma,y(t),t)t-\sigma)$$
$$- L_2(y(t),e(x(t-\sigma),t-\sigma,y(t),t),t)$$

Thus, E's potential payoff changes according to P's position and choice of control at the past - and most recently observed - state and according to E's present state and choice of control.

3.2.2 - The potential value. The remaining payoff, J_r , is the payoff function minus that portion which E knows with certainty, and the potential terminal payoff is the solution of an optimal control problem. Given that an observed phase is a member of the preliminary target set, solution of the optimal control problem requires computation of P's best terminating control function and the associated payoff contribution. This defines the potential terminal payoff on the preliminary target set. From E's point of view, the potential terminal payoff on θ_p replaces the terminal payoff on θ, and the potential payoff is the quantity of significance for E.

Since J_p is defined on S_P and S_E , it is appropriate to address the problem of determinating the upper and lower value of J_p .

Definition. The *potential lower value* $\underline{V}(x_o,\tau_o,y_o,t_o)$, is defined to be

$$\underline{V}(x_o,\tau_o,y_o,t_o) = \sup_{e\in S_E} \inf_{p\in S_P} J_p(x_{o\sigma},\tau_o,y_o,t_o,p,e) \tag{18 a}$$

and the *potential upper value* $\overline{V}(x_o,\tau_o,y_o,t_o)$, is defined to be :

$$\overline{V}(x_o,\tau_o,y_o,t_o) = \inf_{p\in S_P} \sup_{e\in S_E} J_p(x_{o\sigma},\tau_o,y_o,t_o,p,e) \tag{18 b}$$

Assumption 3.1 - Assume there exists $e^o \in S_E$, $p^o \in S_P$ such that, for any initial phase $(x_{o\sigma},\tau_o,y_o,t_o) \in G-\theta_p$, the pair (p^o,e^o) is a potentially terminal admissible pair, and

$$\underline{V}(x_{o\sigma},\tau_o,y_o,t_o) = V^o(x_{o\sigma},\tau_o,y_o,t_o) = \overline{V}(x_{o\sigma},\tau_o,y_o,t_o)$$

$$-\infty < V^o(x_{o\sigma},\tau_o,y_o,t_o) = J_p(x_{o\sigma},\tau_o,y_o,t_o,p^o,e^o) < \infty$$

Thus, we assume that J_p has a saddlepoint with respect to the strategy mappings. The pair (p^o,e^o) will be called an *optimal strategy pair,* and $V^o(x_{o\sigma},\tau_o,y_o,t_o)$ will be called the *potential value.*

Definition. For an optimal strategy pair (p^o,e^o) the corresponding set of observation quadruples, $M_{p^oe^o}$, will be called an *optimal* E-*viewable motion.*

It follows from the preceding assumption that

$$V^o(x_{o\sigma},\tau_o,y_o,t_o) = W(\varphi_{p^oe^o}(t_p),\tau(t_p),\psi_{p^oe^o}(t_p),t_p)$$

$$+ \int_{\tau_o}^{t_p-\sigma} L_1(\varphi_{p^oe^o}(\alpha+\sigma),p^o(\varphi_{p^oe^o}(\alpha+\sigma),\alpha,\psi_{p^oe^o}(\alpha+\sigma),\alpha+\sigma),\alpha)d\alpha$$

$$+ \int_{t_o}^{t_p} L_2(\psi_{p^oe^o}(\alpha),e^o(\varphi_{p^oe^o}(\alpha),\alpha-\sigma,\psi_{p^oe^o}(\alpha),\alpha),\alpha)d\alpha$$

where $t_p = t_p(x_{o\sigma},\tau_o,y_o,t_o,p^o,e^o)$.

3.2.3 - Payoff inequalities for DGWITL. Some fundamental inequalities can be derived from (18) and Assumption 3.1 . For a pair of admissible strategies (p,e) define

$$\hat{J}_p(e) = \inf_{p\in S_P} J_p(p,e)$$

and assume that the infimum occurs at $p_e \in S_P$ where we write p_e to emphasize the dependence of the extremizing strategy on the choice of e under consideration. It follows that

$$J_p(p_e,e) \leqslant J_p(p,e) \qquad \forall p \in S_P$$

Next, define

$$J_o = \sup_e J_p(p_e,e)$$

and assume that the supremum occurs at $e^{o} \in S_E$. Then

$$J_p(p_e, e) \leqslant J_p(p_{e^o}, e^o) \qquad \forall e \in S_E$$

Let $p_e = p_{e^o}$ and $e = e^o$ in (19) to obtain

$$(20) \qquad J_p(p_e, e) \leqslant J_p(p^o, e^o) \leqslant J_p(p, e^o) \qquad \forall p \in S_p,\ e \in S_E$$

While (20) might appear to be the usual saddlepoint inequality, it is a restricted relationship since the RHS allows arbitrary choice of $p \in S_P$, while the LHS allows arbitrary choice of $e \in S_E$ but requires that p be restricted to be p_e . Thus, (20) is intermediate to the development of the saddlepoint inequality. To obtain it one would define

$$\check{J}(p) = \sup_{e \in S_E} J_p(p, e)$$

and assume that $e_p \in S_E$ with

$$(21) \qquad J_p(p, e_p) \geqslant J_p(p, e) \qquad \forall e \in S_E$$

Then, let

$$J^o = \inf_{p \in S_P} \check{J}(p)$$

and assume $p^o \in S_P$, with

$$J_p(p^o, e_{p^o}) \leqslant J_p(p, e_p) \qquad \forall p \in S_P$$

Next, let $e^o = e_{p^o}$ and $p = p^o$ in (21) to obtain

$$(22) \qquad J_p(p^o, e) \leqslant J_p(p^o, e^o) \leqslant J_p(p, e_p) \qquad \forall p \in S_P,\ e \in S_E$$

The following saddlepoint inequality follows from (20), (22) and Assumption 3.1 .

$$J_p(p^o, e) \leqslant J_p(p^o, e^o) \leqslant J_p(p, e^o) \qquad \forall p \in S_P,\ e \in S_E$$

From the viewpoint of DGWITL, the right-hand-side of (20) states that use of e^o by E guarantees that J_p will not decrease if P does not use p^o ; that is, e^o enables E to establish a lower bound on the potential payoff. The left-hand-side states that, if E uses $e \neq e^o$, then P may have a strategy p_e which will result in a payoff less than $J_p(p^o,e^o)$. An assumption that p^o and e^o are unique would lead to strict inequalities in (20).

3.3 - *Conditions of necessity and sufficiency for DGWITL.*

3.3.1 - *Introduction.* In this section, we define for DGWITL a generalization of the control-theoretic concepts of pre-Hamiltonian, Hamiltonian, and the Hamilton-Jacobi equation. We then present a theorem which states conditions under which the potential value satisfies a generalized Hamilton-Jacobi equation. A second theorem provides a sufficiency condition for DGWITL .

3.3.2 - *Some mathematical preliminaries.* For a given $x,\tau,y,t,\lambda_x,\lambda_y,\mu,\beta$, the scalar generalized pre-Halmitonian function $H(\cdot)$ is defined as

$$H(x,\tau,y,t,\lambda_x,\lambda_y,\mu,\beta) = \langle f(x,\tau,\mu),\lambda_x \rangle + \langle g(y,t,\beta),\lambda_y \rangle$$
$$+ L_1(x,\mu,\tau) + L_2(y,\beta,t)$$

where λ_x and λ_y are n-vectors, μ and β are q and m-vectors, respectively, and $\langle\ ,\ \rangle$ denotes the usual inner product on $R^n \times R^n$.

Definition. For a given $x,\tau,y,t,\lambda_x,\lambda_y$, if there exists a unique $\mu \in K_u$, denoted as $k_1(x,\tau,y,t,\lambda_x,\lambda_y)$, and a unique $\beta \in K_v$, denoted as $k_2(x,\tau,y,t,\lambda_x,\lambda_y)$, such that $k_1(\cdot)$ and $k_2(\cdot)$ are continuous in t and τ , Lipschitzian in $(x,y,\lambda_x,\lambda_y)$, and

$$-\infty < \sup_{\beta\in K_v} \inf_{\mu\in K_u} H(x,\tau,y,t,\lambda_x,\lambda_y,\mu,\beta) = H(x,\tau,y,t,\lambda_x,\lambda_y,$$
$$k_1(x,\tau,y,t,\lambda_x,\lambda_y),\ k_2(x,\tau,y,t,\lambda_x,\lambda_y))$$
$$= \inf_{\mu\in K_u} \sup_{\beta\in K_v} H(x,\tau,y,t,\lambda_x,\lambda_y,\mu,\beta) < \infty$$

we define the *generalized Hamiltonian* to be

$$H^o(x,\tau,y,t,\lambda_x,\lambda_y) =$$
$$H(x,\tau,y,t,\lambda_x,\lambda_y,k_1(x,\tau,y,t,\lambda_x,\lambda_y),k_2(x,\tau,y,t,\lambda_x,\lambda_y))$$

Definition. The *generalized Hamilton-Jacobi equation* is defined to be

$$V^o_\tau + V^o_t + H^o(x,\tau,y,t,V^o_x,V^o_y) = 0$$

where V^o_x denotes $\frac{\partial V^o}{\partial x}$ and V^o_y denotes $\frac{\partial V^o}{\partial y}$, the gradients of $V^o(x,\tau,y,t)$ with respect to x and y, V^o_τ denotes $\frac{\partial V^o}{\partial \tau}(x,\tau,y,t)$ and V^o_t denotes $\frac{\partial V^o}{\partial t}(x,\tau,y,t)$.

3.3.3 - *Generalization of the Hamilton-Jacobi theory.* In the theory of differential games and in the theory of optimal control, much attention has been devoted to showing that, under certain conditions, the optimal payoff satisfies a partial differential equation, which is often referred to as the Hamilton-Jacobi equation, main Equation, or Bellman's equation. We now state a theorem which generalizes the results for those problems and applies to differential games with information time lag.

Theorem 3.1 (Necessary Conditions).

Let G_P *and* G_E *be subsets of* $R^n \times R^1$ *to which all motions of* (6) *and* (7) *are restricted, with* $G = G_P \times G_E$. *Let the target set* θ *and the preliminary target set* θ_p *be closed subsets of* G, *and let* G_P *and* G_E *have the following properties :* (i) *the scalar functions* $L_1(\cdot)$ *and* $L_2(\cdot)$ *are continuous in the triples* (x,μ,τ) *and* (y,β,t), *respectively ;* (ii) *the functions* $f(\cdot)$ *and* $g(\cdot)$ *are continuous in* τ *and* t *and Lipschitzian in the pairs* (x,μ), (y,β), *respectively ;* (iii) *for any initial phase pair in* $G-\theta_p$, *there is an optimal* E*-viewable motion of* (8) *and* (9) *which reaches* θ_p *without leaving* G ; (iv) *the regions* $G-\theta_p$ *and* $G-\theta$ *are open ;* (v) *the potential value function* $V^o(x,\tau,y,t)$ *is continuously differentiable in its argument ;* (vi) *for given* $x,\tau,y,t,\lambda_x,\lambda_y$, *there are unique* $\mu = k_1(x,\tau,y,t,\lambda_x,\lambda_y) \in K_u$ *and* $\beta = k_2(x,\tau,y,t,\lambda_x,\lambda_y) \in K_v$ *having the property that*

$$\sup_{\beta\in K_v} \inf_{\mu\in K_u} H(x,\tau,y,t,\lambda_x,\lambda_y,\mu,\beta)$$

$$= H(x,\tau,y,t,\lambda_x,\lambda_y,k_1(x,\tau,y,t,\lambda_x,\lambda_y),k_2(x,\tau,y,t,\lambda_x,\lambda_y))$$

$$= H^o(x,\tau,y,t,\lambda_x,\lambda_y) = \inf_{\mu\in K_u} \sup_{\beta\in K_v} H(x,\tau,y,t,\lambda_x,\lambda_y,\mu,\beta) \tag{23}$$

and (vii) *the functions* $k_1(\cdot)$ *and* $k_2(\cdot)$ *are continuous in* τ *and* t, *and Lipschitzian in* $(x,y,\lambda_x,\lambda_y)$. *Then, the potential value function* $V^o(x,\tau,y,t)$ *satisfies the generalized Hamilton-Jacobi equation in the region* $G-\theta_p$. *That is*

$$V^o_\tau + V^o_t + H^o(x,\tau,y,t,V^o_x,V^o_y) = 0 \tag{24}$$

For the proof of Theorem 3.1, see Section 6.1 .

Lemma 3.1 (Lemma of Caratheodory for DGWITL).

Suppose that a q*-vector function* k_1 *and an* m*-vector function* k_2, *both defined on* $R^n \times R^1 \times R^n \times R^1$, *with* k_1 *and* k_2 *continuous in* τ *and* t *and Lipschitzian in* x *and* y, *are such that, for all quadruples* $(x,\tau,y,t) \in G-\theta_p$,

(i) (k_1,k_2) *is a potentially terminal pair,*

(ii) $k_1(x,\tau,y,t) \in K_u$ and $k_2(x,\tau,y,t) \in K_v$

(iii) $L_1(x,k_1(x,\tau,y,t)) = 0$, $L_2(y,k_2(x,\tau,y,t),t) = 0$

(iv) $L_1(x,\mu,\tau) > 0$ *if* $\mu \in R^q$ *and* $\mu \neq k_1(x,\tau,y,t)$

(v) $L_2(y,\beta,t) < 0$ *if* $\beta \in R^m$ *and* $\beta \neq k_2(x,\tau,y,t)$

and assume that $W \equiv 0$, *that is, the potential terminal payoff is zero. For any* $(x_{o\sigma},\tau_o,y_o,t_o) \in G-\theta_p$, *the potential payoff along motions of* (8) *and* (9) *has the following properties :*

(a) $J_p(x_{o\sigma},\tau_o,y_o,t_o k_1,k_2) \equiv 0$ (the proof is direct)

(b) *The E-observable motion*

$M_{k_1k_2} = \{(\varphi_{k_1k_2}(t;x_{o\sigma},\tau_o,y_o,t_o),\tau(t),\psi_{k_1k_2}(t;x_{o\sigma},\tau_o,y_o,t_o),t) :$

$t \geq t_o\}$ *has the property that any other motion of* (8) *and* (9) *with* p *constrained to be* k_1, *say*

$M_{k_1e} = \{(\varphi_{k_1e}(t;x_{o\sigma},\tau_o,y_o,t_o),\tau(t),\psi_{k_1e}(t;x_{o\sigma},\tau_o,y_o,t_o),t):$

$t \geq t_o\}$, *which connects* $(x_{o\sigma},\tau_o,y_o,t_o)$ *with a phase in* θ_p *and remains entirely in* G, *must result in a negative payoff.*

(c) *The E-observable motion*

$M_{k_1k_2} = \{(\varphi_{k_1k_2}(t;x_{o\sigma},\tau_o,y_o,t_o),\tau(t),\psi_{k_1k_2}(t;x_{o\sigma},\tau_o,y_o,t_o),t) :$

$t \geq t_o\}$ *has the property that any other motion of* (8) *and* (9) *with* e *constrained to be* k_2, *say*

$M_{pk_2} = \{(\varphi_{pk_2}(t;x_{o\sigma},\tau_o,y_o,t_o),\tau(t),\psi_{pk_2}(t;x_{o\sigma},\tau_o,y_o,t_o),t) :$
$t \geq t_o\}$, *which connects* $(x_{o\sigma},\tau_o,y_o,t_o)$ *with a phase in* θ_p *and remains entirely in* G, *must result in a positive payoff.*

Proof of (b). Let M_{k_1e} be a motion connecting $(x_{o\sigma},\tau_o,y_o,t_o)$ with a phase in θ_p and remaining entirely in G, and suppose that $J_p(x_{o\sigma},\tau_o,y_o,t_o,k_1,e) = 0$. From Assumption (v) and the continuity of L, it is clear that, along M_{k_1e}, we must have

$$e(\varphi_{k_1e}(t;x_{o\sigma},\tau_o,y_o,t_o),\tau(t),\psi_{k_1e}(t;x_{o\sigma},\tau_o,y_o,t_o),t)$$
$$= k_2(\varphi_{k_1e}(t;x_{o\sigma},\tau_o,y_o,t_o),\tau(t),\psi_{k_1e}(t;x_{o\sigma},\tau_o,y_o,t_o),t)$$

at every continuity point of e, otherwise we would have $J < 0$. Note that we would obtain the same motion M_{k_1e} if we always let e be defined by this relation, that is $M_{k_1e} = M_{k_1k_2}$; and, since k_1 and k_2 are both continuous in τ and t Lipschitzian in x and y, the motion of (8) and (9) is unique. Hence, these motions are the same.

Theorem 3.2 (Sufficiency Conditions)

Let G *and* θ_p *be given, and let* G *have the following properties : (i)* L_1 *is continuous in the triple* (x,μ,τ) *;* L_2 *is continuous in the triple* (y,β,t) *; (ii)* f *is continuous in* τ *and Lipschitzian in the pair* (x,μ) *;* g *is continuous in* t *and Lipschitzian in the pair* (y,β) *; (iii) for given* $x,\tau,y,t,\lambda_x,\lambda_y$*, there is a unique* $\mu \in K_u$*, denoted by* $k_1(x,\tau,y,t,\lambda_x,\lambda_y)$*, and a unique* $\beta \in K_v$ *denoted by* $k_2(x,\tau,y,t,\lambda_x,\lambda_y)$ *having the property that*

$$\sup_{\beta\in K_v}\ \inf_{\mu\in K_u} H(x,\tau,y,t,\lambda_x,\lambda_y,\mu,\beta)$$
$$= H(x,\tau,y,t,\lambda_x,\lambda_y,k_1(x,\tau,y,t,\lambda_x,\lambda_y),k_2(x,\tau,y,t,\lambda_x,\lambda_y))$$
$$= \inf_{\mu\in K_u}\ \sup_{\beta\in K_v} H(x,\tau,y,t,\lambda_x,\lambda_y,\mu,\beta)$$

(iv) the functions k_1 *and* k_2 *are continuous in* τ *and* t *and Lipschitzian in the quadruple* $(x,y,\lambda_x,\lambda_y)$ *; (v)* $V(x,\tau,y,t)$ *is a scalar function which is continuously differentiable in its*

argument ; (vi) $V(x,\tau,y,t)$ *satisfies the generalized Hamilton-Jacobi equation, that is*

$$V_\tau + V_t + H^o(x,\tau,y,t,V_x,V_y) = 0$$

(vii) *the function* $V(x,\tau,y,t) \equiv W(x,\tau,y,t)$ *on* θ_p ; *and* (viii) *the pair* $(k_1(x,\tau,y,t,V_x,V_y),k_2(x,\tau,y,t,V_x,V_y))$ *is a potentially terminal admissible strategy pair.*

Then,

$$V(x,\tau,y,t) = V^o(x,\tau,y,t) = \sup_{e\in S_E} \inf_{p\in S_P} J_p(x,\tau,y,t,p,e)$$

$$= J_p(x,\tau,y,t,p^o,e^o) = \inf_{p\in S_P} \sup_{e\in S_E} J_p(x,\tau,y,t,p,e)$$

where

$$p^o(x,\tau,y,t) = k_1(x,\tau,y,t,V_x,V_y) \quad \text{and}$$

$$e^o(x,\tau,y,t) = k_2(x,\tau,y,t,V_x,V_y)$$

For proof of Theorem 3.2, see Section 6.2 .

It follows from Theorem 3.2 that any other scalar function $V^*(x,\tau,y,t)$ which satisfies properties (i) through (vii) and produces $p^* = k_1(x,\tau,y,t,V^*_x,V^*_y)$ and $e^* = k_2(x,\tau,y,t,V^*_x,V^*_y)$ which form a potentially terminal admissible pair must be such that $V^* = V = V^o$. Thus, Theorem 3.2 contains an implicit statement about uniqueness of solutions to the generalized Hamilton-Jacobi equation. Although multiple solutions may be found which satisfy the equation and its boundary condition, only one solution can generate an admissible terminal pair. The requirement of termination appears as property (vii) of Theorem 3.2 .

3.5 - *Canonic equations for DGWITL*

3.5.1 - Introduction. The Hamilton-Jacobi equation (HJE) in optimal control theory (Ref.19) and the Main Equation in differential game theory (Ref.1) express a relation between the system velocity vector and the gradient of the payoff along an optimal path in regions where the payoff is smooth. Solution of the equation yields both the optimal payoff and the strategy synthesis associated with optimality. The boundary condition for the equation is provided by the payoff on the terminal surface. Because the Hamilton-Jacobi equation is usually nonlinear, attempts to solve it in closed form are often futile.

As an alternate to closed form or direct solution of the HJE, the method of characteristics (Ref. 20) in the classical theory of partial differential equations can be used to develop an indirect solution to the problem. Instead of obtaining a global expression for the payoff and the feedback control structure, one can write the canonic equations, or the path equations, which describe the time history of the payoff along an optimal path. The method effectively converts the problem of solving a partial differential equation into an equivalent problem of solving a system or ordinary differential equations.

The generalized Hamilton-Jacobi equation for DGWITL contains two temporal parameters, τ and t . In Section 3.5.2 we will convert the equation into a partial differential equation containing a single temporal parameter. Then, in Section 3.5.3 we develop the canonic equations for the potential value of differential games with information time lag.

3.5.2 - *Potential value along* $\tau = t-\sigma$. Although the generalized Hamilton-Jacobi equation is defined over the τt plane, we restrict attention to a locus corresponding to a fixed time lag σ . Let $\tau = t-\sigma$, with σ fixed, and let $\hat{V}^{o}(x,y,t;\sigma)$ denote $V^{o}(x,\tau,y,t)$ along $\tau = t-\sigma$, that is

$$\hat{V}^{o}(x,y,t;\sigma) = V^{o}(x,\tau,y,t)\Big|_{\tau=t-\sigma}$$

Likewise, let

$$\hat{H}^{o}(x,y,t,\hat{V}^{o}_{x},\hat{V}^{o}_{y};\sigma) = H^{o}(x,\tau,y,t,V^{o}_{x},V^{o}_{y})\Big|_{\tau=t-\sigma}$$

Then, along $\tau = t-\sigma$

$$\frac{\partial\hat{V}^{o}}{\partial t} = \frac{\partial V^{o}}{\partial \tau}\frac{d\tau}{dt} + \frac{\partial V^{o}}{\partial t} = \frac{\partial V^{o}}{\partial \tau} + \frac{\partial V^{o}}{\partial t},$$

$$\frac{\partial\hat{V}^{o}}{\partial x} = \frac{\partial V^{o}}{\partial x}(x,\tau,y,t) \qquad \frac{\partial\hat{V}^{o}}{\partial y} = \frac{\partial V^{o}}{\partial y}(x,\tau,y,t)$$

so (24) becomes

$$\frac{\partial\hat{V}^{o}}{\partial t}(x,y,t;\sigma) + \hat{H}^{o}(x,y,t,\hat{V}^{o}_{x}(x,y,t;\sigma),\hat{V}^{o}_{y}(x,y,t;\sigma);\sigma) = 0 \tag{25}$$

subject to

$$\hat{V}^{o}(x,y,t_{p};\sigma) = W(x,t_{p}-\sigma,y,t_{p}) \tag{26}$$

While (24) describes $V^{o}(\cdot)$ over the region of τt plane below $\tau = t$, (25) describes $V^{o}(\cdot)$ on a line corresponding to play with fixed σ.

3.5.3 - *The canonic equations.* We first develop the expression for the Eulerian derivative of $\hat{V}^{o}$ along an optimal path. Let (p,e) be a potentially terminal admissible strategy pair, and let $\varphi_{pe}(\cdot)$ and $\psi_{pe}(\cdot)$ denote the corresponding E-observable trajectories for P and E. Then define

$$\tilde{V}^{o}(t) = \hat{V}^{o}(\varphi_{pe}(t;x_{o\sigma},t_{o}-\sigma,y_{o},t_{o}), \psi_{pe}(t;x_{o\sigma},t_{o}-\sigma,y_{o},t_{o}),t;\sigma) \tag{27}$$

Along an optimal path

$$\frac{d}{dt}\tilde{V}^o(t) = \frac{\partial\hat{V}^o}{\partial x_1}\frac{dx_1}{dt} + \cdots + \frac{\partial\hat{V}^o}{\partial x_n}\frac{dx_n}{dt}$$

$$+ \frac{\partial\hat{V}^o}{\partial y_1}\frac{dy_1}{dt} + \cdots + \frac{\partial\hat{V}^o}{\partial y_n}\frac{dy_n}{dt}$$

$$+ \frac{\partial\hat{V}^o}{\partial t} ,$$

and all partial derivatives are evaluated at the argument given on the right-hand-side of (27). At points on the path corresponding to (p,e)

$$\frac{dx}{dt} = f(\varphi_{pe}(t\ ;\ x_{o\sigma},t_o-\sigma,y_o,t_o) ,$$

$$p(\varphi_{pe}(t\ ;\ x_{o\sigma},t_o-\sigma,y_o,t_o),t_o-\sigma ,$$

$$\psi_{pe}(t\ ;\ x_{o\sigma},t_o-\sigma,y_o,t_o),t),t-\sigma) ,$$

$$\frac{dy}{dt} = g(\psi_{pe}(t\ ;\ x_{o\sigma},t_o-\sigma,y_o,t_o) ,$$

$$e(\varphi_{pe}(t\ ;\ x_{o\sigma},t_o-\sigma,y_o,t_o),t_o-\sigma ,$$

$$\varphi_{pe}(t\ ;\ x_{o\sigma},t_o-\sigma,y_o,t_o),t),t)$$

For simplicity, we will write $\varphi_{pe}(t)$ and $\psi_{pe}(t)$ to denote E's observation at time t of $x(t-\sigma)$ and $y(t)$. Along and optimal path

$$\frac{\partial \hat{V}^o}{\partial t}(\varphi_{p^oe^o}(t),\psi_{p^oe^o}(t),t;\sigma)$$

$$+ < f(\varphi_{p^oe^o}(t),t-\sigma,p^o(\varphi_{p^oe^o}(t),t-\sigma,\psi_{p^oe^o}(t),t),$$

$$\hat{V}^o_x(\varphi_{p^oe^o}(t),\psi_{p^oe^o}(t);\sigma) >$$

$$+ < g(\psi_{p^oe^o}(t),t,e^o(\varphi_{p^oe^o}(t),t-\sigma,\psi_{p^oe^o}(t),t),$$

$$\hat{V}^o_y(\varphi_{p^oe^o}(t),\psi_{p^oe^o}(t),t;\sigma) >$$

$$+ L_1(\varphi_{p^oe^o}(t),t-\sigma,p^o(\varphi_{p^oe^o}(t),t-\sigma,\psi_{p^oe^o}(t),t))$$

$$+ L_2(\psi_{p^oe^o}(t),t,e^o(\varphi_{p^oe^o}(t),t-\sigma,\psi_{p^oe^o}(t),t)) = 0$$

Therefore

$$\frac{d}{dt}\tilde{V}^o(t) = \frac{d}{dt}\hat{V}^o(x,y,t;\sigma)\Bigg|_{\substack{\text{along}\\ \text{an}\\ \text{optimal}\\ \text{path}}} = \frac{d}{dt}V^o(x,t-\sigma,y,t)\Bigg|_{\substack{x=\varphi_{p^oe^o}(t)\\ y=\psi_{p^oe^o}(t)}}$$

$$= - L_1(\varphi_{p^oe^o}(t),t-\sigma,p^o(\varphi_{p^oe^o}(t),t-\sigma,\psi_{p^oe^o}(t),t))$$

$$- L_2(\psi_{p^oe^o}(t),t,e^o(\varphi_{p^oe^o}(t),t-\sigma,\psi_{p^oe^o}(t),t)),$$

or

$$\frac{d\tilde{V}^o}{dt}(t) = -L_1(x,t-\sigma,p^o(x,t-\sigma,y,t)) - L_2(y,t,e^o(x,t-\sigma,y,t)),$$

where x and y are understood to be along the optimal path. Hence, the potential value seen by E changes incrementally by an amount determined by P's loss function at time $t-\sigma$ and E's loss function at time t . If $L_1 \equiv L_2 \equiv 0$ the Eulerian derivative of the potential value is zero, and the system remains on a surface of constant potential value under optimal play.

As the next step in the development of the canonic equations, we seek expressions for the time derivatives of the components of the gradient of $\hat{V}^o(\cdot)$ along an optimal path. The final result will be the set of generalized canonic equations which describe the characteristics of the GHJE along $\tau = t-\sigma$.

For $(x,t-\sigma,y,t) \in G-\theta_p$ we have

$$\begin{aligned} \frac{\partial \hat{V}^o}{\partial t}(x,y,t;\sigma) &+ < f(x,t-\sigma,p^o(x,t-\sigma,y,t)),\hat{V}^o_x(x,y,t;\sigma) > \\ &+ L_1(x,t-\sigma,p^o(x,t-\sigma,y,t)) \\ &+ < g(y,t,e^o(x,t-\sigma,y,t)),\hat{V}^o_y(x,y,t;\sigma) > \\ &+ L_2(y,t,e^o(x,t-\sigma,y,t)) \end{aligned} \tag{28}$$

and we observe that, for points on an E-viewable motion,

$$\begin{aligned} \frac{d}{dt}\frac{\partial \hat{V}^o}{\partial x} &= \frac{\partial^2 \hat{V}^o}{\partial x^2} f(x,t-\sigma,p^o(x,t-\sigma,y,t)) \\ &+ \frac{\partial^2 \hat{V}^o}{\partial x \partial y} g(y,t,e^o(x,t-\sigma,y,t)) + \frac{\partial}{\partial t}\frac{\partial \hat{V}^o}{\partial x} \end{aligned} \tag{29 a}$$

$$\begin{aligned} \frac{d}{dt}\frac{\partial \hat{V}^o}{\partial y} &= \frac{\partial^2 \hat{V}^o}{\partial y \partial x} f(x,t-\sigma,p^o(x,t-\sigma,y,t)) \\ &+ \frac{\partial^2 \hat{V}^o}{\partial y^2} g(y,t,e^o(x,t-\sigma,y,t)) + \frac{\partial}{\partial t}\frac{\partial \hat{V}^o}{\partial y} \end{aligned} \tag{29 b}$$

where all partial derivatives are evaluated at
$x = \varphi_{p^o e^o}(t;x_{o\sigma},t_o-\sigma,y_o,t_o)$, $y = \psi_{p^o e^o}(t;x_{o\sigma},t_o-\sigma,y_o,t_o)$, with

$$\dot{x} = f(\varphi_{p^o e^o}(t),t-\sigma,p^o(\varphi_{p^o e^o}(t),t-\sigma,\psi_{p^o e^o}(t),t)$$

$$\dot{y} = g(\psi_{p^o e^o}(t),t,e^o(\varphi_{p^o e^o}(t),t-\sigma,\psi_{p^o e^o}(t),t)$$

$$\frac{\partial^2 \hat{V}^o}{\partial x^2} = \begin{bmatrix} \frac{\partial^2 \hat{V}^o}{\partial x_1^2} & \frac{\partial^2 \hat{V}^o}{\partial x_1 \partial x_2} & \cdots & \frac{\partial^2 \hat{V}^o}{\partial x_1 \partial x_n} \\ \vdots & & & \\ \frac{\partial^2 \hat{V}^o}{\partial x_n \partial x_1} & & \cdots & \frac{\partial^2 \hat{V}^o}{\partial x_n^2} \end{bmatrix}$$

$$\frac{\partial^2 \hat{V}^o}{\partial y \partial x} = \begin{bmatrix} \frac{\partial^2 \hat{V}^o}{\partial y_1 \partial x_1} & \frac{\partial^2 \hat{V}^o}{\partial y_1 \partial x_2} & \cdots & \frac{\partial^2 \hat{V}^o}{\partial y_1 \partial x_n} \\ \frac{\partial^2 \hat{V}^o}{\partial y_2 \partial x_1} & \frac{\partial^2 \hat{V}^o}{\partial y_2 \partial x_2} & \cdots & \frac{\partial^2 \hat{V}^o}{\partial y_2 \partial x_n} \\ \vdots & & & \\ \frac{\partial^2 \hat{V}^o}{\partial y_n \partial x_1} & & \cdots & \frac{\partial^2 \hat{V}^o}{\partial y_n \partial x_n} \end{bmatrix}$$

$$\frac{\partial^2 \hat{V}^o}{\partial x \partial y} = \begin{bmatrix} \frac{\partial^2 \hat{V}^o}{\partial x_1 \partial y_1} & \frac{\partial^2 \hat{V}^o}{\partial x_1 \partial y_2} & \cdots & \frac{\partial^2 \hat{V}^o}{\partial x_1 \partial y_n} \\ \frac{\partial^2 \hat{V}^o}{\partial x_2 \partial y_1} & \frac{\partial^2 \hat{V}^o}{\partial x_2 \partial y_2} & \cdots & \frac{\partial^2 \hat{V}^o}{\partial x_2 \partial y_n} \\ \cdot & & & \\ \cdot & & & \\ \cdot & & & \\ \frac{\partial^2 \hat{V}^o}{\partial x_n \partial y_1} & & \cdots & \frac{\partial^2 \hat{V}^o}{\partial x_n \partial y_n} \end{bmatrix}$$

$$\frac{\partial^2 \hat{V}^o}{\partial y^2} = \begin{bmatrix} \frac{\partial^2 \hat{V}^o}{\partial y_1^2} & \frac{\partial^2 \hat{V}^o}{\partial y_1 \partial y_2} & & \frac{\partial^2 \hat{V}^o}{\partial y_1 \partial y_n} \\ \cdot & & & \\ \cdot & & & \\ \cdot & & & \\ \frac{\partial^2 \hat{V}^o}{\partial y_n \partial y_1} & & \cdots & \frac{\partial^2 \hat{V}^o}{\partial y_n^2} \end{bmatrix}$$

and

$$\frac{\partial^2 \hat{V}^o}{\partial x^2} = \left[\frac{\partial^2 \hat{V}^o}{\partial x^2}\right]^T, \quad \frac{\partial^2 \hat{V}^o}{\partial y^2} = \left[\frac{\partial^2 \hat{V}^o}{\partial y}\right]^T, \quad \frac{\partial^2 \hat{V}^o}{\partial x \partial y} = \left[\frac{\partial^2 \hat{V}^o}{\partial y \partial x}\right]^T$$

Because (29) contains the Jacobian matrices for $\hat{V}^o$ it is not directly amenable to solution. But, since (28) holds identically in $G-\theta_p$, its derivative with respect to x and y produces the following vector identities in $G-\theta_p$.

$$0 = \frac{\partial}{\partial x}\frac{\partial \hat{V}^o}{\partial t} + \frac{\partial f}{\partial x}\frac{\partial \hat{V}^o}{\partial x} + \frac{\partial^2 \hat{V}^o}{\partial x^2} f + \frac{\partial L_1}{\partial x}$$

$$+ \frac{\partial p^o}{\partial x}\frac{\partial f}{\partial \mu}\frac{\partial \hat{V}^o}{\partial x} + \frac{\partial p^o}{\partial x}\frac{\partial L_1}{\partial \mu}$$

$$+ \frac{\partial e^o}{\partial x}\frac{\partial g}{\partial \beta}\frac{\partial \hat{V}^o}{\partial y} + \frac{\partial e^o}{\partial x}\frac{\partial L_2}{\partial \beta}$$

$$+ \frac{\partial^2 \hat{V}^o}{\partial x \partial y} g \,, \tag{30 a}$$

$$0 = \frac{\partial}{\partial y}\frac{\partial \hat{V}^o}{\partial t} + \frac{\partial g}{\partial y}\frac{\partial \hat{V}^o}{\partial y} + \frac{\partial^2 \hat{V}^o}{\partial y^2} g + \frac{\partial L_2}{\partial y}$$

$$+ \frac{\partial e^o}{\partial y}\frac{\partial g}{\partial \beta}\frac{\partial \hat{V}^o}{\partial y} + \frac{\partial e^o}{\partial y}\frac{\partial L_2}{\partial \beta}$$

$$+ \frac{\partial p^o}{\partial y}\frac{\partial f}{\partial \mu}\frac{\partial \hat{V}^o}{\partial x} + \frac{\partial p^o}{\partial y}\frac{\partial L_1}{\partial \mu}$$

$$+ \frac{\partial^2 \hat{V}^o}{\partial y \partial x} f \tag{30 b}$$

In (30) all partial derivatives of $\hat{V}^o$ are taken at $(x,y,t;\sigma)$; partial derivatives of p^o and e^o are taken at $(x,t-\sigma,y,t)$;

$\frac{\partial f}{\partial \mu}$ denotes $\left.\frac{\partial f}{\partial \mu}(x,t-\sigma,\mu)\right|_{\mu = p^o(x,t-\sigma,y,t)}$

$\frac{\partial L_1}{\partial \mu}$ denotes $\left.\frac{\partial L_1}{\partial \mu}(x,t-\sigma,\mu)\right|_{\mu = p^o(x,t-\sigma,y,t)}$

$\frac{\partial g}{\partial \beta}$ denotes $\left.\frac{\partial g}{\partial \beta}(y,t,\beta)\right|_{\beta = e^{o}(x,t-\sigma,y,t)}$

$\frac{\partial L_2}{\partial \beta}$ denotes $\left.\frac{\partial L_2}{\partial \beta}(y,t,\beta)\right|_{\beta = e^{o}(x,t-\sigma,y,t)}$

$\frac{\partial L_1}{\partial x}$ denotes $\left.\frac{\partial L_1}{\partial x}(x,t-\sigma,\mu)\right|_{\mu = p^{o}(x,t-\sigma,y,t)}$

$\frac{\partial L_2}{\partial y}$ denotes $\left.\frac{\partial L_2}{\partial y}(y,t,\beta)\right|_{\beta = e^{o}(x,t-\sigma,y,t)}$

Since the identities of (30) hold in $G-\theta_p$, they hold along an optimal path in $G-\theta_p$. By letting

$$\lambda_x(t) = \left.\frac{\partial \hat{V}^o}{\partial x}\right|_{\substack{x = \varphi_{p^o e^o}(t) \\ y = \psi_{p^o e^o}(t)}} \qquad \lambda_y(t) = \left.\frac{\partial \hat{V}^o}{\partial y}\right|_{\substack{x = \varphi_{p^o e^o}(t) \\ y = \psi_{p^o e^o}(t)}}$$

we combine (29) and (30) to obtain

$$-\frac{d\lambda_x}{dt} = \left[\frac{\partial f}{\partial x} + \frac{\partial p^o}{\partial x}\frac{\partial f}{\partial \mu}\right]\lambda_x + \frac{\partial e^o}{\partial x}\frac{\partial g}{\partial \beta}\lambda_y$$

$$+ \frac{\partial L_1}{\partial x} + \frac{\partial p^o}{\partial x}\frac{\partial L_1}{\partial \mu} + \frac{\partial e^o}{\partial x}\frac{\partial L_2}{\partial \beta} \tag{31 a}$$

$$-\frac{d\lambda_y}{dt} = \left[\frac{\partial g}{\partial y} + \frac{\partial e^o}{\partial y}\frac{\partial g}{\partial \beta}\right]\lambda_y + \frac{\partial p^o}{\partial y}\frac{\partial f}{\partial \mu}\lambda_x$$

$$+ \frac{\partial L_2}{\partial y} + \frac{\partial e^o}{\partial y}\frac{\partial L_2}{\partial \beta} + \frac{\partial p^o}{\partial y}\frac{\partial L_1}{\partial \mu} \tag{31 b}$$

where the partial derivatives are evaluated in the manner already mentioned. Equation (31) describes the time derivative of the components of the gradient of the potential value along $\tau = t-\sigma$. Next, let

$$\hat{k}_1^o(x,y,t,\hat{V}_x^o,\hat{V}_y^o;\sigma) = k_1(x,\tau,y,t,\hat{V}_x^o,\hat{V}_y^o)\Big|_{\tau = t-\sigma}$$

and

$$\hat{k}_2^o(x,y,t,\hat{V}_x^o,\hat{V}_y^o;\sigma) = k_2(x,\tau,y,t,\hat{V}_x^o,\hat{V}_y^o)\Big|_{\tau = t-\sigma}$$

Equation (31) together with

$$\text{(32 a)} \quad \frac{dx}{dt} = f(x,t-\sigma,\hat{k}_1^o(x,y,t,\hat{V}_x^o,\hat{V}_y^o;\sigma))$$

$$\text{(32 b)} \quad \frac{dy}{dt} = g(y,t,\hat{k}_2^o(x,y,t,\hat{V}_x^o,\hat{V}_y^o;\sigma))$$

form the canonic system for DGWITL†.

To express (31) and (32) in terms of the pre-Hamiltonian we assume, for simplicity, that the constraint sets K_u and K_v are not functions of the spatial coordinates - the other case can be treated as in (Ref. 21). It follows from the smoothness of the solution in $G-\theta_p$ that (31) can be written as

$$-\frac{d\lambda_x}{dt} = \frac{\partial f}{\partial x}\lambda_x + \frac{\partial L_1}{\partial x}$$

$$-\frac{d\lambda_y}{dt} = \frac{\partial g}{\partial y}\lambda_y + \frac{\partial L_2}{\partial y}$$

† Although the potential value need only be continuously differentiable in the derivation of the generalized Hamilton-Jacobi equation, continuity of the second derivatives must be assumed for the derivation of the canonic equations given by (32) .

Let :

$$\hat{H}(x,y,t,\lambda_x,\lambda_y,\mu,\beta;\sigma) = H(x,\tau,y,t,\lambda_x,\lambda_y,\mu,\beta)\Big|_{\tau = t-\sigma}$$

$$= \langle f(x,t-\sigma,\mu),\lambda_x \rangle + \langle g(y,t,\beta),\lambda_y \rangle$$

$$+ L_1(x,t-\sigma,\mu) + L_2(y,t,\beta)$$

Then it follows immediately that

$$\text{(33 a)} \qquad \frac{dx}{dt} = \frac{\partial\hat{H}}{\partial\lambda_x}(x,y,t,\lambda_x,\lambda_y,\mu,\beta;\sigma)$$

$$\text{(33 b)} \qquad \frac{dy}{dt} = \frac{\partial\hat{H}}{\partial\lambda_y}(x,y,t,\lambda_x,\lambda_y,\mu,\beta;\sigma)$$

$$\text{(33 c)} \qquad -\frac{d\lambda_x}{dt} = \frac{\partial\hat{H}}{\partial x}(x,y,t,\lambda_x,\lambda_y,\mu,\beta;\sigma)$$

$$\text{(33 d)} \qquad -\frac{d\lambda_y}{dt} = \frac{\partial\hat{H}}{\partial y}(x,y,t,\lambda_x,\lambda_y,\mu,\beta;\sigma)$$

form the canonic system for the differential game with information time lag when μ and β are evaluated at $\hat{k}_1^o(\cdot)$ and $\hat{k}_2^o(\cdot)$.

Note that, while the structure of (33) appears to be the same as that for the canonic system of differential games without time lag, the systems are distinct. $\hat{H}(\cdot)$ employs P's velocity vector at time $t-\sigma$, and the boundary condition for (33) is determined by the potential terminal payoff on the preliminary target set .

4. Qualitative Aspects of DGWITL

A number of qualitative effects provide additional insight about DGWITL and supplement the analytic detail of the previous section. Some will be discussed here, while others will be treated in the context of the examples in Section 5.

4.1 - *Existence of the potential value.*

Once the potential payoff has been characterized, as in (12) it is important to observe that the formal problem of finding the potential value for a fixed σ is the same as the problem of finding the value in a DG with perfect information. Therefore, the existence theory for DG (Refs. 24,25) may be applied to the problem of determinating the existence of a saddle-point of $J_p(\cdot)$ subject to (8) and (9). This same observation provides another means to obtain the Hamilton-Jacobi equation and the canonic equations along the line defined by $\tau = t-\sigma$, provided that the appropriate adjustments are made in the arguments of the results which apply to games without information time lag.

4.2 - *Data reference plane.*

In general, we have a data tuplet (x,τ,y,t) and we associate (x,τ) with P and (y,t) with E. For fixed spatial variables x and y, the temporal parameters τ and t define the qualitative nature of the game. We refer to the τt plane, as depicted in Figure 4.1, as the *data reference plane*. The line $\tau = t$ partitions the plane into a region for which $\tau \leqslant t$ and a region for which $\tau \geqslant t$. The first region can be associated with DGWITL in which the evader has an information time lag. The second region can be associated with DGWITL in which the pursuer has an information time lag, provided that τ is interpreted to be real time. The line $\tau = t$ corresponds to a DG with full information for both players (DGWFI).

For fixed x and y, the potential value defines a surface above the data reference plane. This surface describes the potential outcome of DGWITL for various data delays and initial times. As we have seen, the potential value along lines for which $\tau = t-\sigma$ corresponds to games with fixed information time lags. The potential value along lines drawn parallel to the τ axis define the *aging data problem*, that is, they reveal how the age of the spatial data affects the potential outcome of the game. This behavior is of interest because E's strategy should - and does - depend on when P was located at a given position x.

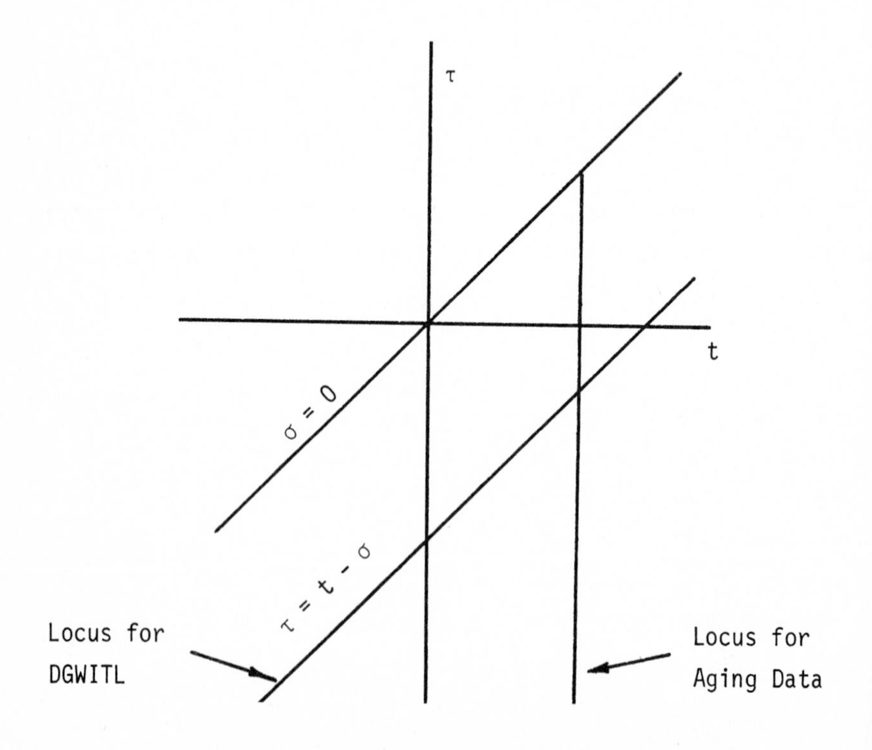

Fig. 4.1 Data Reference Plane

4.3 - *Strategy realization - mixed strategies.*

The potential value is E's bound on the potential outcome of the game. To obtain it, we considered all possible present and future behavior of E, and all possible past, present and future behavior of P, as conditioned on the data known by E. The strategy obtained for E is optimal in the sense that it bounds the outcome of the game from below, and the strategy for P is optimal in the sense that it induces the worst case payoff to E. However, we must observe that this strategy for P is not physically realizable, since it corresponds to E's determination posteriori at time t of what P should have chosen for $\tilde{u}$ at time $t-\sigma$, *given the data observed by* E *at time* t . Needless to say, this data was not available to P at time $t-\sigma$, so the strategy cannot be realized by P in real time†.

For the DGWITL in which E has information time lag, if we consider that, at each time t, the pursuer knows x(t) and y(t), we are lead to the conclusion that synthesis of the DGWFI strategy by P will bound the possible outcomes of the game. However, this bound does not, in general, coincide with E's bound on the potential payoff. Furthermore, P knows that E cannot - in the presence of delay - synthesize the strategy to induce the payoff bound, and E knows that P cannot synthesize the strategy that induces the bound on the potential payoff. It is clear that a form of mixed strategies is required to establish equality of the payoff bounds. At the present time this is an unsolved problem.

---

† See (Refs. 13, 15) for a discussion of P's real time behavior.

4.4 - *Effects of information delay.*

Until recently, the possible quantitative effects of information delay in a differential game were unknown. Within the theory presented in the preceding section it is now possible to determine whether the DGWITL strategy must be synthesized or whether the effects of delay are negligible. The effects of delay vary from problem to problem, but in the next section we will show how the delay leads to unacceptable performance degradation for a basic example. The demonstration serves to underscore the practical importance of the theory of differential games with information time lag.

4.5 - *Comparison of DG and DGWITL.*

The relationship between DG and DGWITL can be seen in Tables 4.1 a, and 4.1 b, where we indicate some of the more obvious features. Other comments concerning the nature of DGWITL can be found in (Refs. 13, 15).

	DG	DGWITL
Dynamics of Interest to E	$\dot{x} = f(x,t,\tilde{u}(t))$ $\dot{y} = g(y,t,\tilde{v}(t))$	$\dot{x} = f(x,t-\sigma,\tilde{u}(t-\sigma))$ $\dot{y} = g(y,t,\tilde{v}(t))$
Payoff of Interest to E	$J(x,y,t,p,e)$	$J_p(x,t-\sigma,y,t,p,e)$
Payoff Bound Under Optimal Play	Game Value for P and E $V^o(x,y,t)$	Game Potential Value for E $V^o(x,t-\sigma,y,t)$

Table 4.1 a - A Comparison of DG and DGWITL

	DG	DGWITL
Conditions for Termination	Target Set, Terminal Time	Preliminary Target Set, Potential Terminal Time
Necessary Condition	Hamilton-Jacobi Equation	Generalized Hamilton-Jacobi-Equation
Boundary Condition	Terminal Payoff on Target Set	Potential Terminal Payoff on Preliminary Target Set†
Strategy Structure	Time Dependent Feedback of $x(t)$ and $y(t)$	Time Dependent Feedback of $x(t-\sigma)$ and $y(t)$ with Parametric Dependence on σ
	Solution for V^{o} Gives p^{o} and e^{o}	Solution for V^{o} Gives p^{o} and e^{o}, but p^{o} is Non-realizable
Player Behavior Considered	Present and Future for P and E	Present and Future for E ; Past, Present and Future for P

Table 4.1 b - A Comparison of DG and DGWITL

† Requires solution of an optimal control problem

5. Examples

5.1 - Introduction

In this section we present two examples to illustrate the details of the theory and to demonstrate the practical importance of information time lag in differential games.

5.2 - Norm-invariant systems and DGWITL

5.2.1 - Problem introduction. Norm-invariant systems have been used in the past to construct a nonlinear problem which can be solved in closed form by optimal control theory (Ref. 22) and the theory of differential games (Ref. 23). Here we re-examine them to demonstrate some salient features of DGWITL.

5.2.2 - The game without time lag. We begin by solving the game in which both players have perfect information. The dynamics are described by :

(34 a) $$\dot{x} = f(x,t) + b\tilde{u}(t)$$

(34 b) $$\dot{y} = g(y,t) + c\tilde{v}(t)$$

where $x,y \in R^n$, b, $c \in R^1$ and $|b| > |c|$. At each time t, the pursuer (P) chooses $\tilde{u}(t) \in R^n$ and the evader (E) chooses $\tilde{v}(t) \in R^n$, with $\|\tilde{u}(t)\| \leq 1$, $\|\tilde{v}(t)\| \leq 1$. The autonomous systems have the norm-invariant property, that is, $< x,f(x,t) > = 0$, $< y,g(y,t) > = 0$ for all (x,t), $(y,t) \in R^n \times R^1$. A motion of (34) beginning at some initial phase in $R^n \times R^1 \times R^n \times R^1$ is terminated when it reaches the target set θ, where

$$\theta = \{(x,t,y,t) : \|x\| = \|y\| + a, \ a > 0\}$$

The payoff made by P to E is determined by $J = \int_t^{t_f} d\alpha$

where t_f is the usual terminal time. Admissible strategies for P and E are taken to be mappings on $R^n \times R^n \times R^1$ which are continuous in t and Lipschitzian in the pair (x,y). The value of the game, $V^o(x,y,t)$ is defined to be

$$V^o(x,y,t) = \max_e \min_p J = \min_p \max_e J$$

The assumption that $|b| > |c|$ is sufficient to guarantee that V^o exists.

To produce the Hamilton-Jacobi equation we form

$$H(x,y,t,\lambda_x,\lambda_y,\mu,\beta) = < \lambda_x, f(x,t) > + < \lambda_x, b\mu >$$
$$+ < \lambda_y, g(y,t) > + < \lambda_y, c\beta > + 1$$

Taking min-max with respect to μ and β produces

$$\mu^o = \frac{-b\lambda_x}{|b| \, \| \lambda_x \|} \qquad \beta^o = \frac{c\lambda_y}{|c| \, \| \lambda_y \|}$$

The Hamiltonian is

$$H^o(x,y,t,\lambda_x,\lambda_y) = H(x,y,t,\lambda_x,\lambda_y,\mu^o,\beta^o)$$

from which we obtain the Hamilton-Jacobi equation

$$V_t^o + < V_x^o, f(x,t) > + < V_y^o, g(y,t) >$$
$$- |b| \, \| V_x^o \| + |c| \, \| V_y^o \| + 1 = 0$$

In this example the Hamilton-Jacobi equation can be solved directly to obtain both the value function and the optimal strategy pair.

Case I : $\|x\| \geqslant \|y\| + a$

$$V^{o}(x,y,t) = \frac{\|x\| - \|y\| - a}{|b| - |c|}$$

$$u^{o}(x,y,t) = \frac{-bx}{|b| \, \|x\|}$$

$$v^{o}(x,y,t) = \frac{-cy}{|c| \, \|y\|}$$

Case II : $\|x\| \leqslant \|y\| + a$

$$V^{o}(x,y,t) = \frac{\|y\| - \|x\| + a}{|b| - |c|}$$

$$u^{o}(x,y,t) = \frac{bx}{|b| \, \|x\|}$$

$$v^{o}(x,y,t) = \frac{cy}{|c| \, \|y\|}$$

5.2.3 - *The game with information time lag.* To formulate the game with information time lag for the evader note that the preliminary target set is given by

$$\begin{aligned} \theta_p = \{(x,\tau,y,t) \; : \; & \tau = t-\sigma, \; \|y\| \leqslant \|x\| \\ & \text{and } \|x\| - \sigma|b| \leqslant \|y\| + a \\ & \text{or } \|y\| \geqslant \|x\| \text{ and } \|x\| + \sigma|b| \geqslant \|y\| + a\} \end{aligned}$$

and σ is fixed (no speedup of data arrival). θ_p contains those quadruples from which termination *could* have occured in the past σ seconds, given $(x(t-\sigma),t-\sigma,y(t),t)$. Upon observation of the data, E computes θ_p to determine whether capture is imminent.

Figure 5.1 illustrates the preliminary target set centered about a fixed $y \in R^2$. The shaded regions represent members of θ_p (i.e. the potentially terminal phases). In these regions the evader's best estimate of when the game will terminate is the current time - the game is potentially over, and the potential value is zero. Note, however, that E cannot be certain that the game is over since E cannot be certain about P's moves from $x(t-\sigma)$ over $t-\sigma$ to t .

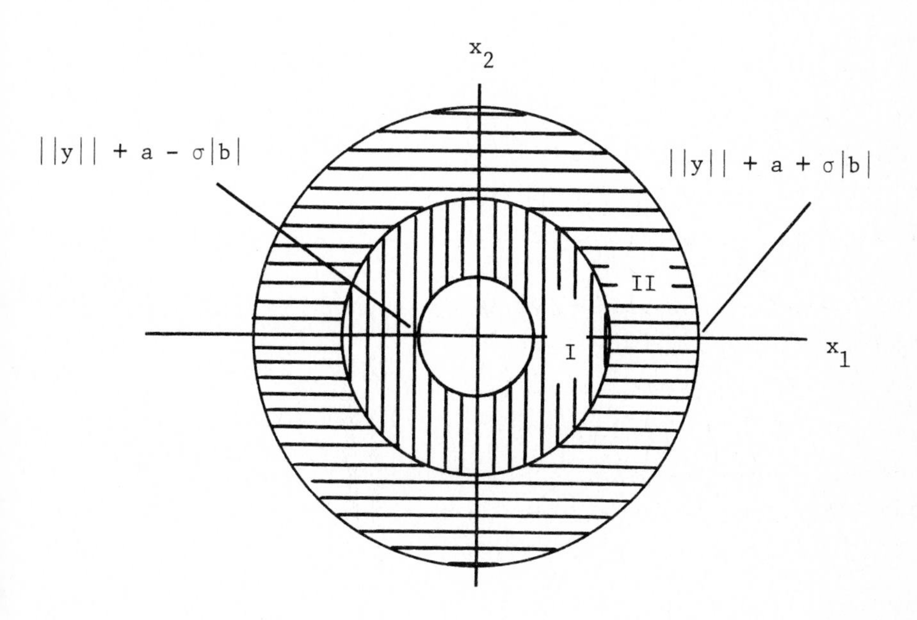

Fig. 5.1 Preliminary Target Set

Due to the particular structure of $J(\cdot)$, the remaining payoff and the potential payoff are given by

$$J_p(x,\tau,y,t) = \int_t^{t_p} d\alpha$$

Here, the potential terminal time is the first time at which an E-observable motion emanating from (x,τ,y,t) penetrates θ_p. Intuitively, if (x,τ,y,t) is near θ_p, then E realizes that the potential terminal time is small. P could have maneuvered from $x(\tau)$ such that $x(t)$ and $y(t)$ are on θ.

To find the potential value, we form the generalized pre-Hamiltonian

$$H(x,\tau,y,t,\lambda_x,\lambda_y,\mu,\beta) = < \lambda_x, f(x,\tau) > + < b\mu, \lambda_x >$$

$$+ < \lambda_y, g(y,t) > + < c\beta, \lambda_y > + 1$$

The min-max of H w.r.t. μ and β, respectively, occurs at

$$\mu^o = \frac{b\lambda_x}{|b| \, \|\lambda_x\|} \qquad \beta^o = \frac{c\lambda_y}{|c| \, \|\lambda_y\|}$$

The generalized Hamiltonian-Jacobi equation is

$$V^o_\tau + V^o_t + < V^o_x, f(x,\tau) > + < V^o_y, g(y,t) >$$

$$+ |c| \, \|V^o_y\| - |b| \, \|V^o_x\| + 1 = 0$$

with boundary condition : $V^o(x,\tau,y,t) = 0$ for (x,τ,y,t) such that

$$\|y\| + a - (\tau-t)|b| \leqslant \|x\| \leqslant \|y\| + a + (\tau-t)|b|$$

and $\tau = \sigma-t$. Application of Theorem 3.2 to this problem confirms that it has the following solution.

Case I : $\|x\| \geqslant \|y\| + a + \sigma|b|$

$$V^o(x,\tau,y,t) = \frac{\|x\| - \|y\| - a - |b| \, (\tau-t)}{|b| - |c|}$$

$$= \frac{\|x\| - \|y\| - a - \sigma|b|}{|b| - |c|}$$

Case II : $\|x\| \leqslant \|y\| + a - \sigma|b|$

$$V^{o}(x,\tau,y,t) = \frac{\|y\| + a - |b|(\tau - t) - \|x\|}{|b| - |c|}$$

$$= \frac{\|y\| + a - \sigma|b| - \|x\|}{|b| - |c|}$$

Figure 5.2 contains a plot of $V^{o}(\cdot)$ vs $\|x\|$. As $\|x\|$ increases, the phase quadruple enters θ_p and eventually passes to the other side of θ_p.

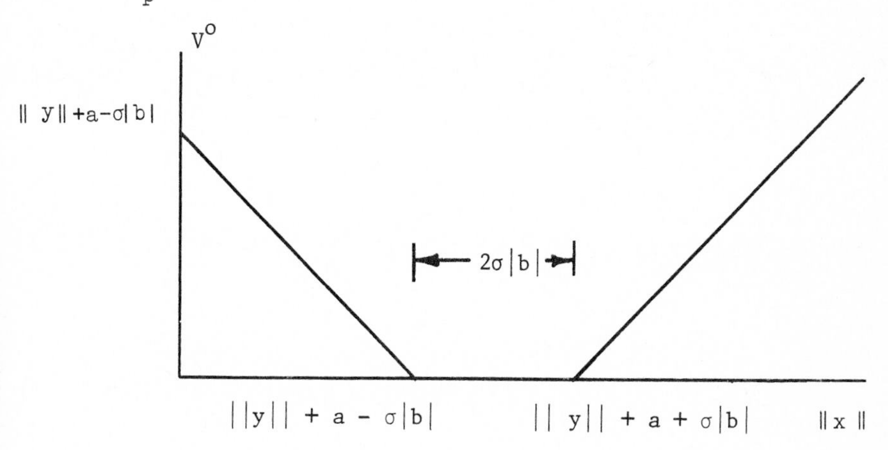

Fig. 5.2 Potential Value for Fixed $\|y\|$

Note that, as the time lag decreases, the uncertainty decreases, and the solution becomes the zero time lag solution. Also, $\theta_p \to \theta$ as $\sigma \to 0$. For this example, σ may be interpreted to broaden the capture surface that appeared in the DG with perfect information.

5.2.4 - The aging data problem. Recall that the study of the behavior of the potential value as the age of the spatial data increases is called the *aging data problem*. One important qualitative feature which we might anticipate in DGWITL is that for sufficiently old data the game should be over (assuming that the duration is not fixed a priori). This expectation is confirmed by this example, which demonstrates the importance of accounting for the time associated with a player's physical position.

For the example, fix $\|x\|$, $\|y\|$ and let τ decrease (σ increase). Figure 5.3 illustrates the behavior of $V^o(\cdot)$ for Region II of θ_p .

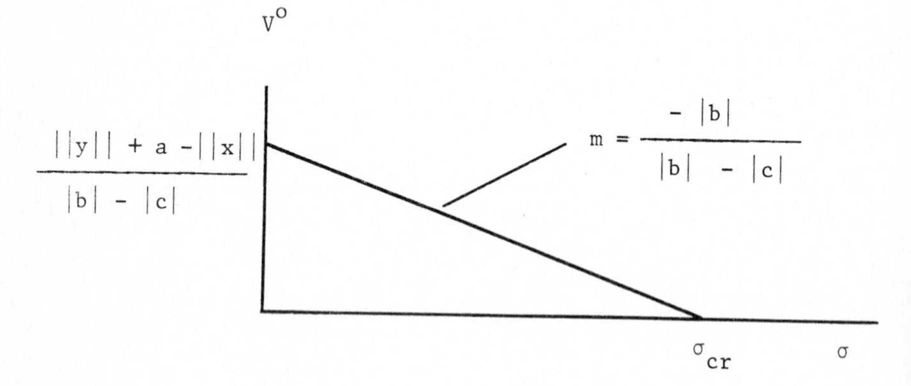

Fig. 5.3 Degradation of Potential Value

Beyond $\sigma_{cr} = \frac{\|y\| + a - \|x\|}{|b|}$ E can only conclude that capture is imminent. Alternately, as σ increases, θ_p enlarges to include the data. At this point E's optimum strategy is simply to contemplate his existence. Note too that if we let $|c|$ vary, the zone of uncertainty does not change because P's ability to effect capture from $x(\tau)$ has not changed, while at the same time $y(t)$ is fixed. However, V^o with $\sigma = 0$ increases as $|c| \to |b|$ because this implies that more time is needed for capture under optimal play.

The important problem of determining the time lag corresponding to the maximum tolerable degradation of potential payoff, as the data lag increases, requires examination of $V^o(\cdot)$ above the τt plane. For this example, Figure 5.3 reveals a linear degradation of potential payoff as the time lag increases. Therefore, for a specified r, $0 \leqslant r \leqslant 1.0$, it is necessary that

$$\sigma \leqslant \frac{\|y\| + a - \|x\|}{|b|}(1-r)$$

so that

$$\frac{V^o(x,t-\sigma,y,t)}{V^o(x,t,y,t)} \geqslant r$$

5.2.5 - *The optimal strategies*. The optimal strategies for the DGWITL are given by the following expressions.

Case I : $\|x\| \geqslant \|y\| + a + \sigma|b|$

$$p^o(x,\tau,y,t) = \frac{-bx(\tau)}{|b|\ \|x(\tau)\|}$$

$$e^o(x,\tau,y,t) = \frac{-cy(t)}{|c|\ \|y(t)\|}$$

Case II : $\|x\| \leqslant \|y\| + a - \sigma|b|$

$$p^o(x,\tau,y,t) = \frac{bx(\tau)}{|b|\ \|x(\tau)\|}$$

$$e^o(x,\tau,y,t) = \frac{cy(t)}{|c|\ \|y(t)\|}$$

Observe that the structure of E's strategy in the DGWITL is identical to the structure of the strategy for $\sigma = 0$. Although the strategies are state-decoupled, the inequalities which define the region of space for use of the strategy are σ dependent. Both the strategy inequalities and the potential value depend on $x(\tau)$ and $y(t)$.

5.2.6 - *Lack of payoff bound coincidence.* Consider the DGWITL for the norm-invariant systems in which $b = 2$, $c = 1$, $a = 20$ and $\sigma = 5$ secs. Let $\|x(\tau)\| = 70$ and $\|y(t)\| = 10$. Then, the potential value for E is $V^o = 30$ secs . Now suppose that P had not exerted control effort over the interval of delay, that is, $\|x(\tau)\| = \|x(t)\| = 70$. From P's point of view, the value is given by the expression for V^o with $\sigma = 0$; this bound on the payoff to E is 40 secs. Thus, P and E compute different estimates of the worst case outcome of the game. Moreover, E's payoff bound (lower) is strictly less than the payoff bound (upper) which can be imposed by P . This lack of coincidence of the payoff bounds is common in differential games with information time lag - suggesting the need for research aimed at implementing mixed strategies in DGWITL. In this particular exemple, mixed strategies are of no avail to E .

5.3 - *Linear systems and DGWITL*

5.3.1 - *Problem introduction.* In this section we examine the well-known differential game in which both players control a linear system and the payoff is quadratic. Versions of this problem were previously solved by function space methods in (Refs. 10, 13). In the context of the theory developed in the preceding sections, the problem will be solved in closed form to obtain a solution expressed in terms of a generalized Riccati matrix . Structural aspects of the DGWITL strategy will be examined.

5.3.2 - *Solution of linear DGWITL.* The systems controlled by P and E are described by†

$$\dot{x} = A_p(t)x(t) + B_p(t)\tilde{u}(t) \; ; \; x(\tau_o) = x_{o\sigma}$$

$$\dot{y} = A_e(t)y(t) + B_e(t)\tilde{v}(t) \; ; \; y(t_o) = y_o$$

---

† $G = R^n \times R^1 \times R^n \times R^1$

with $\tilde{u}(t) \in K_u = R^q$, $\tilde{v}(t) \in K_v = R^m$. The target set is given by

$$\theta = \{(x,t,y,t) : t = T\}$$

Therefore, the preliminary target set is simply

$$\theta_p = \{(x,\tau,y,t) : \tau = T-\sigma,\ t = T\}$$

where $\sigma = \tau_o - t_o$. The terminal time is $t_f = T$, and the potential terminal time is $t_p(x_o,\tau_o,y_o,t_o,p,e) = T$, since the game has specified duration. The remaining payoff function is

$$J_r(x_{o\sigma},\tau_o,y_o,t_o,p,\bar{u},e) = \| x(T) - y(T) \|_F^2$$

$$+ \int_{\tau_o}^{T-\sigma} \| u \|^2_{R_p(\alpha)} d\alpha + \int_{T-\sigma}^{T} \| \bar{u} \|^2_{R_p(\alpha)} d\alpha - \int_{t_o}^{T} \| v \|^2_{R_e(\alpha)} d\alpha$$

where $F > 0$, $R_p(\cdot) > 0$, $R_e(\cdot) > 0$, $\| u \|^2_{R_p(\alpha)}$ denotes $\tilde{u}'(\alpha)R_p(\alpha)\tilde{u}(\alpha)$ and all matrices are of consistent dimension. The potential terminal payoff on θ_p is the solution to the following optimal control problem. Let $(x,\tau,y,t) \in \theta_p$, that is $\tau = T-\sigma$ and $t = T$. Then

$$W(x,T-\sigma,y,T) = \min_{\bar{u}\in\bar{C}_u} \| \bar{x}(T) - y \|_F^2 + \int_{T-\sigma}^{T} \| \bar{u} \|^2_{R_p(\alpha)} d\alpha$$

and $\bar{C}_u$ is the set of piecewise continuous $\bar{u} : [T-\sigma, T] \to R^q$. It can be shown that

$$W(x,T-\sigma,y,T) = [\varphi_p(T,T-\sigma)x-y]'[F^{-1}+SR_p^{-1}S^*]^{-1}[\varphi_p(T,T-\sigma)x-y]$$

where

$$SR_p^{-1}S^* \triangleq \int_{T-\sigma}^{T} \varphi_p(T,\alpha)B_p(\alpha)R_p^{-1}(\alpha)B_p'(\alpha)\varphi_p'(T,\alpha)d\alpha$$

and φ_p is the transition matrix for $A_p(\cdot)$. The optimal (posteriori) terminating control for P is given by

$$\overline{u}(\alpha) = -R_p^{-1}(\alpha)B_p'(\alpha)\varphi_p'(T,\alpha)[F^{-1}+SR_p^{-1}S^*]^{-1}[\varphi_p(T,T-\sigma)x-y] ,$$

$$\alpha \in [T-\sigma,T]$$

Thus, $\overline{u}$ is E's determination, at time T, of the best control that P could have applied over $[T-\sigma,T]$ from $x(T-\sigma)$, with knowledge of $y(T)$. By definition, the potential payoff is

$$J_p(x_{o\sigma},\tau_o,y_o,t_o,p,e) =$$

$$[\varphi_p(T,T-\sigma)x-y]'[F^{-1}+SR_p^{-1}S^*]^{-1}[\varphi_p(T,T-\sigma)x-y]$$

$$+\int_{t_o-\sigma}^{T-\sigma} \|u\|^2_{R_p(\alpha)}d\alpha - \int_{t_o}^{T} \|v\|^2_{R_e(\alpha)}d\alpha$$

where $x = x(T-\sigma)$, $y = y(T)$, $x_{o\sigma} = x(t_o-\sigma)$, $y_o = y(t_o)$. The problem is to determine a saddlepoint of $J_p(\cdot)$ with respect to $p \in S_P$, $e \in S_E$ subject to

$$\dot{x} = A_p(t-\sigma)x + B_p(t-\sigma)\tilde{u}(t-\sigma),\ x(t_o-\sigma) = x_{o\sigma}$$

$$\dot{y} = A_e(t)y + B_e(t)\tilde{v}(t),\ y(t_o) = y_o$$

with $\tilde{u}(\cdot)$ and $\tilde{v}(\cdot)$ specified by p and e. The solution to this problem will be in the form of a payoff bound for E, a strategy for E, and an ideal exploitation strategy for P. This exploitation strategy forms E's posteriori determination at time t of P's best control at time $t-\sigma$, given knowledge of $y(t)$.

To apply the theory of DGWITL to this problem, we first form the generalized pre-Hamiltonian H

$$H(x,\tau,y,t,\lambda_x,\lambda_y,\mu,\beta) = \langle A_p(\tau)x + B_p(\tau)\mu,\lambda_x \rangle$$

$$+ \langle A_e(t)y+B_e(t)\beta,\lambda_y\rangle + \langle\mu,R_p(\tau)\mu\rangle - \langle\beta,R_e(t)\beta\rangle$$

Taking min max of $H(\cdot)$ with respect to $\mu \in K_u$, $\beta \in K_v$ gives

$$k_1(x,\tau,y,t,\lambda_x,\lambda_y) = -\frac{1}{2} R_p^{-1}(\tau)B_p'(\tau)\lambda_x$$

$$k_2(x,\tau,y,t,\lambda_x,\lambda_y) = \frac{1}{2} R_e^{-1}(t)B_e'(t)\lambda_y$$

and the generalized Hamiltonian becomes

$$H^o(x,\tau,y,t,\lambda_x,\lambda_y) = x'A_p'(\tau)\lambda_x - \frac{1}{4} \lambda_x' B_p(\tau)R_p^{-1}(\tau)B_p'(\tau)\lambda_x + \frac{1}{4} \lambda_y' B_e(t)R_e^{-1}(t)B_e'(t)\lambda_y + y'A_e'(t)\lambda_y$$

Then, the generalized Hamilton-Jacobi equation satisfied by the potential value function is

$$V_\tau^o + V_t^o + x'A_p'(\tau)V_x^o + y'A_e'(t)V_y^o - \frac{1}{4} V_x^{o\prime} B_p(\tau)R_p^{-1}(\tau)B_p'(\tau)V_x^o + \frac{1}{4} V_y^{o\prime} B_e(t)R_e^{-1}(t)B_e'(t)V_y^o = 0$$

with boundary condition

$$V^o(x,\tau,y,t)\Big|_{t=T} = [\varphi_p(T,T-\sigma)x-y]'[F^{-1}+SR_p^{-1}S^*]^{-1}[\varphi_p(T,T-\sigma)x-y] ,$$

$$= \begin{bmatrix} x \\ y \end{bmatrix}' \begin{bmatrix} \varphi_p'(T,T-\sigma) \\ \hline -I \end{bmatrix} [F^{-1}+SR_p^{-1}S^*]^{-1}[\varphi_p(T,T-\sigma) \mid -I] \begin{bmatrix} x \\ y \end{bmatrix}$$

where I denotes the identity matrix for R^n. Letting $\sigma \to 0$ and $\tau \to t$ in the above expressions produces the known results for DG without time lag.

Following the procedures for linear optimal control and DG, we next assume that

$$V^o(x,\tau,y,t) = \begin{bmatrix} x \\ y \end{bmatrix}' P(\tau,t) \begin{bmatrix} x \\ y \end{bmatrix}$$

where $P(\cdot)$ is a $2n \times 2n$ symmetric matrix. Then

$$V^o_x(x,\tau,y,t) = 2(P_{11}(\tau,t)x + P_{12}(\tau,t)y)$$

$$V^o_y(x,\tau,y,t) = 2(P_{12}(\tau,t)x + P_{22}(\tau,t)y)$$

and the generalized Hamilton-Jacobi equation becomes

$$0 = \begin{bmatrix} x \\ y \end{bmatrix}' \{ \frac{\partial P}{\partial \tau}(\tau,t) + \frac{\partial P}{\partial t}(\tau,t) + P(\tau,t)A(\tau,t) + A'(\tau,t)P(\tau,t) + P(\tau,t)B(\tau,t)P(\tau,t)\} \begin{bmatrix} x \\ y \end{bmatrix}$$

in $G - \theta_p$ with

$$A(\tau,t) = \left[\begin{array}{c|c} A_p(\tau) & 0 \\ \hline 0 & A_e(t) \end{array}\right]$$

$$B(\tau,t) = \left[\begin{array}{c|c} -B_p(\tau)R_p^{-1}(\tau)B_p'(\tau) & 0 \\ \hline 0 & B_e(t)R_e^{-1}(t)B_e'(t) \end{array}\right]$$

and

$$P(\tau,t) = \left[\begin{array}{cc} P_{11}(\tau,t) & P_{12}(\tau,t) \\ \hline P_{12}(\tau,t) & P_{22}(\tau,t) \end{array}\right]$$

and the potential value is

$$V^o(x,\tau,y,t) = \begin{bmatrix} x \\ y \end{bmatrix}' P(\tau,t) \begin{bmatrix} x \\ y \end{bmatrix}$$

in the neighborhood of $(T-\sigma,T)$ in which the solution to (35) is positive definite.

Note that the generalized Riccati matrix is defined on two temporal parameters, τ and t, corresponding to the spatial data, x and y, observed by E. Likewise, the potential value is defined on both temporal parameters and both spatial variables. This point is emphasized by Figure 5.4, which illustrates the data reference plane for this problem. The shaded region of the plane below the line defined by $\sigma = 0$ is the region over which (35) must be solved. The line defined by $t = T$ is the locus on which the potential terminal payoff provides boundary data for V^o. If the DG with perfect information has already been solved, its solution may be used to provide boundary data long the line defined by $\sigma = 0$.

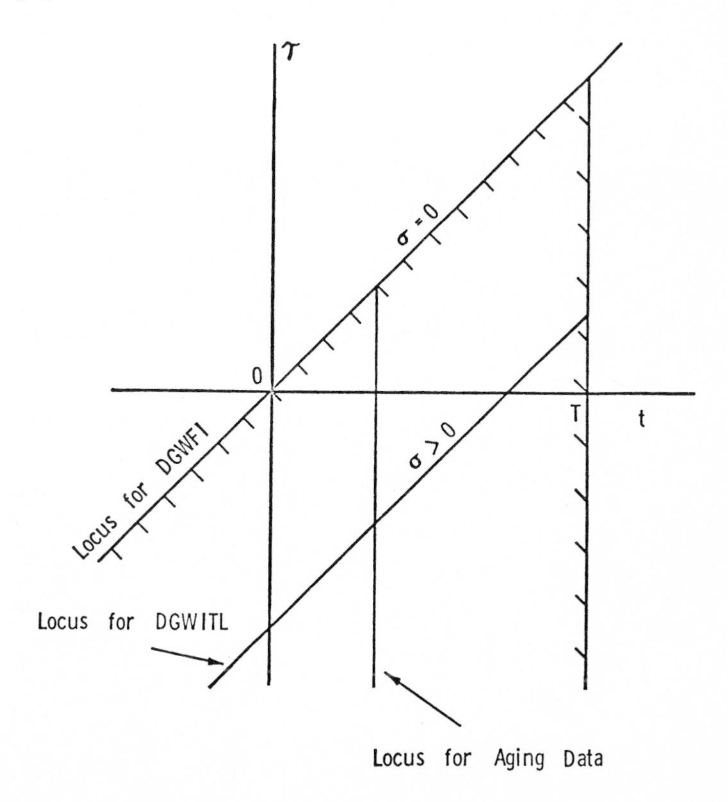

Fig. 5.4 Data Reference Plane

5.3.3 - *Parametric solution of the Riccati equation.* Converting the generalized Hamilton-Jacobi equation over the τt plane into a "delayed argument" partial differential equation along the line given by $\tau = t-\sigma$ requires converting the generalized Riccati equation into a Riccati equation in t, with σ appearing as a parameter in the arguments of the time-varying matrices. For a fixed σ, let

$$\hat{P}(t;\sigma) \triangleq P(\tau,t)\Big|_{\tau=t-\sigma}$$

Then

$$\frac{d\hat{P}}{dt}(t;\sigma) = \frac{\partial P}{\partial \tau}(\tau,t)\Big|_{\tau=t-\sigma} \frac{d\tau}{dt} + \frac{\partial P}{\partial t}(\tau,t)\Big|_{\tau=t-\sigma}$$

It follows directly that the solution to (35) along the line with $\tau = t-\sigma$ must satisfy the following equation.

$$\frac{d\hat{P}}{dt}(t;\sigma) + P(t;\sigma)A(t-\sigma,t) + A'(t-\sigma,t)\hat{P}(t;\sigma) + \hat{P}(t;\sigma)B(t-\sigma,t)\hat{P}(t;\sigma) = 0$$

subject to

$$\hat{P}(T;\sigma) = \begin{bmatrix} \varphi_p'(T,T-\sigma) \\ \hline -I \end{bmatrix} [F^{-1} + SR_p^{-1}S^*]^{-1}[\varphi_p(T,T-\sigma) \mid -I]$$

The solution to the DGWITL can be expressed in terms of $\hat{P}(\cdot;\sigma)$, as given below.

$$V^{o}(x,y,t;\sigma) = \begin{bmatrix} x \\ y \end{bmatrix}' \hat{P}(t;\sigma) \begin{bmatrix} x \\ y \end{bmatrix}$$

$$p^{o}(x,y,t;\sigma) = -R_p^{-1}(t-\sigma)B_p'(t-\sigma)\,[\hat{P}_{11}(t;\sigma)x + \hat{P}_{12}(t;\sigma)y]$$

$$e^{o}(x,y,t;\sigma) = R_e^{-1}(t)B_e'(t)\,[\hat{P}_{12}(t;\sigma)x + \hat{P}_{12}(t;\sigma)y]$$

Figure 5.5 contains the block diagram of the evader's DGWITL optimal strategy, where, for emphasis, we have $x_p = x$, $x_e = y$. Although the feedback structure is linear, the DGWITL strategy is distinct from the strategy for the game without time lag. The important differences are that (i) the strategy incorporates the delayed state vector rather than the actual state vector of the pursuer, and (ii) the strategy requires a gain matrix using the solution to a matrix Riccati equation with delayed arguments.

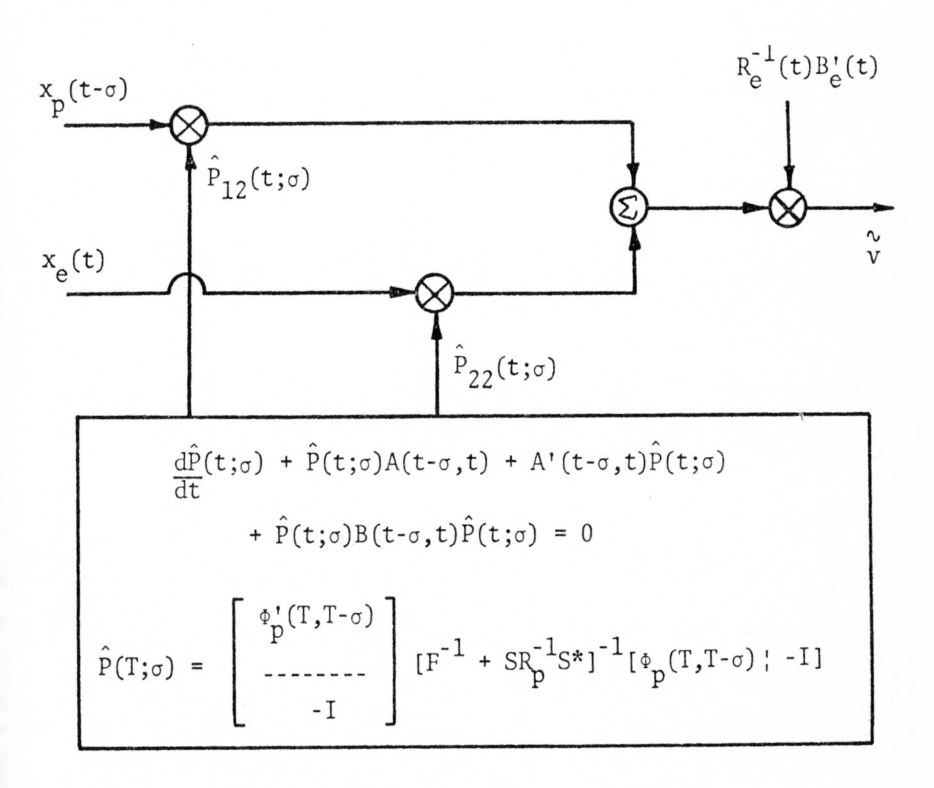

Fig. 5.5 Evader DGWITL Strategy

5.3.4 - *Predictor decomposition of the evader strategy.* The evader strategy can be decomposed by means of a simple transformation to reveal that part of the strategy which is a predictor. Let $\tilde{P}$ be defined by

$$\tilde{P}(t;\sigma) = \left[\begin{array}{c|c} \varphi_p'(t-\sigma,t) & 0 \\ \hline 0 & -I \end{array}\right] \hat{P}(t;\sigma) \left[\begin{array}{c|c} \varphi_p(t-\sigma,t) & 0 \\ \hline 0 & -I \end{array}\right]$$

Then,

$$\begin{bmatrix} p^o(\cdot) \\ e^o(\cdot) \end{bmatrix} = \left[\begin{array}{c|c} -R_p^{-1}(t-\sigma)B_p'(t-\sigma)\varphi_p'(t,t-\sigma) & 0 \\ \hline & R_e^{-1}(t)B_e'(t) \end{array}\right] \tilde{P}(t;\sigma) \left[\begin{array}{c} \varphi_p(t,t-\sigma)x(t-\sigma) \\ \hline y(t) \end{array}\right]$$

and $\tilde{P}(\cdot;\sigma)$ satisfies the Riccati equation in the diagram of Figure 5.6, and $\tilde{x}_p(t) = \varphi_p(t,t-\sigma)x(t-\sigma)$. Note that $\tilde{x}_p(t)$ is simply that portion of $x_p(t)$ that is due to the autonomous motion of P's system over the delay interval. It is clear from this representation of the evader strategy that the effect of the information time lag is to create both a simple predictor and a σ-dependent Riccati gain in the strategy.

5.4 - *Strategy comparison for DGWITL*

5.4.1 - *Introduction.* In practice, an information delay might not be detected by E. E would then be led to synthesize the perfect information strategy in ignorance of the fact that $x(t-\sigma)$ was being observed instead of $x(t)$. Or, an aware E may simply conjecture that the delay is of no consequence. We call the strategy which uses the delayed state in the perfect information strategy a "direct insertion feedback" strategy (DIFBK). Another possibility is that E may predict - in the manner already discussed - P's present state and continue to use the Riccati gain

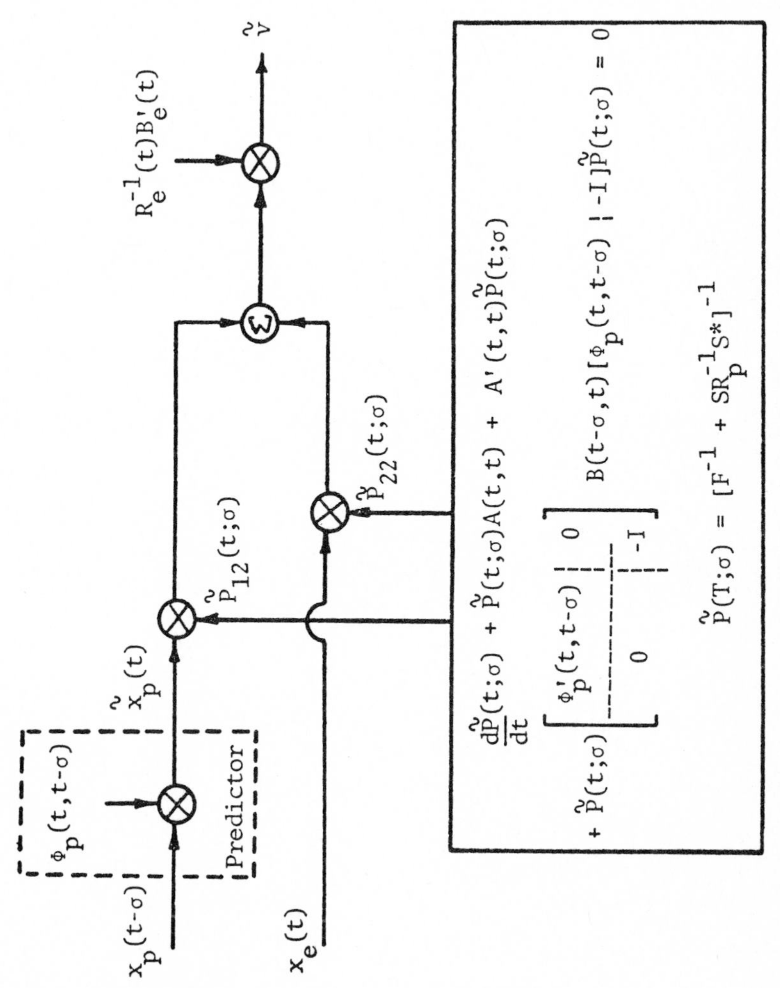

Fig. 5.6 Predictor Decomposition of Evader DGWITL Strategy

corresponding to the perfect information game. We call this strategy a "predictor feedback" strategy (PRFBK).

Both of these suboptimal strategies may be synthesized in Figure 5.6 by (i) setting $\sigma = 0$ in the predictor and in the Riccati equation to form the DIFBK strategy, and (ii) setting $\sigma = 0$ in only the Riccati equation to form the PRFBK strategy.

In applying game theory we would like to know whether it is important to take into account information delays, or whether one of the suboptimal strategies described here is adequate. Although the answer to these questions depends on the nature of each differential game we provide, in the next section, a quantitative illustration of the effect of information delays, with the intention of giving insight about the approach that can be taken for the study of other examples and demonstrating the importance of such consideration in practical problems.

5.4.2 - *A scalar example.* In this section we will use a scalar version of the linear regulator results to demonstrate that even small information lags can lead to serious performance degradation if the delay is ignored and a DIFBK strategy or a PRFBK strategy is used. We will also show that use of the DGWITL strategy effectively compensates for the information lag and results in marked performance improvement.

The equations of motion are

$$\dot{x}_p = a_p x_p + b_p \tilde{u}(t) \tag{37}$$

$$\dot{x}_e = a_e x_e + b_e \tilde{v}(t) \tag{38}$$

and all parameters are scalar constants. The remaining payoff for E is

$$J_r(\cdot) = |x_p(T) - x_e(T)|^2 + \frac{1}{c_p}\int_{\tau}^{T-\sigma} \tilde{u}^2(\alpha)d\alpha$$

$$(39) \qquad - \frac{1}{c_e}\int_{t}^{T} \tilde{v}^2(\alpha)d\alpha + \frac{1}{c_p}\int_{T-\sigma}^{T} \bar{u}^2(\alpha)d\alpha$$

where $\sigma = \tau - t$, $c_p > 0$, $c_e > 0$ and T is fixed. The potential terminal payoff, $W(\cdot)$, is the solution to the optimal control problem given by

$$W(x_p(T-\sigma), T-\sigma, x_e(T), T) = \min_{\bar{u} \in \bar{C}_u} \{|\bar{x}_p(T) - x_e(T)|^2 + \frac{1}{c_p}\int_{T-\sigma}^{T} \bar{u}^2(\alpha)d\alpha\}$$

subject to $\dot{x}_p = a_p x_p + b_p \bar{u}(t)$, $T-\sigma \leqslant t \leqslant T$, $x_p(T-\sigma)$ given .

The solution can easily be shown to be :

$$W(x_p(T-\sigma), T-\sigma, x_e(T), T) = \frac{\left[e^{a_p} x_p(T-\sigma) - x_e(T)\right]^2}{1 + \frac{c_p b_p^2}{2a_p}[e^{2a_p\sigma} - 1]}$$

Therefore, the potential payoff to E is given by

$$J_p = \frac{\left[e^{a_p\sigma} x_p(T-\sigma) - x_e(T)\right]^2}{1 + \frac{c_p b_p^2}{2a_p}[e^{2a_p\sigma} - 1]} + \frac{1}{c_p}\int_{\tau}^{T-\sigma} \tilde{u}^2(\alpha)d\alpha - \frac{1}{c_e}\int_{t}^{T} \tilde{v}^2(\alpha)d\alpha$$

where $\tilde{u}(\cdot)$ and $\tilde{v}(\cdot)$ are specified by strategies p and e .

Performing the analysis described in the preceding sections leads to the following expressions for the optimal strategies and the potential value.

$$p^{o}(x_p,\tau,x_e,t) = \frac{-c_p b_p [\, e^{a_p(T-\tau)} x_p - e^{a_e(T-t)} x_e \,]\, e^{a_p(T-\tau)}}{G(\tau,t)}$$

$$e^{o}(x_p,\tau,x_e,t) = \frac{-c_e b_e [\, e^{a_p(T-\tau)} x_p - e^{a_e(T-t)} x_e \,]\, e^{a_e(T-t)}}{G(\tau,t)}$$

$$V^{o}(x_p,\tau,x_e,t) = \frac{[\, e^{a_p(T-\tau)} x_p - e^{a_e(T-t)} x_e \,]^2}{G(\tau,t)}$$

where

$$G(\tau,t) = 1 + \frac{c_p b_p^2}{2a_p}[\, e^{2a_p(T-\tau)} - 1 \,] - \frac{c_e b_e^2}{2a_e}[\, e^{2a_e(T-t)} - 1 \,] \tag{40}$$

Along the line defined by $\tau = t-\sigma$, the solution is given by

$$p^{o}(x_p,x_e,t;\sigma) = \frac{-c_p b_p [\, e^{a_p(T-t+\sigma)} x_p(t-\sigma) - e^{a_e(T-t)} x_e(t) \,]\, e^{a_p(T-t+\sigma)}}{\hat{G}(t;\sigma)} \tag{41}$$

$$e^{o}(x_p,x_e,t;\sigma) = \frac{-c_e b_e [\, e^{a_p(T-t+\sigma)} x_p(t-\sigma) - e^{a_e(T-t)} x_e(t) \,]\, e^{a_e(T-t)}}{\hat{G}(t;\sigma)} \tag{42}$$

$$V^{o}(x_p,x_e,t;\sigma) = \frac{[\, e^{a_p(T-t+\sigma)} x_p(t-\sigma) - e^{a_e(T-t)} x_e(t) \,]^2}{\hat{G}(t;\sigma)} \tag{43}$$

and

$$\hat{G}(t,\sigma) = 1 + \frac{c_p b_p^2}{2a_p}[\, e^{2a_p(T-t+\sigma)} - 1 \,] - \frac{c_e b_e^2}{2a_e}[\, e^{2a_e(T-t)} - 1 \,] \tag{44}$$

5.4.3 - *Performance simulations*. The importance of properly compensating for information time lag in a differential game can be demonstrated by simulating the performance of the DGWITL strategy, the direct insertion feedback strategy, and the predictor feedback strategy. Recall that the direct insertion feedback strategy ignores the information delay and inserts the delayed information directly into the perfect information strategy, while the predictor feedback strategy incorporates the predictor portion of the DGWITL strategy but uses an incorrect Riccati gain. These three strategies were generated from (42) and used in a digital simulation of (37), (38) and (39). Because the pursuer strategy given by (41) is not physically realizable in real time, we simulated a pursuer using a perfect information strategy, as given by (41) with $\sigma = 0$ †

The following parameters were used in the simulations.

$$a_p = -0.2 \qquad a_e = -0.4$$
$$b_p = 0.8 \qquad b_e = 1.2$$
$$c_p = 1.0 \qquad c_e = 1.0$$
$$T = 5.0 \qquad 0 \leqslant \sigma \leqslant 5.0$$
$$t_o \leqslant 5.0$$

Once the parameters for (37), (38) and (39) have been chosen, there are two ways to conduct a simulation of the strategies. The first method fixes the spatial data $(x_p(t_o-\sigma),\ x_e(t_o))$ and examines a family of games for various choices of t_o and σ. The second method fixes $(x_p(t_o),\ x_e(t_o))$ and, likewise, examines games for t_o and σ of interest. Both methods require assumptions which specify the behavior of the players over $[t_o-\sigma, t_o]$.

† Recall that $p^o(x_p,x_e,t;\sigma)$ is E's posteriori determination at time t of P's best decision at time $t-\sigma$, given $y(t)$.

In using the first method it is necessary to generate a "history" for E and simulate P's response to obtain a data record for $x_p(\alpha)$, $t_o-\sigma \leqslant \alpha \leqslant t_o$. The initializing path for E was taken to be the autonomous motion of (37), with terminal boundary condition given by $x_e(t_o)$. P was assumed to be using the perfect information strategy. With this method it is possible to examine performance for fixed E-viewed data - each choice of t_o and σ defines a different value for $x_p(t_o)$.

The second method also requires generation of histories over $[t_o-\sigma, t_o]$, but with terminal boundary conditions for both $x_p(t_o)$ and $x_e(t_o)$. Each choice of t_o and σ leads to a different $x_p(t_o-\sigma)$ depending on the assumed behavior for P.

The first simulation method was used to provide a comparison of the three strategies for fixed initial spatial data, and the second method was used to examine the effect of information delay on the potential payoff to the evader. In using the second method we assumed that the player histories over $[t_o-\sigma, t_o]$ were due to autonomous motion terminating at the prescribed $x_p(t_o)$ and $x_e(t_o)$.

Each method has a distinct physical interpretation. The first method reveals how a player's behavior depends on the time associated with the observed spatial date, e.g. what E does depends on when P was located at the position prescribed by the spatial data. The second method indicates how the lack of information about the present position of P degrades the potential payoff and increases the uncertainty about the outcome, as the delay increases.

Figures 5.7 - 5.13 summarize the simulation results for fixed E-observed data, with $x_p(t_o-\sigma) = 10.0$, $x_e(t_o) = 20.0$. Figure 5.7 contains the potential value, as computed from (43). Figures 5.8-5.10 illustrate the simulated J_{TL}, J_{DI}, and J_{PR} - the remaining payoffs when the optimal DGWITL strategy, the direct insertion strategy, and the predictor strategy are used, respectively, by E. Examination of the curves for a given t_o and σ shows that the DGWITL strategy bounds the payoff from below by the potential value. (It follows from (39) that V^o is always positive). Since P cannot realize the optimal exploitation strategy, J_{TL} is actually greater than V^o. On the other hand, the direct insertion feedback strategy and the predictor feedback strategy permit large negative payoffs. It is important to note that serious performance degradation occurs when the time lag is ignored, even for data lags that are small relative to the time constants of the systems, e.g. for $t_o = 0.0$ and $\sigma = 1.0$ (the time constants are 2.5 secs. and 5.0 secs.). Figure 5.11 shows the relative performance of the strategies for $t_o = 0.0$. η_{DI} and η_{PR} denote the loss (relative to J_{TL}) in payoff incurred by the insertion strategy and the predictor strategy, respectively, normalized by the maximum loss incurred by the insertion strategy for $0 \leqslant \sigma \leqslant 5.0$.

The boundary conditions for DGWITL are such that each of the strategies perform the same for games with $t_o = 5.0$ and for games with $\sigma = 0.0$, as verified by the simulations.

Since these simulations are for fixed $x_p(t_o-\sigma)$, $x_e(t_o)$, the data in Figures 5.7 - 5.13 is useful for comparing the performance of the strategies on a game-by game basis, that is, for each pair of t_o and σ. However, the curves are not appropriate for examining the effect of the information delay on the outcome of the game for fixed initial $x_p(t_o)$ and $x_e(t_o)$, (this comparison will be made in Figure 5.16). Thus, in Figure 5.10, the information delay appears to improve the performance of E. The improvement is real - in the sense that the potential value for a game beginning at $t_o = 4.0$ secs., with $x_p(t_o-\sigma) = 10.0$, $x_e(t_o) = 20.0$, and $\sigma = 4.0$ secs., is greater than the potential value for the same data with $\sigma = 3.0$ secs. Two games are being compared, and, because the systems are stable and the initial coordinates were both positive, the natural motion of (37) over $[t_o-\sigma, t_o]$ works to P's disadvantage - in the sense that if P was inactive over $[t_o-\sigma, t_o]$ the actual separation at time t_o would be greater than $x_p(t_o-\sigma) - x_e(t_o)$. Hence, the initial conditions for the game with σ = 4.0 secs. are more favorable to E than the initial conditions for the game with σ = 3.0 secs.

Figure 5.12 shows the behavior of J_{TL}, J_{DI}, and J_{PR} for the same spatial data as in the previous figures, but with t_o = - 5.0 secs.For this game of longer duration, non-negligible payoff degradation accompanies use of the direct insertion feedback strategy and the predictor feedback strategy, even for relatively small information delays. Figure 5.13 contains the trajectories that result from use of the DGWITL strategy and the DIFBK strategy by E, for the game with σ = 2.0 secs. Although the terminal miss for DIFBK is better, E's net performance is worse because the DIFBK strategy erroneously uses excessive energy in the terminal phase of play.

Figure 5.14 shows the remaining payoffs for a game of 5.0 secs. duration, but with different initial spatial coordinates - $x_p(t_o-\sigma) = 20.0$, $x_e(t_o) = 40.0$. The detrimental effects of ignoring the information delay in synthesizing the DIFBK strategy are very apparent. Plots of J_{TL}, J_{DI}, and J_{PR} in Figure 5.15 are for the same spatial data as in Fig. 5.14, but with t_o = - 5.0 secs. In this case both the suboptimal strategies lead to unacceptable performance. On the other hand, use of the DGWITL strategy results in performance that is much less sensitive to the information delay.

It is clear from the examples in Figs. 5.7 - 5.15 that ignoring information delay in synthesizing a strategy for a differential game can result in serious performance degradation, even for small delays. Another interesting aspect of a DGWITL is the effect of information delay on the potential value, for fixed

$x_p(t_o)$, $x_e(t_o)$. In Figure 5.16 we have the plot of $V^*(\sigma)$, the potential value for games with $x_p(t_o) = 10.0$, $x_e(t_o) = 20.0$, $t_o = 0.0$ secs., and both player's initial history generated by autonomous motion. Observe that $V^*(0)$ corresponds to the value of a game with perfect information. The plot of $V^*(\sigma)$ reveals degradation of the potential payoff as the information delay increases. Figure 5.17 illustrates the magnitude of the per cent change in potential payoff as the delay increases, indicating that 10% degradation occurs at σ less than one-half the smallest time constant.

5.4.4 - *Conjugate points*. In general, the existence of a solution to a linear differential game depends on the occurence of a conjugate point in the solution of the Riccati equation for the payoff gain matrix. For this example, it is easily verified that the conjugate point of the Riccati gain matrix corresponding to (43) is solely determined by the zero crossing of $\hat{G}(t;)$ in the neighborhood of T, and ultimately on the relative value of the parameters in (37), (38) and (39).

To determine the effect of information delay on the existence of solutions to the linear DGWITL we examine (40) and seek the locus of conjugate points in the τt plane. Letting $\tau_{cr}(t)$ denote the value of $\tau \leqslant T$ at which (40) has a zero crossing, we are able to locate the conjugate points for fixed system parameters. The results of this analysis are given in Figure 5.18, which shows $\tau_{cr}(t)$ for various system configurations, with $b_p = 0.8$, $c_p = c_e = 1.0$, $a_p = -.2$, $a_e = -.4$, and $T - 5.0$. The area under the curve defined by $\tau_{ct}(\cdot)$ is the region of the τt plane for which the Riccati gain matrix and $\hat{G}(\cdot;\sigma)$ are positive. Lines drawn parallel to and below the line defined by $\tau = t-\sigma$ correspond to DGWITL. Note that for $b_e = 1.35$ the line with $\sigma = 0$ does not intersect $\tau_{cr}(\cdot)$, so neither the game with perfect information or the DGWITL have conjugate points. For $b_e = 1.4$ the differential game with full information (DGWFI) has a conjugate point at $t = 3.5$ secs . A DGWITL with $0 \leqslant \sigma \leqslant 0.7$ also has a conjugate point. The time at which it occurs is farther from T than the time associated with the DGWFI conjugate point. Hence, the information delay effectively extends the half-interval about T for which the solution to the game exists. For $\sigma > 0.7$ the DGWITL does not have a conjugate point. Thus, if sufficient information delay is present, the conjugate point vanishes. For $b_e = 1.45$ the DGWFI has a conjugate point at $t = 3.8$. The DGWITL also has a conjugate point at $\hat{t}(\sigma)$, where, for $\sigma = 3.0$, $\hat{t} = 1.25$ secs . These three choices of b_e demonstrate that the information delay effectively lengthens the solution interval about T, and, depending on the parameters, may even extend the interval to $-\infty$.

5.4.5 - *Remarks* . By means of a relatively simple example we have shown that information time lag can lead to serious performance degradation. We expect that similar phenomena can occur in more complex and realistic problems. Hence, it is apparent that the overall synthesis procedure in dynamic conflict should include measures to detect information delays and determine their significance by means of the DGWITL theory. Such efforts will allow the system designer to (i) assess the potential payoff degradation ; (ii) establish acceptable levels of information delay ; (iii) evaluate suboptimal strategies ; and (iv) determine whether deliberate introduction of information delay can achieve an overall economy of design.

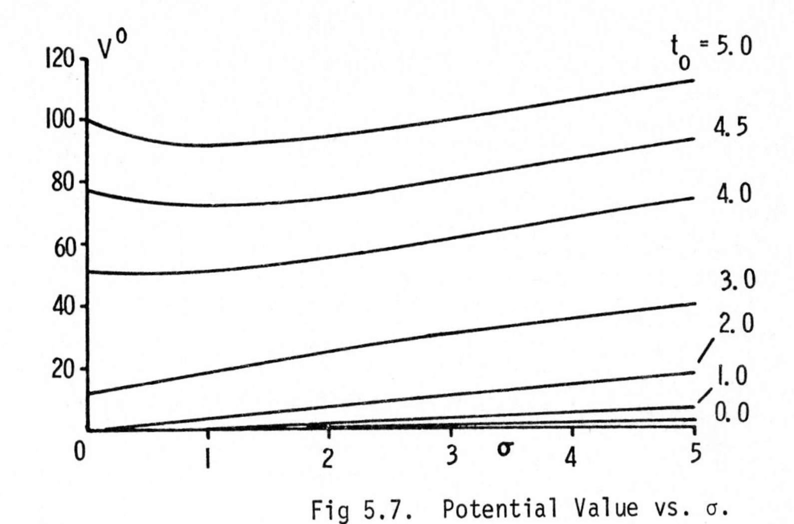

Fig 5.7. Potential Value vs. σ.

Fig. 5.8. J_{TL} vs. σ.

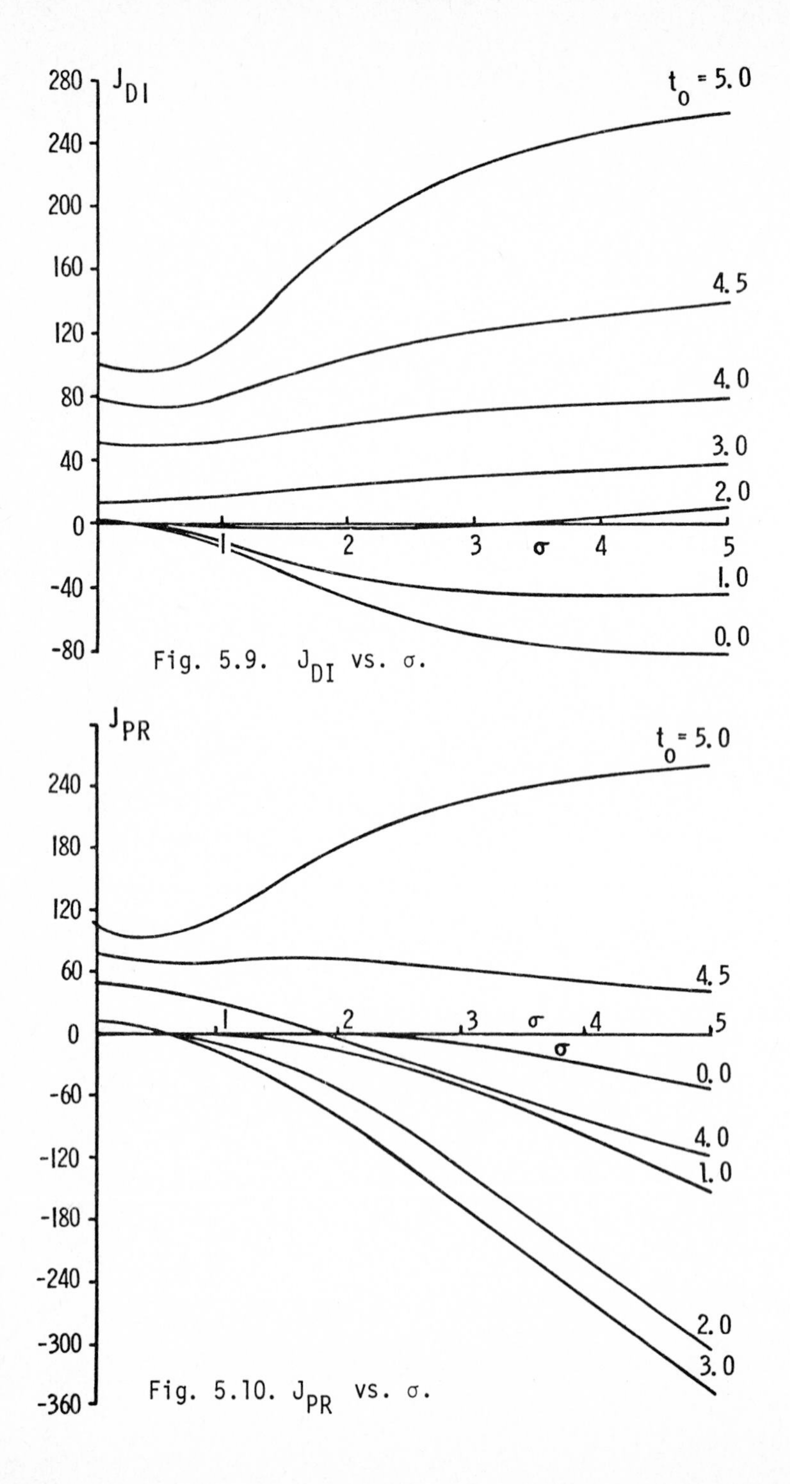

Fig. 5.9. J_{DI} vs. σ.

Fig. 5.10. J_{PR} vs. σ.

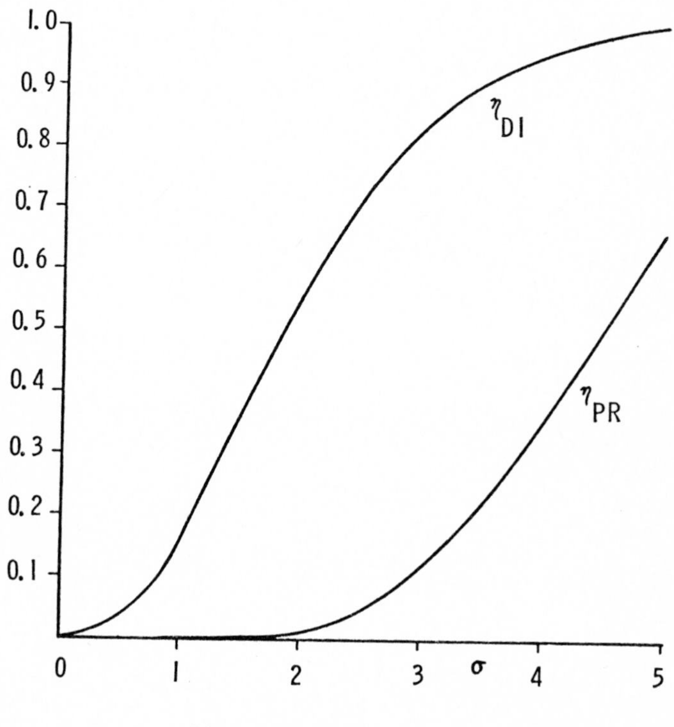

Fig. 5.11. η_{DI} and η_{PR} vs. σ.

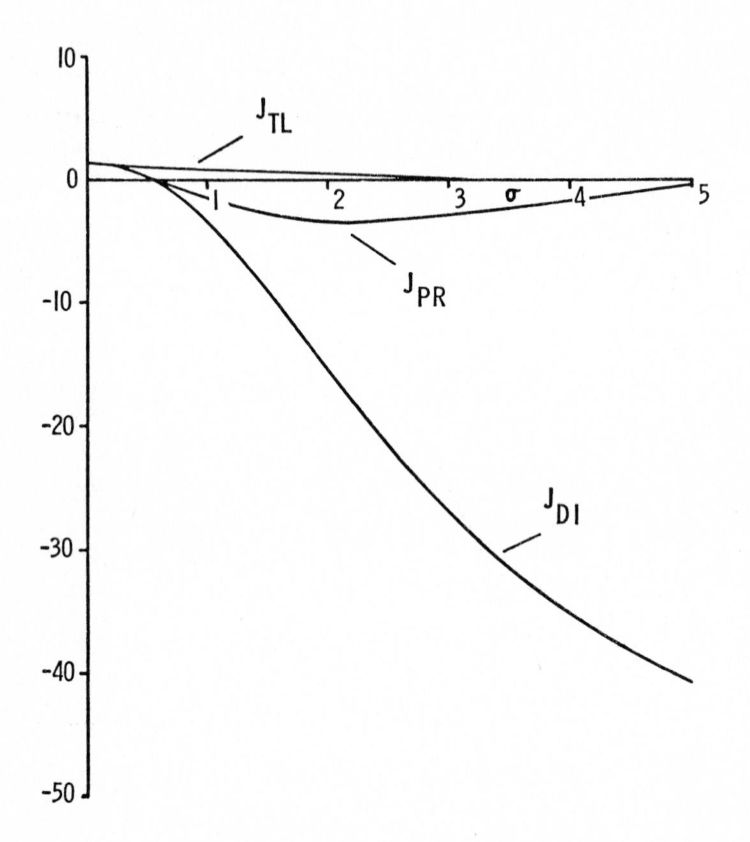

Fig. 5.12. J_{TL}, J_{DI}, and J_{PR} vs. σ.

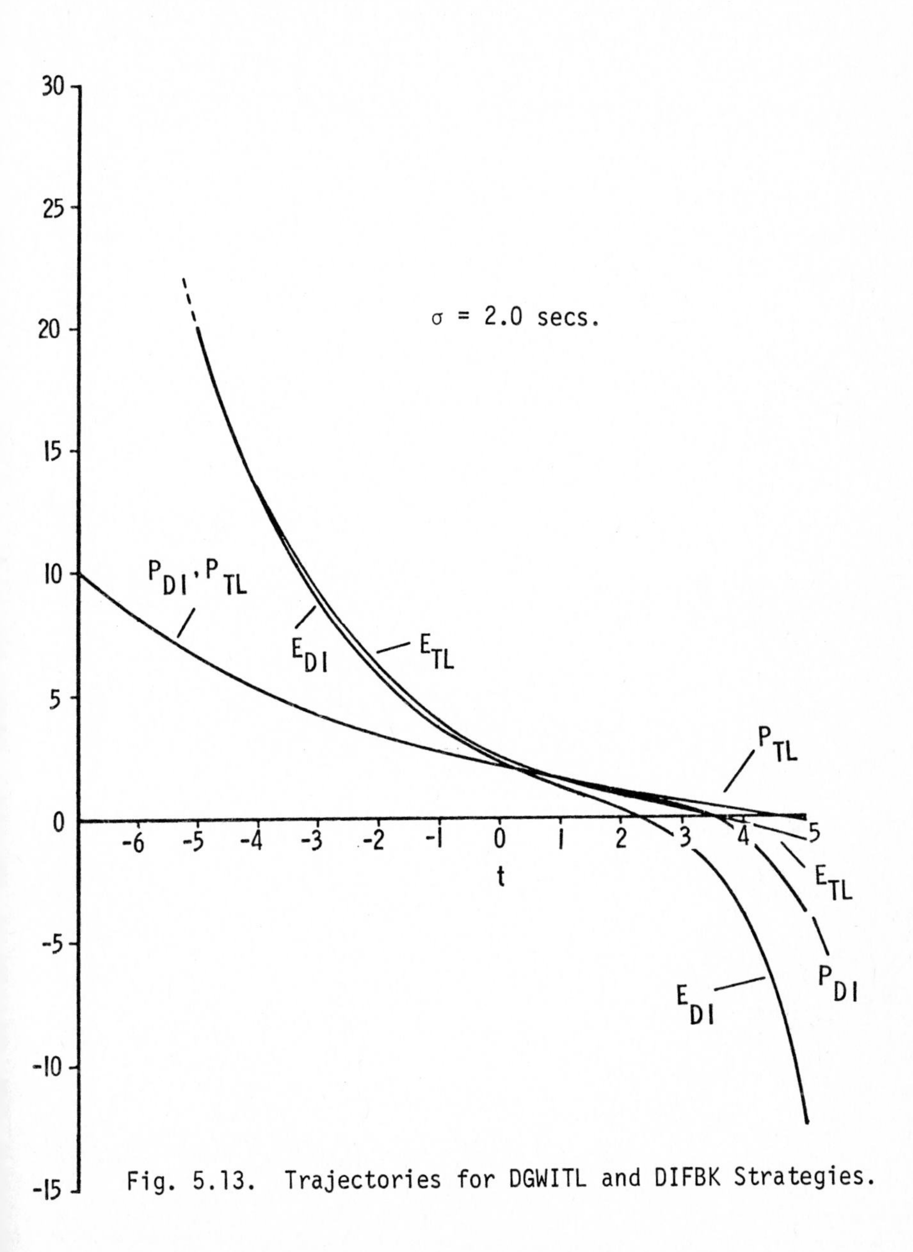

Fig. 5.13. Trajectories for DGWITL and DIFBK Strategies.

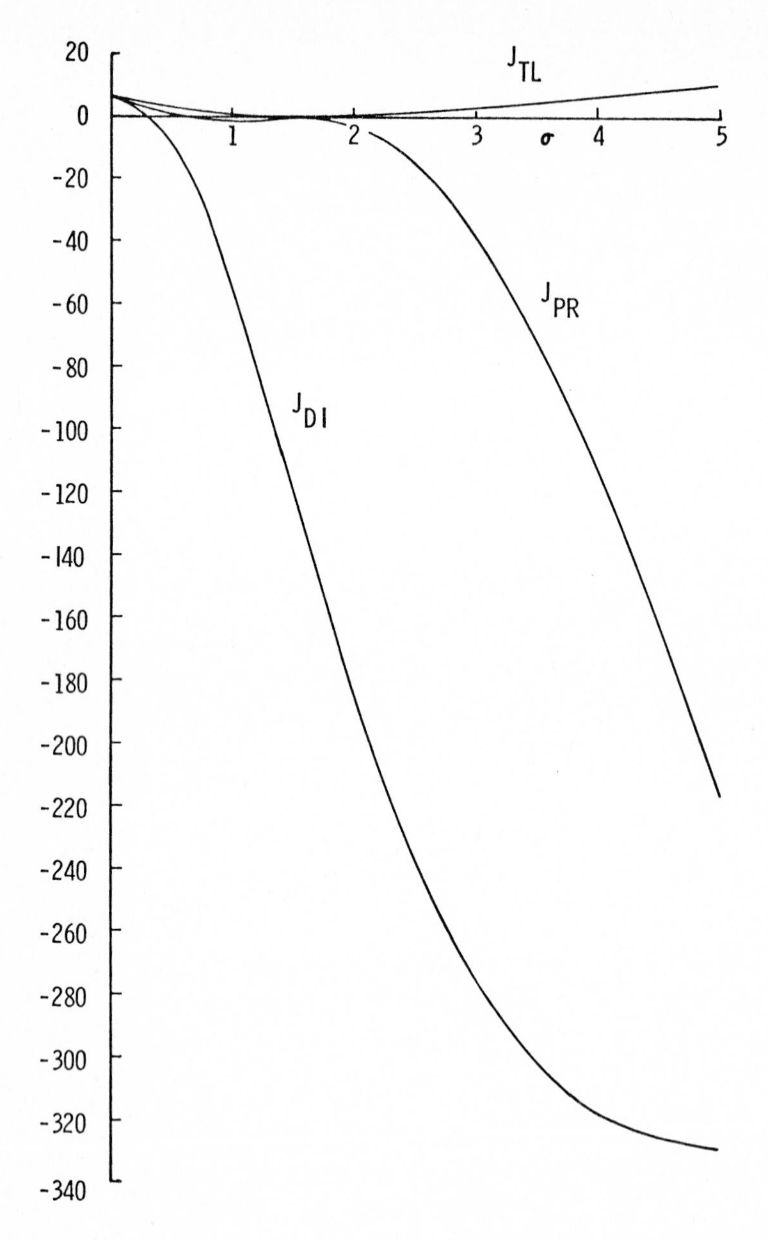

Fig. 5.14. J_{TL}, J_{DI}, and J_{PR} vs. σ.

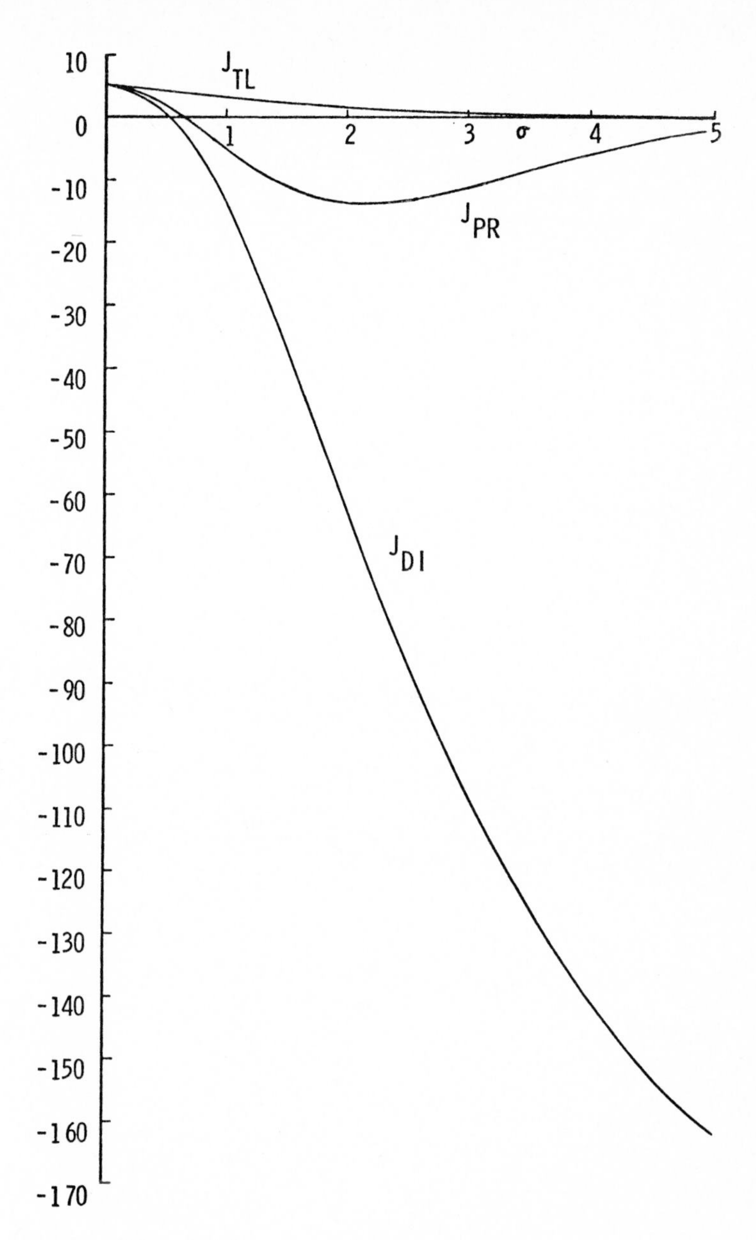

Fig. 5.15. J_{TL}, J_{DI}, and J_{PR} vs. σ.

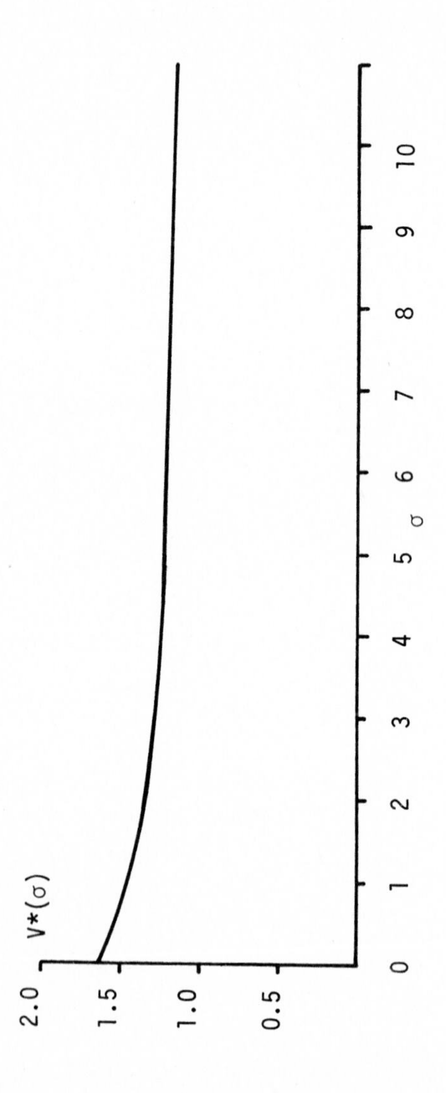

Fig. 5.16. $V^*(\sigma)$ vs. σ .

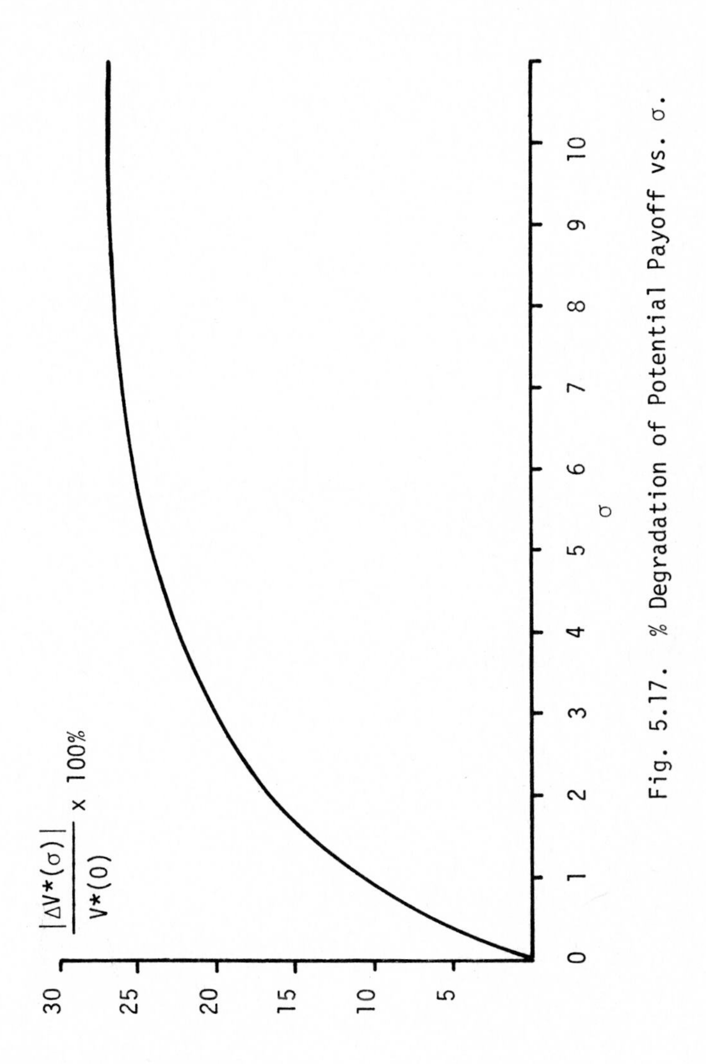

Fig. 5.17. % Degradation of Potential Payoff vs. σ.

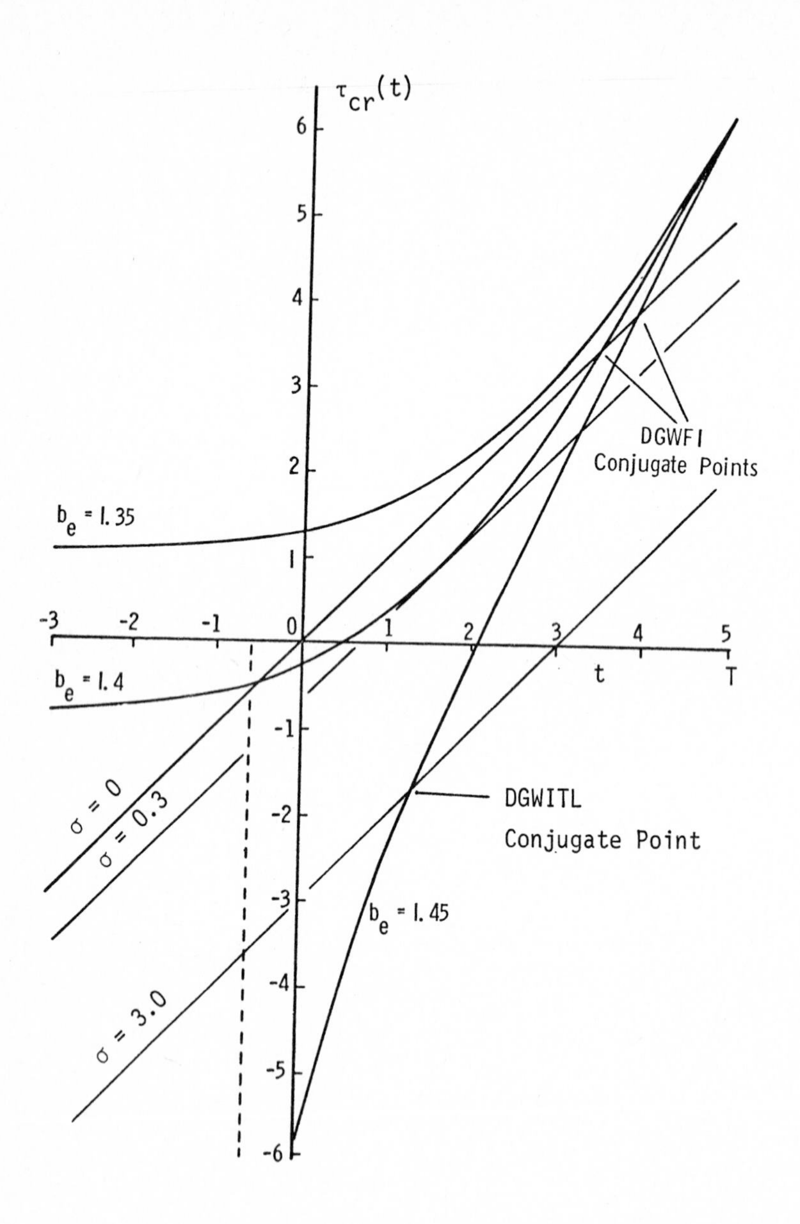

Fig. 5.18. τ_{cr} vs. t.

6. Appendix

6.1 - Proof of Theorem 3.1.

To prove Theorem 3.1, let (p,e) be a potentially terminal admissible pair such that $p(x,\tau,y,t)$ and $e(x,y,\tau,t)$ coincide with $p^o(x,\tau,y,t)$, $e^o(x,y,\tau,t)$ for $t \geqslant t^*$, where $t_o \leqslant t^* \leqslant t_p(x_{o\sigma},\tau_o,y_o,t_o,p^o,e^o)$. Let $\tilde{S}_P$ and $\tilde{S}_E$ denote the subsets of S_P and S_E which contain all p and e defined in the above manner. Since $p^o \in \tilde{S}_P$ and $e^o \in \tilde{S}_E$, extremization w.r.t. strategies in $\tilde{S}_P$ and $\tilde{S}_E$ is equivalent to extremization w.r.t. strategies in S_P and S_E. We then write

$$J_p(x_{o\sigma},\tau_o,y_o,t_o,p,e) =$$

$$\int_{\tau_o}^{\tau(t^*)} L_1(\varphi_{pe}(\alpha+\sigma),p(\varphi_{pe}(\alpha+\sigma),\alpha,\psi_{pe}(\alpha+\sigma),\alpha+\sigma,\alpha)d\alpha$$

$$+\int_{t_o}^{t^*} L_2(\psi_{pe}(\alpha),e(\varphi_{pe}(\alpha),\alpha-\sigma,\psi_{pe}(\alpha),\alpha)d\alpha$$

$$+\int_{\tau(t^*)}^{t_p-\sigma} L_1(\varphi_{pe}(\alpha+\sigma),p(\varphi_{pe}(\alpha+\sigma),\alpha,\psi_{pe}(\alpha+\sigma),\alpha+\sigma),\alpha)d\alpha$$

$$+\int_{t^*}^{t_p} L_2(\psi_{pe}(\alpha),e(\varphi_{pe}(\alpha),\alpha-\sigma,\psi_{pe}(\alpha),\alpha),\alpha)d\alpha$$

$$+ W(\varphi_{pe}(t_p;x_{o\sigma},\tau_o,y_o,t_o),\tau(t_p),\psi_{pe}(t_p;x_{o\sigma},\tau_o,y_o,t_o),t_p)$$

and, because of the nature of p and e,

$$J_p(x_{o\sigma},\tau_o,y_o,t_o,p,e) =$$

$$\int_{\tau_o}^{\tau(t^*)} L_1(\varphi_{pe}(\alpha+\sigma),p(\varphi_{pe}(\alpha+\sigma),\alpha,\psi_{pe}(\alpha+\sigma),\alpha+\sigma),\alpha)d\alpha$$

$$+\int_{t_o}^{t^*} L_2(\psi_{pe}(\alpha),e(\varphi_{pe}(\alpha),\alpha-\sigma,\psi_{pe}(\alpha),\alpha),\alpha)d\alpha$$

$$+ V^o(\varphi_{pe}(t^*;x_{o\sigma},\tau_o,y_o,t_o),\tau(t^*),\psi_{pe}(t^*;x_{o\sigma},\tau_o,y_o,t_o),t^*)$$

with $\varphi_{pe}(\alpha) = \varphi_{pe}(\alpha;x_{o\sigma},\tau_o,y_o,t_o)$

and $\psi_{pe}(\alpha) = \psi_{pe}(\alpha;x_{o\sigma},\tau_o,y_o,t_o)$

It follows from the assumptions on the structure of f, g, p and e, that φ_{pe} and ψ_{pe} are unique and continuous in t. From this observation and the assumed structure of L_1, L_2, p, e, it follows that

$$\tilde{L}_1(\alpha) = L_1(\varphi_{pe}(\alpha+\sigma),p(\varphi_{pe}(\alpha+\sigma),\alpha,\psi_{pe}(\alpha+\sigma),\alpha+\sigma),\alpha)$$

is a continuous function of α for $:\tau_o = t_o-\sigma \leqslant \alpha \leqslant \tau(t^*) = t^*-\sigma$

Likewise,

$$\tilde{L}_2(\alpha) = L_2(\psi_{pe}(\alpha),e(\varphi_{pe}(\alpha),\alpha-\sigma,\psi_{pe}(\alpha),\alpha),\alpha)$$

is a continuous function of α for $t_o \leqslant \alpha \leqslant t^*$. Next, apply the mean value theorem for integrals to obtain

$$\int_{\tau_o}^{\tau(t^*)} \tilde{L}_1(\alpha)d\alpha = [\,\tau(t^*)-\tau_o\,]\tilde{L}_1(\tau_o+\theta_1(\tau(t^*)-\tau_o))$$

$$= [\,t^*-t_o\,]\tilde{L}_1(t_o-\sigma+\theta_1(t^*-t_o))$$

where $0 \leqslant \theta_1 \leqslant 1$. Likewise,

$$\int_{t_o}^{t^*} \tilde{L}_2(\alpha)d\alpha = [t^*-t_o]\ \tilde{L}_2(t_o + \theta_2(t^*-t_o))$$

where $0 \leqslant \theta_2 \leqslant 1$.

Since

$$\tilde{f}(\alpha) = f(\varphi_{pe}(\alpha;x_{o\sigma},\tau_o,y_o,t_o),p(\varphi_{pe}(\alpha;x_{o\sigma},\tau_o,y_o,t_o),\alpha-\sigma,$$
$$\psi_{pe}(\alpha;x_{o\sigma},\tau_o,y_o,t_o),\alpha),\alpha-\sigma)$$

is a continuous function of α for $t_o \leqslant \alpha \leqslant t^*$ we write

$$\varphi_{pe}(t^*,x_{o\sigma},\tau_o,y_o,t_o) = x_{o\sigma}+[t^*-t_o]\tilde{f}(t_o+\theta_3(t^*-t_o)),$$

where $0 \leqslant \theta_3 \leqslant 1$. Likewise,

$$\tilde{g}(\alpha) = g(\psi_{pe}(\alpha;x_{o\sigma},\tau_o,y_o,t_o),e(\varphi_{pe}(\alpha;x_{o\sigma},\tau_o,y_o,t_o),\alpha-\sigma,$$
$$\psi_{pe}(\alpha;x_{o\sigma},\tau_o,y_o,t_o),\alpha),\alpha)$$

is continuous in α, so

$$\psi_{pe}(t^*;x_{o\sigma},\tau_o,y_o,t_o) = y_o +[t^*-t_o]\tilde{g}(t_o+\theta_4(t^*-t_o))$$

where $0 \leqslant \theta_4 \leqslant 1$. Then

$$V^o(\varphi_{pe}(t^*;x_{o\sigma},\tau_o,y_o,t_o),\tau(t^*),\psi_{pe}(t^*;x_{o\sigma},\tau_o,y_o,t_o),t^*)$$
$$= V^o(x_{o\sigma}+[t^*-t_o]\tilde{f}(t_o + \theta_3(t^*-t_o)),\tau(t^*),$$
$$y_o + [t^*-t_o]\tilde{g}(t_o + \theta_4(t^*-t_o)),t^*)$$

With the assumption that V^o is continuously differentiable in its argument, we apply the mean value theorem of differential calculus in the following manner :

$$V^{o}(\hat{z}) - V^{o}(z_{o}) = < \nabla V^{o}(z_{o} + \theta_{5}(\hat{z}-z_{o})), \hat{z}-z_{o} >$$

where $0 \leqslant \theta_{5} \leqslant 1$, $\hat{z} = (x,\tau,y,t)$ and ∇ denotes the gradient operator. Hence

$$V^{o}(x_{o\sigma}+[t^{*}-t_{o}]\tilde{f}(t_{o}+\theta_{3}(t^{*}-t_{o})),$$

$$\tau(t^{*}), y_{o}+[t^{*}-t_{o}]\tilde{g}(t_{o}+\theta_{4}(t^{*}-t_{o})), t^{*})$$

$$= V^{o}(x_{o\sigma},\tau_{o},y_{o},t_{o})+(t^{*}-t_{o})V^{o}_{\tau}(x_{o\sigma}+\theta_{5}(t^{*}-t_{o})\tilde{f}(t_{o}+\theta_{3}(t^{*}-t_{o})),$$

$$\tau_{o}+\theta_{5}(t^{*}-t_{o}), y_{o}+\theta_{5}(t^{*}-t_{o})\tilde{g}(t_{o}+\theta_{4}(t^{*}-t_{o})), t_{o}+\theta_{5}(t^{*}-t_{o}))$$

$$+ (t^{*}-t_{o})V^{o}_{t}(x_{o\sigma}+\theta_{5}(t^{*}-t_{o})\tilde{f}(t_{o}+\theta_{3}(t^{*}-t_{o})),$$

$$\tau_{o}+\theta_{5}(t^{*}-t_{o}), y_{o}+\theta_{5}(t^{*}-t_{o})\tilde{g}(t_{o}+\theta_{4}(t^{*}-t_{o})), t_{o}+\theta_{5}(t^{*}-t_{o}))$$

$$+ < V^{o}_{x}(x_{o\sigma}+\theta_{5}(t^{*}-t_{o})\tilde{f}(t_{o}+\theta_{3}(t^{*}-t_{o})), \tau_{o}+\theta_{5}(t^{*}-t_{o}), y_{o}+\theta_{5}(t^{*}-t_{o})$$

$$\tilde{g}(t_{o}+\theta_{4}(t^{*}-t_{o})), t_{o}+\theta_{5}(t^{*}-t_{o})), (t^{*}-t_{o})\tilde{f}(t_{o}+\theta_{3}(t^{*}-t_{o})) >$$

$$+ < V^{o}_{y}(x_{o\sigma}+\theta_{5}(t^{*}-t_{o})\tilde{f}(t_{o}+\theta_{3}(t^{*}-t_{o})), \tau_{o}+\theta_{5}(t^{*}-t_{o}), y_{o}+\theta_{5}(t^{*}-t_{o})$$

$$\tilde{g}(t_{o}+\theta_{4}(t^{*}-t_{o})), t_{o}+\theta_{5}(t^{*}-t_{o})), (t^{*}-t_{o})\tilde{g}(t_{o}+\theta_{4}(t^{*}-t_{o})) >$$

And it follows that :

$$J_p(x_{o\sigma},\tau_o,y_o,t_o,p,e) =$$

$$[t^*-t_o][V^o_\tau(x_{o\sigma}+\theta_5(t^*-t_o)\tilde{f}(t_o+\theta_3(t^*-t_o)),\tau_o+\theta_5(t^*-t_o),$$

$$y_o+\theta_5(t^*-t_o)\tilde{g}(t_o+\theta_4(t^*-t_o)),t_o+\theta_5(t^*-t_o))$$

$$+V^o_t(x_{o\sigma}+\theta_5(t^*-t_o)\tilde{f}(t_o+\theta_3(t^*-t_o)),\tau_o+\theta_5(t^*-t_o),y_o+\theta_5(t^*-t_o)$$

$$\tilde{g}(t_o+\theta_4(t^*-t_o)),t_o+\theta_5(t^*-t_o))$$

$$+<V^o_x(x_{o\sigma}+\theta_5(t^*-t_o)\tilde{f}(t_o+\theta_3(t^*-t_o)),\tau_o+\theta_5(t^*-t_o),y_o+\theta_5(t^*-t_o)$$

$$\tilde{g}(t_o+\theta_4(t^*-t_o)),t_o+\theta_5(t^*-t_o)),f(t_o+\theta_3(t^*-t_o))>$$

$$+<V^o_y(x_{o\sigma}+\theta_5(t^*-t_o)\tilde{f}(t_o+\theta_3(t^*-t_o)),\tau_o+\theta_5(t^*-t_o),y_o+\theta_5(t^*-t_o)$$

$$\tilde{g}(t_o+\theta_4(t^*-t_o)),t_o+\theta_5(t^*-t_o)),g(t_o+\theta_4(t^*-t_o))$$

$$+\tilde{L}_1(t_o-\sigma+\theta_1(t^*-t_o))+\tilde{L}_2(t_o+\theta_2(t^*-t_o)]+V^o(x_{o\sigma},\tau_o,y_o,t_o),$$

where V^o_x, V^o_y, V^o_τ, V^o_t are gradients of V^o w.r.t. x,y,τ,t respectively.

Since

$$\sup_{e\in S_E}\inf_{p\in S_P} J_p(x_{o\sigma},\tau_o,y_o,t_o,p,e) = \sup_{e\in\tilde{S}_E}\inf_{p\in\tilde{S}_P} J_p(x_{o\sigma},\tau_o,y_o,t_o,p,e)$$

and

$$\sup_{e\in S_E}\inf_{p\in S_P} J_p(x_{o\sigma},\tau_o,y_o,t_o,p,e) = V^o(x_{o\sigma},\tau_o,y_o,t_o)$$

it follows that

$$0 \equiv \sup_{e\in\tilde{S}_E} \inf_{p\in\tilde{S}_P} [t^*-t_o][< V_x^o(x_{o\sigma}+\theta_5(t^*-t_o)$$

$$\tilde{f}(t_o+\theta_3(t^*-t_o)),\tau_o+\theta_5(t^*-t_o),y_o+\theta_5(t^*-t_o)$$

$$\tilde{g}(t_o+\theta_4(t^*-t_o)),t_o+\theta_5(t^*-t_o)),\tilde{f}(t_o+\theta_3(t^*-t_o)) >$$

$$+ < V_y^o(x_{o\sigma}+\theta_5(t^*-t_o)\tilde{f}(t_o+\theta_3(t^*-t_o)),\tau_o+\theta_5(t^*-t_o),y_o+\theta_5(t^*-t_o)$$

$$\tilde{g}(t_o+\theta_4(t^*-t_o)),t_o+\theta_5(t^*-t_o)),\tilde{g}(t_o+\theta_4(t^*-t_o)) >$$

$$+ V_\tau^o(x_{o\sigma}+\theta_5(t^*-t_o)\tilde{f}(t_o+\theta_3(t^*-t_o)),\tau_o+\theta_5(t^*-t_o),y_o+\theta_5(t^*-t_o)$$

$$\tilde{g}(t_o+\theta_4(t^*-t_o)),t_o+\theta_5(t^*-t_o))$$

$$+ V_t^o(x_{o\sigma}+\theta_5(t^*-t_o)\tilde{f}(t_o+\theta_3(t^*-t_o)),\tau_o+\theta_5(t^*-t_o),y_o+\theta_5(t^*-t_o)$$

$$\tilde{g}(t_o+\theta_4(t^*-t_o)),t_o+\theta_5(t^*-t_o))$$

$$+ \tilde{L}_1(t_o-\sigma+\theta_1(t^*-t_o)) + \tilde{L}_2(t_o+\theta_2(t^*-t_o))]$$

Cancel $[t^*-t_o]$, let $t^* \to t_o+$ and note that

$$\tilde{f}(t_o+\theta_3(t^*-t_o)) \to f(\varphi_{pe}(t_o;x_{o\sigma},\tau_o,y_o,t_o),p(\varphi_{pe}(t_o;x_{o\sigma},\tau_o,y_o,t_o),$$
$$t_o-\sigma,\psi_{pe}(t_o;x_{o\sigma},\tau_o,y_o,t_o),t_o),t_o-\sigma)$$
$$= f(x_{o\sigma},p(x_{o\sigma},\tau_o,y_o,t_o),\tau_o),$$

$$\tilde{g}(t_o+\theta_4(t^*-t_o)) \to g(\psi_{pe}(t_o;x_{o\sigma},\tau_o,y_o,t_o),e(\varphi_{pe}(t_o;x_{o\sigma},\tau_o,y_o,t_o),$$
$$t_o-\sigma,\psi_{pe}(t_o;x_{o\sigma},\tau_o,y_o,t_o),t_o)$$
$$= g(y_o,e(x_{o\sigma},\tau_o,y_o,t_o),t_o)$$

and observe that in the limit, extremization w.r.t. $e \in \tilde{S}_E$ and $p \in \tilde{S}_P$ is equivalent to extremization w.r.t. $p(x_{o\sigma},\tau_o,y_o,t_o) \in K_u$ and $e(x_{o\sigma},\tau_o,y_o,t_o) \in K_v$. Hence

$$
\begin{aligned}
0 = \sup_{e(x_{o\sigma},\tau_o,y_o,t_o) \in K_v} \ \inf_{p(x_{o\sigma},\tau_o,y_o,t_o) \in K_u} \ & [< V_x^o(x_{o\sigma},\tau_o,y_o,t_o), f(x_{o\sigma},p(x_{o\sigma},\tau_o,y_o,t_o),\tau_o) > \\
& + < V_y^o(x_{o\sigma},\tau_o,y_o,t_o), g(y_o,e(x_{o\sigma},\tau_o,y_o,t_o), t_o) > \\
& + L_1(x_{o\sigma},p(x_{o\sigma},\tau_o,y_o,t_o)\ \tau_o) \\
& + L_2(y_o,e(x_{o\sigma},\tau_o,y_o,t_o),t_o)] \\
& + V_\tau^o(x_{o\sigma},\tau_o,y_o,t_o) + V_t^o(x_{o\sigma},\tau_o,y_o,t_o)
\end{aligned} \tag{45}
$$

Now since any $(x,\tau,y,t) \in G-\theta_p$ may be used for $(x_{o\sigma},\tau_o,t_o,t_o)$ it is evident that (45) must be satisfied pointwise in $G-\theta_p$. That is,

$$
\begin{aligned}
0 = & V_\tau^o(x,\tau,y,t) + V_t^o(x,\tau,y,t) \\
& + \sup_{\beta \in K_v} \ \inf_{\mu \in K_u} \ H(x,\tau,y,t,V_x^o,V_y^o,\mu,\beta)
\end{aligned}
$$

Repetition of these steps using p^o and e^o would lead directly to

$$
\begin{aligned}
0 = & V_\tau^o(x,\tau,y,t) + V_t^o(x,\tau,y,t) \\
& + < V_x^o(x,\tau,y,t), f(x,p^o(x,\tau,y,t),\tau) > \\
& + < V_y^o(x,\tau,y,t), g(y,e^o(x,\tau,y,t),t) > \\
& + L_1(x,p^o(x,\tau,y,t),\tau) + L_2(y,e^o(x,\tau,y,t),t)
\end{aligned}
$$

Therefore

$$V^o_\tau(x,\tau,y,t) + V^o_t(x,\tau,y,t)$$

$$+ H(x,\tau,y,t,V^o_x,V^o_y,p^o(x,\tau,y,t),e^o(x,\tau,y,t))$$

$$\equiv V^o_\tau(x,\tau,y,t) + V^o_t(x,\tau,y,t)$$

$$+ \sup_{\beta\in K_v} \inf_{\mu\in K_u} H(x,\tau,y,t,V^o_x,V^o_y,\mu,\beta)$$

But, according to the assumptions of Theorem 3.1, there exist unique

$$\mu = k_1(x,\tau,y,t,V^o_x,V^o_y) \quad \text{and} \quad \beta = k_2(x,\tau,y,t,V^o_x,V^o_y)$$

such that

$$\sup_{\beta\in K_v} \inf_{\mu\in K_u} H(x,\tau,y,t,V^o_x,V^o_y,\mu,\beta)$$

$$= H(x,\tau,y,t,V^o_x,V^o_y, k_1(x,\tau,y,t,V^o_x,V^o_y).k_2(x,\tau,y,t,V^o_x,V^o_y)).$$

Therefore, we conclude that :

$$p^o(x,\tau,y,t) = k_1(x,\tau,y,t,V^o_x(x,\tau,y,t),V^o_y(x,\tau,y,t))$$

and

$$e^o(x,\tau,y,t) = k_2(x,\tau,y,t,V^o_x(x,\tau,y,t),V^o_y(x,\tau,y,t))$$

Moreover,

$$H(x,\tau,y,t,V^o_x,V^o_y,k_1(x,\tau,y,t,V^o_x,V^o_y),k_2(x,\tau,y,t,V^o_x,V^o_y))$$

$$\triangleq H^o(x,\tau,y,t,V^o_x,V^o_y)$$

so that

$$V^o_\tau(x,\tau,y,t) + V^o_t(x,\tau,y,t) + H^o(x,\tau,y,t,V^o_x,V^o_y) = 0 \qquad \text{Q.E.D.}$$

6.2 - *Proof of Theorem 3.2*

Apply the generalized Lemma of Caratheodory to the quantity

$$V_\tau + V_t + H^o(x,\tau,y,t,V_x,V_y)$$

to obtain

$$J_p(x,\tau,y,t,p^o,e) \leqslant V(x,\tau,y,t)$$

$$= J_p(x,\tau,y,t,p^o,e^o) \leqslant J_p(x,\tau,y,t,p,e^o)$$

for all admissible p,e . Then

$$\sup_{e\in S_E} \inf_{p\in S_P} J(x,\tau,y,t,p,e) \leqslant \sup_{e\in S_E} J_p(x,\tau,y,t,p^o,e)$$

$$= V(x,\tau,y,t)$$

$$= \inf_{p\in S_P} J_p(x,\tau,y,t,p,e^o)$$

$$\leqslant \inf_{p\in S_P} \sup_{e\in S_E} J_p(x,\tau,y,t,p,e)$$

Under the existence assumption, we have

$$V^o(x,\tau,y,t) = \sup_{e\in S_E} \inf_{p\in S_P} J_p(x,\tau,y,t,p,e)$$

$$= \inf_{p\in S_P} \sup_{e\in S_E} J_p(x,\tau,y,t,p,e)$$

Therefore, we conclude that

$$V^o(x,\tau,y,t) = V(x,\tau,y,t) \qquad \text{Q.E.D.}$$

References

(1) R. ISAACS, Differential Games, J. Wiley and Sons, New York, 1965

(2) L.E. DUBINS, "A Discrete Evasion Game", Institute of Air Weapons Research, Technical Note n°2, University of Chicago

(3) R. ISAACS and S. KARLIN, "A Game of Aiming and Evasion", Rand Corporation Report RM-1316, 1954

(4) H. SCARF and L. SHAPLEY, "Games with Information Lag", Rand Corporation Report RM-1320, 1954

(5) R. ISAACS, "A Game of Aiming and Evasion : General Discussion and the Marksman's Strategies", Rand Corporation Report RM-1385, 1954

(6) H. SCARF and L. SHAPLEY, "Games with Partial Information", *Contributions to the Theory of Games*, Vol.III, Annals of Mathematics Study n°39, Princeton University Press, 1957

(7) S. KARLIN, "An Infinite Move Game with a Lag", *Contributions to the Theory of Games*, Vol.III, Annals of Mathematics Study n°39, Princeton University Press, 1957

(8) L.E. DUBINS, "A Discrete Evasion Game", *Contributions to the Theory of Games*, Vol.III, Annals of Mathematics Study n°39, Princeton University Press, 1957

(9) R. ISAACS, "The Problem of Aiming and Evasion", Naval Res. Logistics Quarterly, Vol.2, N°s 1 and 2, June 1955

(10) M.D. CILETTI, "On a Class of Deterministic Differential Games with Imperfect Information", Dept. of Elect. Engineering T.R. EE679, Univerity of Notre Dame, Notre Dame, Ind., Dec. 1967

(11) M.D. CILETTI, "Functional Analysis and Imperfect Information Differential Game", Proc. of Eleventh Midwest Symposium on Circuit Theory, Univ. of Notre Dame, Notre Dame, Ind., 1968

(12) M.D. CILETTI, "A Differential Game with Information Time Lag", Proc. of the First International Conference on the Theory and Applications of Differential Games, Univ. of Massachusetts, Amherst, Mass., 1969

(13) M.D. CILETTI, "Results in the Theory of Linear Differential Games with Information Time Lag", Journal of Optimization Theory and Applications, Vol.5, n°5, 1970, pp.347-362

(14) M.D. CILETTI, "Open-Loop Nash Equilibrium Strategies for an N-Person, Nonzero-Sum Differential Game with Information Time Lag", in *Differential Games and Related Topics*, ed. by H.W. KUHN and G.P. SZEGO, North Holland Publishing Company, Amsterdam, 1971

(15) M.D. CILETTI, "New Results in the Theory of Differential Games with Information Time Lag", Journal of Optimization Theory and Applications, Vol.8, n°4, 1971, pp.287-315

(16) B.N. SOKOLOV and F.L. CHERNOUS'KO, "Differential Games with Information Lag", Journal of Applied Mathematics and Mechanics, Vol.34, n°5, 1970,pp.779-785

(17) F.L. CHERNOUS'KO, "Differential Games with Information Delay", Soviet Physics-Doklady, Vol.14, n°10, 1970 pp. 952-954

(18) L.A. PETROSJAN, "Differential Games with Incomplete Information", Soviet Mathematics-Doklady, Vol.11, n°6, 1970, pp.1524-1527

(19) M. ATHANS and P.L. FALB, *Optimal Control - An Introduction to Theory and Applications*, McGraw Hill, New York, 1966

(20) COURANT-HILBERT, *Methods of Mathematical Physics*, Vol.II, Interscience, 1962

(21) A. BLAQUIERE, F. GERARD and G. LEITMANN, *Quantitative and Qualitative Games*, Academic Press, New York, 1969

(22) M. ATHANS, P.L. FALB and R.T. LACOSS, "Time, Fuel-, and Energy- Optimal Control of Nonlinear Norm-Invariant Systems", IEEE Transactions on Automatic Control, Vol. AC-8, 1963, pp.196-202

(23) M.D. CILETTI, "Norm-Invariant Systems - Differential Games", IEEE Transactions on Automatic Control, Vol. AC-13, n°6, 1968, p.735

(24) P. VARAIYA and J. LIN, "Existence of Saddle Points in Differential Games", SIAM J. Control, Vol.7, n°1, 1969, pp.141-157

(25) A. FRIEDMAN, "On the Definition of Differential Games and the Existence of Value and of Saddle Points", Journal of Differential Equations, Vol.7, n°1, 1970, pp.69-91

Part. II

Non Zero-Sum Differential Games

GEOMETRY OF PARETO EQUILIBRIA IN N-PERSON DIFFERENTIAL GAMES.

A. BLAQUIERE, L. JURICEK and K.E. WIESE*

While in optimal control problems and in two-person zero-sum games the performance of a system is measured by a scalar-valued criterion function, in nonzero-sum games several such criterion functions are simultaneously considered, one for each player. In other words, in nonzero-sum games the performance criterion is a vector-valued function. Many new concepts arise. At the outset, we must consider explicitely whether or not the players are permitted to communicate and to cooperate.

Two kinds of nonzero-sum games have been widely discussed in the literature, namely cooperative and non cooperative games. Associated with these two well shaped specimens are the concepts of Pareto and Nash equilibria, respectively. Between them one finds the whole gamut of partly cooperative games, with different kinds and different degrees of cooperation. Some of them can be entered upon through the concept of coalition, which will be discussed in the next chapter.

The difficulty in making explicit assumptions about communication and collusion among the players appears to stem from the variety of rules which may be found in empirical situations ; for instance in Economics one finds the whole gamut from no rules at

* Laboratoire d'Automatique Théorique, Université de Paris 7, France.

all, through moral sanctions, to elaborate legal codes as in the antitrust laws. Clearly, coalition formation is a sociological phenomenon ; the literature on this subject, and common observation, suggest that one important aspect of the phenomenon are the restrictions society places upon coalition formation and coalition changes. The exact nature of these restrictions and sanctions seems to depend upon the specific situation, its history, the general mores of the society, the legal structure,...

Coalition constraints can be introduced in the rule of the game as follows : let the players be $J_1, \ldots J_N$, and denote by P the collection of non-empty subsets of $J_1, \ldots J_N$. Let P be a non-empty subset of P, and $L : G \to P$, where G is an open non-empty set in Euclidean state space E^n. For each $x \in G$, $L(x)$ is called the coalition constraint at point x, and a group C of players is called a coalition at point x if $C \in L(x)$. Indeed, if $L(x) = (\{ J_1 \}, \ldots \{ J_N \})$, $x \in G$, there is no possibility of cooperation between the players. On the other hand, if $L(x) = \{ J_1, \ldots J_N \}$, $x \in G$, the players $J_1, \ldots J_N$ can but cooperate. The latter case will be discussed in this chapter. It will provide us with a geometric framework convenient for dealing with the more complex problem of coalitions.

1. An Introductory Example

The definition of a Pareto equilibrium is best illustrated via the consideration of an introductory static optimization problem.

The problem of curve fitting permits of two somewhat different interpretations. In the first place we may ask for the equation of a curve of prescribed type which passes rigorously through each point of a given set. On the other hand, we may weaken these requirements and ask for some simpler curve whose equation contains too few parameters to permit it to be passed exactly through each given point, but which comes "as close as possible" to each point. For instance, given a set of points, a straight line coming as close as possible to each point may very well be more useful than some complicated curve passing exactly through each point. This will certainly be the case with experimental data which theoretically should fall along a straight line, but which fail to do so because of errors of observation.

As we shall see this problem can be formulated as an optimization problem with a vector-valued criterion function. For instance, let us suppose that we wish to fit a straight line $x_2 = a+bx_1$ to the N points $(x_1^1, x_2^1),\dots(x_1^N, x_2^N)$. Let $W_k(a,b) \triangleq (x_2^k - a - bx_1^k)^2$, $k = 1,\dots N$. (a^*, b^*) is said to be a *noninferior solution*, or a *Pareto equilibrium*, if the inequalities system

$$W_k(a,b) \leqslant W_k(a^*,b^*) \qquad \text{for} \quad k = 1,\dots N$$

$$W_\ell(a,b) < W_\ell(a^*,b^*) \qquad \text{for at least one } \ell \in \{1,\dots N\}$$

admits no solution.

In other words, we wish to minimize the N components of the vector-valued criterion

$$W(a,b) \triangleq (W_1(a,b),\dots W_N(a,b))$$

and (a^*,b^*) is a Pareto equilibrium† if any deviation from (a^*,b^*) cannot result in simultaneous improvement of $W_1\dots W_N$;

---

† In some problems, for instance when the performance index is a pay off, the goal is to maximize the components of the vector-valued criterion, then the inequalities in the definition of a Pareto equilibrium should be reversed.

that is, if

$$(W_k(a,b) \leqslant W_k(a^*,b^*) \quad \text{for} \quad k = 1,\dots N) \quad \Rightarrow$$

$$(W_k(a,b) = W_k(a^*,b^*) \quad \text{for} \quad k = 1,\dots N)$$

The meaning of "as close as possible" is almost universally taken to be the least-square criterion, and the process of applying this criterion is known as the *method of least squares* ; that is, the usual choice of (a,b) is $(\bar{a},\bar{b})$ such that

$$\sum_{k=1}^{N} W_k(\bar{a},\bar{b}) \leqslant \sum_{k=1}^{N} W_k(a,b) \quad \text{for all } (a,b) \in R^2$$

One can easily prove that $(\bar{a},\bar{b})$ is a Pareto equilibrium . Indeed, suppose that there exists (a,b) such that

$$W_k(a,b) \leqslant W_k(\bar{a},\bar{b}) \quad \text{for} \quad k = 1,\dots N$$

$$W_\ell(a,b) < W_\ell(\bar{a},\bar{b}) \quad \text{for at least one } \ell \in \{ 1,\dots N \}$$

By summation over $k = 1,\dots N$ in both sides, one obtains

$$\sum_{k=1}^{N} W_k(a,b) < \sum_{k=1}^{N} W_k(\bar{a},\bar{b})$$

which is in contradiction with the definition of $(\bar{a},\bar{b})$, and proves the property.

Let us note that, in general, a Pareto equilibrium is not unique. We shall leave it to the reader to verify that, for N=3 in the above example, if $x_1^1 \neq x_1^2 \neq x_1^3$ and if there is no straight line through the three given points, then the set of Pareto equilibria in the a-b plane is the triangular area whose sides belong to the straight lines $W_1(a,b) = 0$, $W_2(a,b) = 0$ and $W_3(a,b) = 0$, respectively . The set of all Pareto equilibria is called the *Pareto-optimal set* for the problem.

The non-uniqueness of a Pareto equilibrium is an important feature of the problem. One may question whether one can define an "optimal" Pareto equilibrium in the Pareto-optimal set, where "optimal" is meant in the sense that certain reasonable desiderata are satisfied. An extensive literature exists on this point; see for instance Refs (14) and (15). Here we shall but illustrate this problem by discussing the above example more completely. First let us note that if, instead of applying the least-square criterion, we minimize $\sum_{k=1}^{N} c_k W_k(a,b)$, for any given strictly positive constants $c_1,\dots c_N$, we also find a Pareto equilibrium and, indeed, this Pareto equilibrium depends on the choice of $c_1,\dots c_N$. Now, if points (x_1^1, x_2^1), (x_1^2, x_2^2) and (x_1^3, x_2^3) are experimental data, and if the accuracy is the same for all the measurements, with uncorrelated errors, the least-square criterion is warranted by the symmetry of the problem. In other words, $c_1 = c_2 = c_3 = 1$ is a "realistic" choice. On the contrary, if the accuracy is not the same for all the measurements, it makes sense to use different weights c_1, c_2, c_3 for the different points. Indeed, the better the accuracy of a measurement, the greater should be the corresponding constant. A more detailed discussion of this relation would require an analysis of the measurement errors.

The above solution concepts are by no means limited to the static case. They carry over directly to the dynamic case treated in the next paragraphs.

2. State Equations and Strategies

We shall be interested in the dynamics of a system controlled by N players $J_1,\dots J_N$. We shall assume that the state of the system or, as we shall say, the *state of the game*, at any instant of time is a point z in (N+n)-dimensional Euclidean space E^{N+n}. Let $z \triangleq (x_o,x)$, $x_o \triangleq (x_{o1},\dots x_{oN}) \in E^N$, $x \triangleq (x_1,\dots x_n) \in G \subseteq E^n$, where G is an open set in E^n. We shall be concerned with a system whose dynamical behavior is governed by a vector differential equation, the *state equation*

$$(1) \qquad \frac{dz}{dt} = F(z,u_1,\ldots u_N)$$

$$F(z,u_1,\ldots u_N) \triangleq (f_o(x,u_1,\ldots u_N),\ f(x,u_1,\ldots u_N)),$$

$$f_o \triangleq (f_{o1},\ldots f_{oN}),\quad f \triangleq (f_1,\ldots f_n),$$ where $u_1,\ldots u_N$ are *control vectors* in Euclidean spaces $E^{r_1},\ldots E^{r_N}$, respectively ; namely, $u_k = (u_{k1},\ldots u_{kr_k})$, $k = 1,\ldots N$. We shall suppose that $u_1 \in U_1 \subseteq E^{r_1},\ldots u_N \in U_N \subseteq E^{r_N}$, where $U_1,\ldots U_N$ are open sets in $E^{r_1},\ldots E^{r_N}$, respectively. We shall take $x_n \equiv t$ so that $f_n\ (x,u_1,\ldots u_N) \equiv 1$, and we shall suppose that f_o and f are of class C^1 on $G \times U_1 \times \ldots \times U_N$.

Players $J_1,\ldots J_N$ make their decisions through choosing control vectors $u_1,\ldots u_N$, respectively, at each instant. These choices are governed by functions of x defined on a domain $X \subset G$, $p_1 : X \to E^{r_1},\ldots p_N : X \to E^{r_N}$, which $J_1,\ldots J_N$ select from prescribed sets of functions $S_1,\ldots S_N$, respectively. Like u_k, p_k has r_k components ; that is, $p_k = (p_{k1},\ldots p_{kr_k})$, $k = 1,\ldots N$.

We shall require that S_k, $k = 1,\ldots N$, satisfy the following condition

(i) For any $p_k', p_k'' \in S_k$, $k = 1,\ldots N$, and for any $x^j \in X$ functions $p_k''' : X \to E^{r_k}$, $k = 1,\ldots N$ defined by

$$p_k'''(x) = p_k'(x) \quad \text{for} \quad x_n \leqslant x_n^j$$

$$p_k'''(x) = p_k''(x) \quad \text{for} \quad x_n > x_n^j$$

are strategies ; that is $p_k''' \in S_k$, $k = 1,\ldots N$, and furthermore, we shall constrain the values of the strategies by

(ii) $(p_1(x),\dots p_N(x)) \in K(x) \subseteq U_1 \times \dots \times U_N$

for all $x \in X$.

$K(x)$ is a *constraint set* which may depend on x .

We shall call members $p \triangleq (p_1,\dots p_N)$ of $S \triangleq S_1 \times \dots \times S_N$ *strategy* N-*tuples*. For given strategy N-tuple p, we shall agree to replace Eq.(1) by

$$\frac{dz}{dt} = F(z,p(z)) \tag{2}$$

where $p(z) \triangleq (p_1(x),\dots p_N(x))$

Let $u \triangleq (u_1,\dots u_N)$. If $\tilde{z} : t \mapsto z = \tilde{z}(t)$ is a solution of Eq.(2) defined on $[t_i, t_j]$, then u is given by function of t, $\tilde{u}$ namely $\tilde{u} : t \mapsto u = \tilde{u}(t)$, $\tilde{u}(t) = p(\tilde{z}(t))$, $t \in [t_i, t_j]$.

3. Play, Terminating Play, Playable Strategy N-Tuple

Target Θ is a given set of points in $E^N \times G$. We shall assume that $\Theta \triangleq E^N \times \theta$, where θ is an $(n-1)$-dimensional surface belonging to the boundary ∂X of X, defined by a single equation $m(x) = 0$; that is $x \in \theta \Rightarrow m(x) = 0$ where function m is of class C^1 and grad $m(x) \neq 0$ on a neighborhood of θ in G . We shall suppose that θ is open in the induced topology† .

For any given strategy N-tuple $p \in S$, the graph $\{(t,z) : z = \tilde{z}(t),\ t \in [t_i, t_j]\}$ of a solution $\tilde{z}$ of Eq. (2) such that $\tilde{z}(t_j) \in E^N \times (X \cup \theta)$ will be called a *path generated by* p . We shall call a *play* the evolution of state z along a path. A *terminating play* is a play whose end state belongs to Θ . The corresponding path will be called a *terminating path* .

We shall say that strategy N-tuple $p \in S$ is playable at state $z^i \in E^N \times X$ if it generates a terminating play from z^i. Since

---

† induced by G on the set $\{ x : x \in G,\ m(x) = 0\}$

the right hand side of Eq.(2) does not contain $x_{o1}, \dots x_{oN}$, one can see easily that if p is playable at state (x_o^i, x^i), $x^i \in X$, then p is playable at state (C, x^i), no matter what the real N - tuple C is . Accordingly, we shall say that p is *playable at* $x^i \in X$ if p is playable at (C, x^i) for some C .

4. Payoff, Pareto Optimality

We shall suppose that, at the end of a terminating play given by $\tilde{z}(t) \triangleq (\tilde{x}_o(t), \tilde{x}(t))$, $t \in [t_i, t_f]$, with $\tilde{x}(t_i) = x^i \in X$, generated by strategy N-tuple p, playable at x^i, each player receives a payoff of the form

$$V_k(x^i, \theta; p) \triangleq \int_{t_i}^{t_f} f_{ok}(\tilde{x}(t), p_1(\tilde{x}(t)), \dots p_N(\tilde{x}(t)))dt$$

$$k = 1, \dots N$$

Furthermore, we let

$$V_k(x^i, \theta; p) = 0 \qquad \forall\, x^i \in \theta, \qquad \forall\, p \in S,$$

$$k = 1, \dots N$$

We shall say that a strategy N-tuple p^* is *Pareto optimal on* X if

(i) p^* is playable at point x, for all $x \in X$,

(ii) $V_k(x, \theta; p^*) \triangleq V_k^*(x)$, $k = 1, \dots N$, is defined for all $x \in X$,

(iii) $(V_k(x, \theta; p) \geqslant V_k^*(x), \quad \forall\, k = 1, \dots N) \Rightarrow$

$(V_k(x, \theta; p) = V_k^*(x), \quad \forall\, k = 1, \dots N)$

We shall assume that there exists a N-tuple p^*, Pareto optimal on X .

5. Pareto Surfaces

Since $V_k^*(x)$, $k = 1,\dots N$, is defined on $X \cup \theta$, we can define a *Pareto surface* $\Sigma(C)$ by

$$(3) \qquad \Sigma(C) \triangleq \{ z : x \in X \cup \theta, \quad x_{ok}+V_k^*(x) = C_k, \ k = 1,\dots N \}$$

where $C_1,\dots C_N$ are constant parameters and $C \triangleq (C_1,\dots C_N)$, and a set $A/\Sigma(C)$, namely

$$(4) \qquad A/\Sigma(C) \triangleq \{ z : x \in X \cup \theta, \ x_{ok}+V_k^*(x) > C_k, \ k = 1,\dots N \}$$

Let $y_1 \triangleq (x_{o1},x),\dots y_N \triangleq (x_{oN},x)$. We shall also define a surface $\Sigma_k(C)$, $k = 1,\dots N$, by

$$(5) \qquad \Sigma_k(C) \triangleq \{ y_k : x \in X \cup \theta, \ x_{ok}+V_k^*(x) = C_k \}$$

and a set $A/\Sigma_k(C)$, $k = 1,\dots N$, by

$$(6) \qquad A/\Sigma_k(C) \triangleq \{ y_k : x \in X \cup \theta, \ x_{ok}+V_k^*(x) > C_k \}$$

As the values of parameters $C_1,\dots C_N$ are varied, equations (3) and (5) define a N-parameter family and N one-parameter families of surfaces, respectively, namely $\{\Sigma(C)\}$ and $\{\Sigma_k(C)\}$, $k = 1,\dots N$.

6. A Fundamental Property of Pareto Surfaces

In the following, we shall denote the projection of a set S, $S \subseteq E^{N+n}$, on $\mathcal{E}_k \triangleq \{x_{ok}\} \times E^n$, $k = 1,\dots N$, by $P_k(S)$.

Let $\Pi^*(C)$ be a terminating path generated by p^*, given by $z^* : t \mapsto z^*(t)$, $t \in [t_o,t_f]$, $z^*(t_o) = z^o \in E^N \times X$, $z^*(t_f) = z^f \triangleq (C,x^f) \in \Theta$. Let $\Pi_k^*(C)$ be its projection on $\mathcal{E}_k$, represented by $y_k^* : t \mapsto y_k^*(t)$, $t \in [t_o, t_f]$, $k = 1,\dots N$, and

π^* its projection on $X \cup \theta$, represented by $x^* : t \mapsto x^*(t)$, $t \in [t_o, t_f]$

From the definitions of $\Pi^*(C)$ and $\Sigma(C)$ one deduces directly the following

Lemma 1. $\Pi^*(C) \subset \Sigma(C)$

7. Contingent of a Set

We shall say that a vector η is *tangent* to a set S, $S \subset E^m$ at a point $s \in \overline{S}$, if the following condition is fulfilled :

There exists an infinite sequence of vectors
$\{ \eta^\nu : \nu = 1, 2, \ldots \ell \ldots \text{ and } \eta^\nu \to \eta \text{ as } \nu \to \infty \}$
and an infinite sequence of strictly positive numbers
$\{ \varepsilon^\nu : \nu = 1, 2, \ldots \ell \ldots \text{ and } \varepsilon^\nu \to 0 \text{ as } \nu \to \infty \}$
such that $s + \varepsilon^\nu \eta^\nu \in S$, $\nu = 1, 2, \ldots \ell \ldots$

The *contingent* $C(S,s)$ of S at point s is defined by
$C(S,s) \triangleq \{ s + \eta : \eta \text{ is tangent to } S \text{ at point } s \}$

Properties of the contingent of S at a point $s \in \partial S$ will be called *local properties* of ∂S .

8. Some Local Properties of a Pareto Surface

Let
$Y \triangleq \{(\zeta_1, \ldots \zeta_N) : \zeta_k \triangleq (\xi_{ok}, \xi), \xi_{ok} \in R, k = 1, \ldots N, \xi \in E^n \}$
and let P be the onto and one-one mapping $E^{N+n} \to Y$, that associates with each element $\eta \triangleq (\xi_{o1}, \ldots \xi_{oN}, \xi) \in E^{N+n}$ the N-tuple $P(\eta) \triangleq ((\xi_{o1}, \xi), \ldots (\xi_{oN}, \xi))$. Note that $P(\eta) = (P_1(\eta), \ldots P_N(\eta))$
We shall denote by P^{-1} the inverse of P ; that is,
$P^{-1}((\xi_{o1}, \xi), \ldots (\xi_{oN}, \xi)) = (\xi_{o1}, \ldots \xi_{oN}, \xi)$.

Assumption 1. From now on, we shall assume that p^* agrees on $\Delta \cap X$, where Δ is a neighborhood of π^* in G, with a function-say p^Δ - of class C^1 on Δ ; and that π^* does not reach θ tangentially.

Then, one can prove easily that V_k^*, $k = 1,\dots N$, is of class C^1 on $\Delta \cap X$.

Some local properties of a Pareto surface $\Sigma(C)$ are embedded in

Lemma 2. *If* $z + \eta \in C(A/\Sigma(C),z)$, $z = z^*(t)$, $t \in [t_o, t_f)$, *then* $y_k + \zeta_k \in C(A/\Sigma_k(C), y_k)$, $y_k = y_k^*(t)$, $\zeta_k \triangleq P_k(\eta)$, $k = 1,\dots N$

If $z + \eta \in C(A/\Sigma(C), z)$, there exists an infinite sequence of vectors

$\{\eta^\nu : \nu = 1, 2,\dots \ell \dots \text{and } \eta^\nu \to \eta \text{ as } \nu \to \infty\}$

and an infinite sequence of strictly positive numbers

$\{\varepsilon^\nu : \nu = 1, 2,\dots \ell \dots \text{and } \varepsilon^\nu \to 0 \text{ as } \nu \to \infty\}$

such that

$y_k + \varepsilon^\nu \zeta_k^\nu \in A/\Sigma_k(C)$, $\forall\, k = 1,\dots N$ and $\forall\, \nu = 1,2,\dots \ell \dots$

where $y_k = P_k(z)$ and $\zeta_k^\nu = P_k(\eta^\nu)$

$$\zeta_k^\nu \to \zeta_k = P_k(\eta) \text{ as } \eta^\nu \to \eta \text{ and, accordingly,}$$

$y_k + \zeta_k \in C(A/\Sigma_k(C), y_k)$, $k = 1,\dots N$, which concludes the proof of Lemma 2.

Lemma 3. *If* $y_k + \zeta_k \in C(A/\Sigma_k(C), y_k)$, $y_k = y_k^*(t)$, $t \in [t_o, t_f)$, $k = 1,\dots N$, *and* $(\zeta_1,\dots\zeta_N) \in Y$, *then* $z + \eta \in C(A/\Sigma(C), z)$, $z = z^*(t)$, *where* $\eta = P^{-1}(\zeta_1,\dots \zeta_N)$.

Let $\Phi_k(y_k) \triangleq x_{ok} + V_k^*(x)$, $k = 1,\dots N$, $x \in X \cup \theta$.

$\text{grad}\, \Phi_k(y_k) \triangleq \left(1, \frac{\partial V_k^*(x)}{\partial x_1}, \dots \frac{\partial V_k^*(x)}{\partial x_n}\right)$ is defined on some neighborhood of $y_k^*(t)$, $t \in [t_o, t_f)$, in E_k, $k = 1,\dots N$. Hence the tangent plane $T(\Sigma_k(C), y_k)$ of $\Sigma_k(C)$ at point $y_k = y_k^*(t)$, $k = 1,..N$, is defined; that is, $y_k^*(t)$, $t \in [t_o, t_f)$, is a regular interior[†] point of $\Sigma_k(C)$ and $T(\Sigma_k(C), y_k) \perp \text{grad}\, \Phi_k(y_k)$; $k = 1,..N$.

One can easily verify that

(i) $C(A/\Sigma_k(C), y_k)$, $y_k = y_k^*(t)$, $t \in [t_o, t_f)$, is one of the closed half spaces determined by $T(\Sigma_k(C), y_k)$, $k = 1, .. N$; and

(ii) If $y_k + \zeta_k \in \overset{\circ}{C}(A/\Sigma_k(C), y_k)$, $y_k = y_k^*(t)$, $t \in [t_o, t_f)$, then there exists a positive number α_k such that $\varepsilon \in (0, \alpha_k) \Rightarrow y_k + \varepsilon\, \zeta_k \in A/\Sigma_k(C)$, $k = 1,.. N$; and

(iii) If $y_k + \zeta_k \in \overset{\circ}{C}(A/\Sigma_k(C), y_k)$, $y_k = y_k^*(t)$, $t \in [t_o, t_f)$, then there exists an open ball $B_k(y_k + \zeta_k)$ in E_k which belongs to $\overset{\circ}{C}(A/\Sigma_k(C), y_k)$ and which has the property that for every point $y_k + \zeta_k'$ in $B_k(y_k + \zeta_k)$ there exists a positive number α_k (independant of ζ_k') such that for all $\varepsilon \in (0, \alpha_k)$ point $y_k + \varepsilon\, \zeta_k'$ belongs to $A/\Sigma_k(C)$.

Now let $y_k + \zeta_k \in C(A/\Sigma_k(C), y_k)$, $y_k = y_k^*(t)$, $t \in [t_o, t_f)$, $k = 1, \dots N$, and suppose that $(\zeta_1, \dots \zeta_N) \in Y$. From property (i) above, it follows that, for every $k = 1,\dots N$, there exists an infinite sequence of vectors $\{ \zeta_k^\nu : \nu = 1, 2, \dots \ell \dots$ and $\zeta_k^\nu \to \zeta_k$ as $\nu \to \infty \}$ such that $y_k + \zeta_k^\nu \in \overset{\circ}{C}(A/\Sigma_k(C), y_k)$, and $(\zeta_1^\nu, \dots\dots \zeta_N^\nu) \in Y$[††], $\nu = 1, 2,\dots \ell \dots$; and from property (ii)

† It is an interior point of $\Sigma_k(C)$ since $x^*(t) \in X$, and X is a domain.

†† For instance, for $\zeta_k = (\xi_{ok}, \xi)$, $k=1,..N$, let $\zeta_k^\nu = (\xi_{ok}^\nu, \xi)$ with $\xi_{ok}^\nu > \xi_{ok}$, $k=1,..N$.

above, it follows that, for every $k = 1, 2,..N$, and for every $\nu = 1, 2, \ldots \ell \ldots$, there exists a strictly positive number α_k^ν such that $\varepsilon \in (0,\alpha_k^\nu) \Rightarrow y_k + \varepsilon \zeta_k^\nu \in A/\Sigma_k(C)$.

Let $\alpha^\nu \triangleq \min_{k=1,..N} \alpha_k^\nu$, $\nu = 1, 2,.. \ell \ldots$, then $\varepsilon \in (0,\alpha^\nu) \Rightarrow y_k + \varepsilon \zeta_k^\nu \in A/\Sigma_k(C)$, $k = 1,\ldots N$. Accordingly there exists an infinite sequence of strictly positive numbers $\{ \varepsilon^\nu : \varepsilon^\nu \in (0,\alpha^\nu),\ \nu = 1, 2,\ldots\ell\ldots \text{and } \varepsilon^\nu \to 0 \text{ as } \nu \to \infty \}$ such that $y_k + \varepsilon^\nu \zeta_k^\nu \in A/\Sigma_k(C)$, $\forall\ k = 1,\ldots,N$, and since $(\zeta_1^\nu,\ldots,\zeta_N^\nu) \in Y$, and $(\zeta_1,\ldots,\zeta_N) \in Y$, there exists an infinite sequence of vectors

$\{ \eta^\nu : \nu = 1, 2,.. \ell .. \text{and } \eta^\nu \to \eta \text{ as } \nu \to \infty \}$ such that $z + \varepsilon^\nu \eta^\nu \in A/\Sigma(C)$, $z = z^*(t)$, indeed $\eta^\nu = P^{-1}(\zeta_1^\nu,\ldots\zeta_N^\nu)$ and $\eta = P^{-1}(\zeta_1,\ldots\zeta_N)$.

It follows that $z + \eta \in C(A/\Sigma(C),z)$, $z = z^*(t)$, which concludes the proof of Lemma 3.

From Lemmas 2 and 3, and from the convexity of $C(A/\Sigma_k(C), y_k)$, $y_k = y_k^*(t)$, $t \in [t_o, t_f)$, $k = 1,\ldots N$ according to property (i) above, one deduces directly

Lemma 4. *The contingent of* $A/\Sigma(C)$ *at point* $z = z^*(t), t \in [t_o,t_f)$, *is convex.*

From Lemmas 2 and 3, and from property (iii) above, one can also deduce easily

Lemma 5. *At point* $z = z^*(t)$, $t \in [t_o,t_f)$, $\overset{\circ}{C}(A/\Sigma(C),z) \neq \phi$, *and if* $B(z+\eta)$ *is an open ball with center* $z + \eta$ *which belongs to* $\overset{\circ}{C}(A/\Sigma(C), z)$, *then* $\exists\ \alpha > 0$, $\forall\ \varepsilon \in (0,\alpha)$, $\forall\ z + \eta' \in B(z + \eta)$, $z + \varepsilon \eta' \in A/\Sigma(C)$.

9. Separability of Contingent

We shall say that an (m-1)-dimensional hyperplane $I(s,m-1)$ containing a point $s \in \overline{S}$, $S \subset E^m$, is a *separating plane* of S, if every point of S liesin one of the closed half spaces, $D(S,s,m-1)$, determined by $I(s,m-1)$.

Lemma 6. *If* $I(z, N+n-1)$ *is a separating plane of* $C(A/\Sigma(C), z)$, $z = z^*(t)$, $t \in [t_o, t_f)$, *and if* $y_k + \zeta_k \in C(\Sigma_k(C), y_k)$, $y_k = y_k^*(t)$, $k = 1,\ldots N$, $(\zeta_1,\ldots\zeta_N) \in Y$, *then* $z + \eta \in I(z, N+n-1) \cap C(A/\Sigma(C), z)$ *where* $\eta = P^{-1}(\zeta_1,\ldots\zeta_N)$.

One can readily verify that $C(\Sigma_k(C), y_k)$, $y_k = y_k^*(t)$, $t \in [t_o, t_f)$, is the tangent plane $T(\Sigma_k(C), y_k)$ of $\Sigma_k(C)$ at point $y_k = y_k^*(t)$. Hence $y_k - \zeta_k \in C(\Sigma_k(C), y_k)$, $k = 1,\ldots N$. Since $C(\Sigma_k(C), y_k) \subset C(A/\Sigma_k(C), y_k)$, $k = 1, \ldots N$, $y_k + \zeta_k$ and $y_k - \zeta_k$ belong to $C(A/\Sigma_k(C), y_k)$ at point $y_k = y_k^*(t)$, $k = 1,..N$; and it follows from Lemma 3 that $z + \eta$ and $z - \eta$ belong to $C(A/\Sigma(C), z)$ at point $z = z^*(t)$. At last, since $C(A/\Sigma(C), z)$, $z = z^*(t)$, is a subset of one of the closed half spaces determined by $I(z, N+n-1)$, $z + \eta$ and $z - \eta$ both belong to $I(z, N+n-1)$, which concludes the proof of Lemma 6 .

Lemma 7. *If* $I(z, N+n-1)$ *is a separating plane of* $C(A/\Sigma(C), z)$, $z = z^*(t)$, $t \in [t_o, t_f)$, *then for any vector* $\vec{n}$ *which is normal*[†] *to* $I(z, N+n-1)$, *such that* $z + \vec{n} \in D(C(A/\Sigma(C), z), z, N+n-1)$, *at point* $z = z^*(t)$, *we have* $\vec{n} = a_1 \text{ grad } \Phi_1(z) +\ldots+ a_N \text{ grad } \Phi_N(z)$ *where* $a_1 \ldots a_N$ *are non negative constants, and*

$$\text{grad } \Phi_k(z) \triangleq (0,\ldots 0,1,0,\ldots 0, \frac{V_k^*(x)}{\partial x_1} \cdots \frac{V_k^*(x)}{\partial x_n}) \in E^{N+n}$$

where 1 *is the k-th component of* grad $\Phi_k(z)$, *and* $k = 1,\ldots,N$.

† including the null vector as a special case

Consider the n linearly independent vectors

$$\eta^1 \triangleq (\xi^1_{o1}, \dots \xi^1_{oN}, 1, 0, \dots 0)$$

$$\eta^2 \triangleq (\xi^2_{o1}, \dots \xi^2_{oN}, 0, 1, \dots 0)$$

$$\eta^n \triangleq (\xi^n_{o1}, \dots \xi^n_{oN}, 0, 0, \dots 1)$$

where the components ξ^{ν}_{ok}, $k = 1, \dots N$, $\nu = 1, \dots N$ are such that

$$P_k(z^*(t) + \eta^{\nu}) \in C(\Sigma_k(C), y^*_k(t)), \ t \in [t_o, t_f), \quad k = 1, \dots N, \quad \nu = 1, \dots n$$

From Lemma 6 it follows that, if $I(z^*(t), N+n-1)$ is a separating plane of $C(A/\Sigma(C), z^*(t))$, then

$$z^*(t) + \eta^{\nu} \in I(z^*(t), N+n-1), \qquad \nu = 1, \dots n \tag{7}$$

Let $\vec{N} \triangleq a_1 \text{ grad } \Phi_1(z) + \dots + a_N \text{ grad } \Phi_N(z) =$

$$\left(a_1, \dots a_N, a_1 \frac{\partial V^*_1(x)}{\partial x_1} + \dots + a_N \frac{\partial V^*_N(x)}{\partial x_1}, \dots a_1 \frac{\partial V^*_1(x)}{\partial x_n} + \dots + a_N \frac{\partial V^*_N(x)}{\partial x_n}\right) \tag{8}$$

where $a_1 \dots a_N$ are arbitrary constants .

One can readily verify that $\vec{N}.\eta^{\nu} = 0$, $\nu = 1, \dots n$.

On the other hand it follows from (7) that every vector $\vec{n}$ which is normal to a separating plane of $C(A/\Sigma(C), z^*(t))$ also satisfies $\vec{n}.\eta^{\nu} = 0$, $\nu = 1, .. n$. Let $N \triangleq \{\vec{n} : \vec{n} \perp I(z^*(t), N+n-1)$, $I(z^*(t), N+n-1)$ separating plane of $C(A/\Sigma(C), z^*(t))\}$

$$P \triangleq \{ \vec{N} : \vec{N} = a_1 \text{ grad } \Phi_1(z) + \dots + a_N \text{ grad } \Phi_N(z), \ z = z^*(t), \quad a_k \in R, \ k = 1, \dots N \}$$

Since $\eta^1, \dots \eta^n$ are linearly independent vectors, and since P

is a N-dimensional plane in E^{N+n}, we have $N \subseteq P$. Accordingly we shall let $\vec{n} = \vec{N}$.

Now let us prove that if $z + \vec{n} \in D(C(A/\Sigma(C), z), z, N+n-1)$ at $z = z^*(t)$, then $a_1 \ldots a_N$ are non negative constants. Let $z^* = (x^*_{o1}, \ldots x^*_{oN}, x^*)$ and consider the set $\Omega(z^*(t)$, $t \in [t_o, t_f)$,namely

$$\Omega(z^*(t)) \triangleq \{ z : x_{ok} > x^*_{ok}(t), k = 1,..N, x = x^*(t) \}$$

One can see easily that

$$C(\Omega(z^*(t)), z^*(t)) = \overline{\Omega}(z^*(t)) = \{ z : x_{ok} \geqslant x^*_{ok}(t),$$

$$k = 1,..N, x = x^*(t) \}$$

and since $\Omega(z^*(t)) \subset A/\Sigma(C)$, we have

$$C(\Omega(z^*(t)), z^*(t)) \subset C(A/\Sigma(C), z^*(t)),$$

or equivalently

$$\overline{\Omega}(z^*(t)) \subset C(A/\Sigma(C), z^*(t)) .$$

Now, if $I(z^*(t), N+n-1)$ is a separating plane of $C(A/\Sigma(C), z^*(t))$, it is a separating plane of $\overline{\Omega}(z^*(t))$ and furthermore, $D(C(A/\Sigma(C), z^*(t)), z^*(t), N+n-1) = D(\overline{\Omega}(z^*(t), z^*(t), N+n-1)$.

Let

$$\vec{i}_k \triangleq (0,\ldots 0, 1, 0,\ldots 0) \in E^{N+n} \tag{9}$$

where 1 is the k-th component of $\vec{i}_k$; that is, the component on the x_{ok}-axis, and $k = 1,\ldots N$. Since $z^*(t) + \vec{i}_k$ belongs to $\overline{\Omega}(z^*(t))$, it also belongs to $D(\overline{\Omega}(z^*(t)), z^*(t), N+n-1)$ and hence it belongs to $D(C(A/\Sigma(C), z^*(t)), z^*(t), N+n-1)$. Accordingly if $z^*(t) + \vec{n} \in D(C(A/\Sigma(C), z^*(t)), z^*(t), N+n-1)$, we have[†] $\vec{i}_k \cdot \vec{n} \geqslant 0$, $k = 1, \ldots N$; and from (9) and (8) with $\vec{n} = \vec{N}$, we

---

[†] Since $\vec{n} \perp I(z^*(t), N+n-1)$

have $\vec{i}_k \cdot \vec{n} = a_k$, $k = 1, \ldots N$, which concludes the proof of Lemma 7 .

10. Variational Equations and their Adjoints

We shall associate with Eq.(2) the *variational equations*

$$(10) \qquad \frac{d\xi_{ok}}{dt} = \sum_{\alpha=1}^{n} \left(\frac{\partial f_{ok}}{\partial x_\alpha} + \sum_{\beta=1}^{N} \sum_{\gamma=1}^{r_\beta} \frac{\partial f_{ok}}{\partial u_{\beta\gamma}} \frac{\partial p^*_{\beta\gamma}}{\partial x_\alpha} \right) \xi_\alpha$$

$$k = 1, \ldots N$$

$$(11) \qquad \frac{d\xi_\nu}{dt} = \sum_{\alpha=1}^{n} \frac{\partial f_\nu}{\partial x_\alpha} + \sum_{\beta=1}^{N} \sum_{\gamma=1}^{r_\beta} \frac{\partial f_\nu}{\partial u_{\beta\gamma}} \frac{\partial p^*_{\beta\gamma}}{\partial x_\alpha}) \xi_\alpha$$

$$\nu = 1, \ldots n$$

where partial derivatives are evaluated for $z = z^*(t)$ and $u_k = p^*_k(z^*(t))$, $k = 1,\ldots N$, $t \in [t_o, t_f)$.

Let $\eta \triangleq (\xi_o, \xi)$, $\xi_o \triangleq (\xi_{o1}, \ldots \xi_{oN})$, $\xi \triangleq (\xi_1, \ldots \xi_n)$. For given initial condition η' at $t' \in [t_o, t_f)$, there exists a unique solution $\tilde{\eta} : t \mapsto \eta = \tilde{\eta}(t)$ of Eqs.(10)(11) defined and continuous on $[t', t_f]$ †, such that $\tilde{\eta}(t') = \eta'$. Furthermore, $\tilde{\eta}(t)$ is nonzero for all $t \in [t', t_f]$ provided η' is nonzero. This solution defines a nonsingular linear transformation $A(t',t)$ such that $\tilde{\eta}(t) = A(t', t)\, \eta'$, $t \in [t', t_f]$. For $t = t''$, we write $\tilde{\eta}(t'')=\eta''$ so that $\eta'' = A(t', t'')\, \eta'$, $t_o \leqslant t' \leqslant t'' \leqslant t_f$. Since transformation $A(t', t'')$ is nonsingular, an inverse transformation $A^{-1}(t', t'')$ is defined such that

$$(12) \qquad \eta' = A^{-1}(t', t'')\, \eta'', \quad t_o \leqslant t' \leqslant t'' \leqslant t_f$$

We shall put $A^{-1}(t', t'') = A(t'', t')$, so that (12) reads

$$(13) \qquad \eta' = A(t'', t')\, \eta'', \quad t_o \leqslant t' \leqslant t'' \leqslant t_f$$

---

† Because of the assumption A.1 of paragraph 8 concerning p^*

The equations adjoint to variational equations (10)(11) are

$$(14) \qquad \frac{d\lambda_{ok}}{dt} = 0, \qquad k = 1, \ldots N$$

$$(15) \qquad \frac{d\lambda_\nu}{dt} = -\sum_{\alpha=1}^{N} \left(\frac{\partial f_o}{\partial x_\nu} + \sum_{\beta=1}^{N} \sum_{\gamma=1}^{r_\beta} \frac{\partial f_{o\alpha}}{\partial u_{\beta\gamma}} \frac{\partial p^*_{\beta\gamma}}{\partial x_\nu} \right) \lambda_{o\alpha} - \sum_{\alpha=1}^{n} \left(\frac{\partial f_\alpha}{\partial x_\nu} + \sum_{\beta=1}^{N} \sum_{\gamma=1}^{r_\beta} \frac{\partial f_\alpha}{\partial u_{\beta\gamma}} \frac{\partial p^*_{\beta\gamma}}{\partial x_\nu} \right) \lambda_\alpha$$

$$\nu = 1, \ldots, n$$

Let $\psi \triangleq (\lambda_o, \lambda)$, $\lambda_o \triangleq (\lambda_{o1}, \ldots \lambda_{oN})$, $\lambda \triangleq (\lambda_1 \ldots \lambda_n)$. For given initial condition ψ' at $t' \in [t_o, t_f)$, there exists a unique solution $\tilde{\psi} : t \mapsto \psi = \tilde{\psi}(t)$ of Eqs. (14)(15) defined and continuous on $[t', t_f]$, such that $\tilde{\psi}(t') = \psi'$. Furthermore, $\tilde{\psi}(t)$ is nonzero for all $t \in [t', t_f]$ provided ψ' is nonzero.

As a consequence of (10)(11) and (14)(15), we have

$$\tilde{\psi}(t).\tilde{\eta}(t) = \text{constant} \qquad \forall\, t \in [t_o, t_f]$$

and from (14), we have

$$(16) \qquad \tilde{\lambda}_{ok}(t) = \text{constant} \qquad \forall\, t \in [t_o, t_f]$$

11. Transformation of a Contingent

Lemma 8. $z^*(t') + \eta' \in C(A/\Sigma(C), z^*(t')) \Rightarrow z^*(t'') + \eta'' \in C(A/\Sigma(C), z^*(t''))$, $\eta'' = A(t', t'')\eta'$, $t' \in [t_o, t_f]$, $t'' \in [t_o, t_f)$.

Let $z^*(t') + \eta' \in C(A/\Sigma(C), z^*(t'))$.

Then there exist η^ν and ε^ν ($\varepsilon^\nu > 0$), $\nu = 1, 2, \ldots \ell \ldots$ such that $z^*(t') + \varepsilon^\nu \eta^\nu \in A/\Sigma(C)$ for $\nu = 1, 2, \ldots \ell \ldots$ and $\eta^\nu \to \eta'$ and $\varepsilon^\nu \to 0$ as $\nu \to \infty$. From the definition of a Pareto surface, it follows that there exists a Pareto surface $\Sigma(C_\nu)$ through $z^*(t') + \varepsilon^\nu\eta^\nu$, and that $\Sigma(C_\nu) \subset A/\Sigma(C)$.

From the Assumptions 1 of paragraph 8 concerning p^* and π^*, and from the fact that $z^*(t'')$ is an interior point of $\Sigma(C)$, it follows that, for ε^ν sufficiently small, there exists a path $\Pi^*(C_\nu)$ generated by p^* , given by function z^ν , such that $z^\nu(t')=z^*(t')+\varepsilon^\nu\eta^\nu$. Then $z^\nu(t'') = z^*(t'') + \varepsilon^\nu A(t',t'')\eta^\nu + o(\varepsilon^\nu)$. It follows from Lemma 1 that $\Pi^*(C_\nu) \subset \Sigma(C_\nu)$. Hence

$$z^*(t'') + \varepsilon^\nu(A(t',t'')\eta^\nu + \frac{o(\varepsilon^\nu)}{\varepsilon^\nu}) \in A/\Sigma(C) \text{ , and since}$$

$$A(t',t'')\eta^\nu + \frac{o(\varepsilon^\nu)}{\varepsilon^\nu} \to \eta'' = A(t',t'')\eta' \quad \text{as } \nu \to \infty$$

we conclude that

$$z^*(t'') + \eta'' \in C(A/\Sigma(C), z^*(t'')).$$

Lemma 8 has the following

Corollary 1. $C(A/\Sigma(C), z^*(t_f)) \subset A(t,t_f)\ C(A/\Sigma(C), z^*(t))$ $t \in [t_o,t_f)$, *where* $A(t,t_f)\ C(A/\Sigma(C),z^*(t)) \triangleq$

$$\{z^*(t_f) + \eta^f : \eta^f = A(t,t_f)\eta \text{ , } z^*(t) + \eta \in C(A/\Sigma(C), z^*(t))\ \}$$

Indeed, from Lemma 8 we have

$$z^*(t_f) + \eta^f \in C(A/\Sigma(C), z^*(t_f)) \Rightarrow z^*(t) + \eta \in C(A/\Sigma(C), z^*(t)),$$

$$\eta = A(t_f,t)\eta^f$$

and hence Corollary 1 is established .

Lemma 9. $A(t,t_f)C(A/\Sigma(C), z^*(t))$, $t \in [t_o,t_f)$ *is independent of* t.

Let $t'' \neq t'$, $t', t'' \in [t_o,t_f)$. From Lemma 8 we deduce $C(A/\Sigma(C), z^*(t'')) = A(t',t'')C(A/\Sigma(C), z^*(t'))$, or equivalently $C(A/\Sigma(C), z^*(t')) = A(t'',t')C(A/\Sigma(C), z^*(t''))$.

Accordingly we have

$$\begin{aligned} A(t',t_f)C(A/\Sigma(C), z^*(t')) &= A(t',t_f)A(t'',t')C(A/\Sigma(C), z^*(t'')) \\ &= A(t'',t_f)C(A/\Sigma(C), z^*(t'')) \end{aligned}$$

which concludes the proof.

Now in view of Lemma 9, we can define the cone $\Gamma(A/\Sigma(C), z^*(t_f))$, namely

$$\Gamma(A/\Sigma(C), z^*(t_f)) \triangleq A(t,t_f)C(A/\Sigma(C), z^*(t))$$

where t, $t \in [t_o,t_f)$, is arbitrary.

Lemma 10. $\Gamma(A/\Sigma(C), z^*(t_f))$ *is convex.*

This lemma is a direct consequence of Lemma 4 together with the fact that transformation $A(t,t_f)$ is linear and non singular.

12. Separability of Local Cones

Assumption 2. From now on we shall assume that for all $x \in \pi^*$, and for all u, $u \triangleq (u_1,\dots u_N) \in K(z)$, where $K(z) \triangleq K(x)$, there exist strategies $\alpha_1(x,u,\cdot) \in S_1,\dots\alpha_N(x,u,\cdot) \in S_N$ † which are of class C^1 on some neighborhood of x in G, and which satisfy the conditions $\alpha_1(x,u,x) = u_1,\dots\alpha_N(x,u,x) = u_N$.

We shall denote by $\alpha(z,u,\cdot)$ this strategy N-tuple ; namely $\alpha(z,u,\cdot) \triangleq (\alpha_1(x,u,\cdot),\dots\alpha_N(x,u,\cdot)) \in S$.

We shall be concerned with the vectors of the set

$$R(t) \triangleq \{ v : v \triangleq A(\tau,t)F(z^*(\tau),u), u \in K(z^*(\tau)), \tau \in [t_o,t_f) \},$$

$t \in [t_o,t_f]$, that is, we shall consider the set L(t) of linear combinations, having non-negative coefficients, of a finite number of vectors v_r , $r \in N$, $v_r \in R(t)$, and we shall define the cone $K(z^*(t))$, namely

$$K(z^*(t)) \triangleq \{ z^*(t) + \rho : \rho \in L(t) \}, \qquad t \in [t_o,t_f]$$

Let us also define the cone $V(z^*(t))$, namely

$$V(z^*(t)) \triangleq \{ z^*(t) + \gamma F(z^*(t),u) : \gamma \geq 0, u \in K(z^*(t)) \}$$
$$t \in [t_o,t_f)$$

---

† This amounts to replacing the earlier notation p_k for a strategy by the new one $p_k(\cdot)$, k = 1, 2,.. N .

Lemma 11. $K(z^*(t))$, $t \in [t_o, t_f]$, *is convex*

For $t = t_o$, we have $R(t) = \phi$ and hence $L(t) = \phi$ and $K(z^*(t)) = \phi$; so, $K(z^*(t))$ is convex. For $t \neq t_o$ the convexity of $K(z^*(t))$ follows from its definition together with the convexity of $L(t)$.

Lemma 12. $A(t,t_f)V(z^*(t)) \subset K(z^*(t_f))$, $t \in [t_o,t_f)$,
where $A(t,t_f)V(z^*(t)) \triangleq$
$\{z^*(t_f) + \eta^f : \eta^f = A(t,t_f)\eta,\ z^*(t) + \eta \in V(z^*(t))\}$

This lemma is a direct consequence of the definitions of $K(z^*(t))$ and $V(z^*(t))$.

Lemma 13. $K(z^*(t)) \cap \overset{\circ}{C}(A/\Sigma(C),\ z^*(t)) = \phi$, $t \in (t_o,t_f)$

Let us select certain instants τ_1, $\tau_2, \ldots \tau_s$ which satisfy the inequality $t_o \leqslant \tau_1 \leqslant \tau_2 \leqslant \ldots \leqslant \tau_s < t < t_f$. Let us further select arbitrary non-negative numbers β_1, $\beta_2, \ldots$ β_s and arbitrary (not necessarily distinct) points u^1 , $u^2, \ldots$ u^s of $K(z^*(\tau_1))$, $K(z^*(\tau_2)), \ldots$ $K(z^*(\tau_s))$, respectively.

Let $\varphi \triangleq \sum_{r=1}^{s} \beta_r v_r \in L(t)$, where $v_r \triangleq A(\tau_r,t)F(z^*(\tau_r),\ u^r)$,
$r = 1, 2, \ldots s$, and let $\beta_o = 0$, and

$$I(\tau_r) \triangleq \{z : \tau_r + (\beta_o + \ldots + \beta_{r-1})\varepsilon \leqslant x_n < \tau_r + (\beta_o + \ldots + \beta_r)\varepsilon\}$$

$r = 1, 2, \ldots s$, $\varepsilon > 0$. Then, consider strategy N-tuple $a \triangleq (a_1, \ldots a_N) \in S$ such that

$$a(z) = \alpha(z^*(\tau_r),\ u^r, z) \text{ for all } z \in I(\tau_r) \text{ and}$$

$$a(z) = p^*(z) \text{ for all } z \notin \bigcup_{r=1,..s} I(\tau_r)$$

From Assumption 2 above and from the fact that $z^*(t)$ is an interior point of $\Sigma(C)$, it follows that, for ε sufficiently small, there exists a path $\Pi(C')$ generated by strategy N-tuple a, given by $\tilde{z} : t \mapsto \tilde{z}(t)$, $t \in [t_o, t + (\beta_1 + \ldots + \beta_s)\varepsilon]$, such that $\tilde{z}(t_o) = z^o \triangleq z^*(t_o)$. Then $\tilde{z}(t + (\beta_1 + \ldots + \beta_s)\varepsilon) = z^*(t) + \varepsilon\varphi + o(\varepsilon)$

From the definition of φ and of $K(z^*(t))$, we have $z^*(t) + \varphi \in K(z^*(t))$. Now if $z^*(t) + \varphi \in \overset{o}{C}(A/\Sigma(C), z^*(t))$, it follows from Lemma 5 that, for ε sufficiently small, we have $z^*(t) + \varepsilon(\varphi + \frac{o(\varepsilon)}{\varepsilon}) \in A/\Sigma(C)$.

Accordingly there exists a Pareto surface $\Sigma(C'')$ through $z^*(t) + \varepsilon\varphi + o(\varepsilon)$, and $\Sigma(C'') \subset A/\Sigma(C)$. It follows that there exists a terminating path $\Pi^*(C'')$ generated by p^* from state $z^*(t) + \varepsilon\varphi + o(\varepsilon)$, and from Lemma 1 we have $\Pi^*(C'') \subset \Sigma(C'')$. Furthermore, from condition (i) of paragraph 1, $\Pi(C') \cup \Pi^*(C'')$ is a terminating path emanating from z^o, and since $\Pi^*(C'') \subset A/\Sigma(C)$, its terminating point belongs to $\Theta \cap (A/\Sigma(C))$. In other words, for each player J_k , $k = 1,\ldots N$, the cost associated with $\Pi(C') \cup \Pi^*(C'')$ is greater than $V_k^*(x^o)$ which contradicts the fact that p^* is Pareto optimal on X .

It follows that $z^*(t) + \varphi \notin \overset{o}{C}(A/\Sigma(C), z^*(t))$, and since $\tau_1, \tau_2, \ldots \tau_s$ in the half-open interval $[t_o, t)$, and $u^1, u^2 \ldots u^s$ in the sets $K(z^*(\tau_1)), K(z^*(\tau_2)), \ldots K(z^*(\tau_s))$, respectively, and the non-negative numbers $\beta_1, \beta_2, \ldots \beta_s$, are arbitrary, we can conclude that

$$z^*(t) + \varphi \in K(z^*(t)) \;\Rightarrow\; z^*(t) + \varphi \notin \overset{o}{C}(A/\Sigma(C), z^*(t))$$

and hence Lemma 13 is established.

Lemma 14. $K(z^*(t_f)) \cap \overset{o}{\Gamma}(A/\Sigma(C), z^*(t_f)) = \phi$

Let $z^*(t_f) + \varphi^f \in K(z^*(t_f))$. From the definition of $K(z^*(t_f))$, it follows that there exists a finite set $\{\tau_1, \tau_2, \ldots \tau_s\}$ where $t_o \leqslant \tau_1 \leqslant \tau_2 \leqslant \ldots \leqslant \tau_s < t_f$, and a set $\{u^1, u^2, \ldots, u^s\}$ where $u^1 \in K(z^*(\tau_1))$, $u^2 \in K(z^*(\tau_2)), \ldots$ $u^s \in K(z^*(\tau_s))$, and non-negative numbers $\beta_1, \beta_2, \ldots \beta_s$, such that

$$\varphi^f = \sum_{r=1}^{s} \beta_r A(\tau_r, t_f) F(z^*(\tau_r), u^r)$$

Consider an arbitrary time t, $t \in (\tau_s, t_f)$, and let

$$\varphi \triangleq \sum_{r=1}^{s} \beta_r A(\tau_r, t) F(z^*(\tau_r), u^r)$$

Then

$$\varphi = A(t_f, t)\, \varphi^f \quad \text{and} \quad z^*(t) + \varphi \in K(z^*(t)) \tag{17}$$

Let us suppose that

$$z^*(t_f) + \varphi^f \in \overset{\circ}{\Gamma}(A/\Sigma(C),\, z^*(t_f))$$

or, equivalently, in view of the definition of $\Gamma(A/\Sigma(C), z^*(t_f))$,

$$z^*(t_f) + \varphi^f \in \overset{\circ}{\overbrace{A(t, t_f) C(A/\Sigma(C),\, z^*(t))}}$$

Since $A(t, t_f)$ is linear and non-singular we have

$$\overset{\circ}{\overbrace{A(t, t_f) C(A/\Sigma(C),\, z^*(t))}} = A(t, t_f) \overset{\circ}{C}(A/\Sigma(C),\, z^*(t)).$$

Hence $z^*(t_f) + \varphi^f \in A(t, t_f) \overset{\circ}{C}(A/\Sigma(C),\, z^*(t))$, which implies that

$$z^*(t) + \varphi \in \overset{\circ}{C}(A/\Sigma(C),\, z^*(t)) \tag{18}$$

Since (17) and (18) contradict Lemma 13, Lemma 14 is established.

From Lemmas 10, 11 and 14, we deduce the following

Corollary 2. $K(z^*(t_f))$ *and* $\Gamma(A/\Sigma(C),\, z^*(t_f))$ *are separated in* E^{N+n} .

That is, there exists an (N+n-1)-dimensional hyperplane through $z^*(t_f)$ such that $K(z^*(t_f))$ is entirely contained in one (closed) half-space defined by this hyperplane, and $\Gamma(A/\Sigma(C),\, z^*(t_f))$ is entirely contained in the other. In the following we shall consider such an hyperplane that we shall denote by $I(z^*(t_f), N+n-1)$, and we shall say that $K(z^*(t_f))$ and

$\Gamma(A/\Sigma(C), z^*(t_f))$ are *separated by* $I(z^*(t_f), N+n-1)$.

From Corollary 1, Lemma 12, Corollary 2 and the definitions of $\Gamma(A/\Sigma(C), z^*(t_f))$ and $I(z^*(t_f), N+n-1)$, we have

Corollary 3. $C(A/\Sigma(C), z^*(t_f))$ *and* $A(t,t_f)V(z^*(t))$, $t \in [t_o,t_f)$ *are separated by* $I(z^*(t_f), N+n-1)$.

Let

$$I(z^*(t), N+n-1) \triangleq A(t_f,t)I(z^*(t_f), N+n-1), \quad t \in [t_o,t_f)$$

From Corollary 2, Lemma 12 and the definitions of $\Gamma(A/\Sigma(C), z^*(t_f))$ and $I(z^*(t), N+n-1)$, one readily deduces the following

Lemma 15. $V(z^*(t))$ *and* $C(A/\Sigma(C), z^*(t))$, $t \in [t_o,t_f)$, *are separated by* $I(z^*(t), N+n-1)$.

Now let η^* be a non-zero vector tangent to $\Pi^*(C)$ at point $z^*(t)$, $t \in [t_o,t_f)$. Let ζ^*_k be its projection on E_k, $k = 1,..N$. Since ζ^*_k is tangent to $\Pi^*_k(C)$ at point $y^*_k(t)$, and $\Pi^*_k(C) \subset \Sigma_k(C)$, $k = 1,...N$, we have†

$$y^*_k(t) + \zeta^*_k \in C(\Sigma_k(C), y^*_k(t)), \; k = 1, \dots N$$

Then, by invoking Lemmas 6 and 15 we obtain

Lemma 16. *If* η^* *is a non-zero vector tangent to* $\Pi^*(C)$ *at point* $z^*(t)$, $t \in [t_o,t_f)$, *then*

$$z^*(t) + \eta^* \in I(z^*(t), N+n-1) \cap C(A/\Sigma(C), z^*(t))$$

Likewise, let η^f be a non-zero tangent to $\Theta \cap \Sigma(C)$ at point $z^*(t_f)$. Then, it follows from its definition that it is tangent to $\Sigma(C)$ at point $z^*(t_f)$, and hence $z^*(t_f) + \eta^f \in C(\Sigma(C),z^*(t_f))$. One can prove easily†† that $C(\Sigma(C), z^*(t_f)) \subset C(A/\Sigma(C),z^*(t_f))$.

---

† Of course $(\zeta^*_1, \dots \zeta^*_N) \in Y$

†† See Appendix

Accordingly

$$(19) \qquad z^*(t_f) + \eta^f \in C(A/\Sigma(C),\ z^*(t_f))$$

Since $\Theta \cap \Sigma(C) = \{ z : x_{ok} = C_k \ , k = 1,\ldots N,\ x \in \theta \}$, $\Theta \cap \Sigma(C)$ may be deduced from θ by translation parallel to $C_1\vec{w}_1 + \ldots + C_N\vec{w}_N$, where $\vec{w}_k$ is the unit vector parallel to the x_{ok}-axis, $k = 1,\ldots N$; then it follows from the definition of θ that $-\eta^f$ is tangent to $\Theta \cap \Sigma(C)$ and, accordingly,

$$(20) \qquad z^*(t_f) - \eta^f \in C(A/\Sigma(C),\ z^*(t_f))$$

Now, from Corollary 3, $I(z^*(t_f), N+n-1)$ is a separating plane of $C(A/\Sigma(C),\ z^*(t_f))$. Accordingly, as a direct consequence of (19) and (20), we have the following

Lemma 17. *If* η^f *is a non-zero vector tangent to* $\Theta \cap \Sigma(C)$ *at point* $z^*(t_f)$*, then* $z^*(t_f) + \eta^f \in I(z^*(t_f), N+n-1)$.

Remark. Note that η^f has a null projection on the x_{ok}-axis, $k = 1,\ldots N$. Hence $\eta^f = \zeta^f$, where ζ^f is a non-zero vector tangent to θ at point $x^*(t_f)$.

13. A Maximum Principle

From Lemma 7, Corollaries 2 and 3, and Lemmas 15-17, one can deduce easily the following

Theorem 1. *If* $\Pi^*(C)$ *is a path generated by Pareto optimal strategy* N*-tuple* p^**, represented by* $z^* : t \mapsto z^*(t),\ t \in [t_o, t_f\,]$, *and if Assumptions 1 and 2 are satisfied (for the problem stated above), then there exists a non-zero continuous solution*

$$\tilde{\psi} = (\tilde{\lambda}_{o1},\ldots\tilde{\lambda}_{oN},\ \tilde{\lambda}_1,\ldots\tilde{\lambda}_n) : t \mapsto \tilde{\psi}(t), \quad t \in [t_o,\ t_f\,]$$

of adjoint equations (14)(15) *such that*

$$\left.\begin{array}{ll}
(a) & \max_{u \in K(z^*(t))} \tilde{\psi}(t).F(z^*(t),u) \\
 & \qquad = \tilde{\psi}(t).F(z^*(t),p^*(z^*(t))) \\
(b) & \tilde{\psi}(t).F(z^*(t),p^*(z^*(t))) = 0 \\
(c) & \tilde{\lambda}_{o1}(t),\ldots\tilde{\lambda}_{oN}(t) \text{ are non negative constants}
\end{array}\right\} \text{for all } t \in [t_o,t_f)$$

(d) $(\tilde{\lambda}_1(t_f),\ldots\tilde{\lambda}_n(t_f))$ *is normal to* θ

By using Assumption 2 one can prove easily that conditions (a), (b) and (c) of Theorem 1 hold on $[t_o,t_f]$ by continuity.

Theorem 1 can be easily extended to the case where, at the end of a terminating play, each player receives a payoff of the form

$$V_k(x^i,\theta;p) \triangleq \int_{t_i}^{t_f} f_{ok}(\tilde{x}(t),p_1(\tilde{x}(t)),\ldots p_N(\tilde{x}(t)))dt + w_k(\tilde{x}(t_f))$$

$k = 1,\ldots N$

where functions $w_k : G \to R$, $k = 1, \ldots N$, are of class C^2 on G .

Then, condition (d) of Theorem 1 is replaced by

$$(d)' \quad \tilde{\lambda}(t_f) = \sum_{k=1}^{N} \lambda_{ok} \text{ grad } w_k(x^*(t_f)) + c \text{ grad } m(x^*(t_f))$$

where $x^*(t_f)$ is the projection of $z^*(t_f)$ on G, and $c \in R$.

14. Jump Condition

The definitions and properties of discontinuity manifolds are similar to the ones discussed in Ref.(4), in the case of two-person zero-sum games. The behavior of optimal paths in $E^N \times G$, near such surfaces is established in Ref.(13). Here we shall not reexamine the assumptions of Ref.(13) which are similar to the ones of Ref.(4). We shall suppose that there exists a decomposition $\{X_1, X_2, .. X_k\}$ of X, and a strategy N-tuple p^* optimal on X, satisfying these assumptions, and we shall recall the jump condition, at a point where an optimal path crosses a transition manifold $M_{h\ell} \triangleq \overline{X}_h \cap \overline{X}_\ell$. We shall let $N_{h\ell} \triangleq E^N \times M_{h\ell}$. We shall suppose that $M_{h\ell}$ is defined by a single equation $m_{h\ell}(x) = 0$; that is

$$x \in M_{h\ell} \Rightarrow m_{h\ell}(x) = 0$$

where function $m_{h\ell} : G \to R$ is of class C^1 and grad $m_{h\ell}(x) \neq 0$ on a neighborhood of $M_{h\ell}$ in G .

Let $z^*(t_j) \in N_{h\ell}$, $t_j \in (t_o, t_f)$, and

$$\widetilde{\psi}^-(t_j) \triangleq \lim_{\substack{t \to t_j \\ t < t_j}} \widetilde{\psi}(t) = \sum_{k=1}^{N} \lambda_{ok}^- N_k^-$$

$$\widetilde{\psi}^+(t_j) \triangleq \lim_{\substack{t \to t_j \\ t > t_j}} \widetilde{\psi}(t) = \sum_{k=1}^{N} \lambda_{ok}^+ N_k^+$$

where

$$\lambda_{ok}^- \triangleq \widetilde{\lambda}_{ok}(t_j - 0), \qquad \lambda_{ok}^+ \triangleq \widetilde{\lambda}_{ok}(t_j + 0)$$

$$N_k^- \triangleq (\delta_1^k, \dots \delta_N^k , \text{grad } V_k^-(x^*(t_j)))$$

$$N_k^+ \triangleq (\delta_1^k, \dots \delta_N^k , \text{grad } V_k^+(x^*(t_j)))$$

where $\delta_\nu^k = 0$ for $\nu \neq k$, $\delta_k^k = 1$, and

$$\text{grad } V_k^-(x^*(t_j)) \triangleq \lim_{\substack{t \to t_j \\ t < t_j}} \text{grad } V_k^*(x^*(t))$$

$$\text{grad } V_k^+(x^*(t_j)) \triangleq \lim_{\substack{t \to t_j \\ t > t_j}} \text{grad } V_k^*(x^*(t))$$

The jump condition can be written

$$N_k^- = N_k^+ + a_k \text{ grad } m_{h\ell}(z^*(t_j))$$

$a_k \in R$, $k = 1, \ldots N$; $m_{h\ell}(z) \triangleq m_{h\ell}(x)$

with

$$- \sum_{k=1}^{N} \lambda_{ok}^- a_k = \frac{\sum_{k=1}^{N} \lambda_{ok}^- f_{ok}^-(z^*(t_j), p^*(z^*(t_j)))}{\text{grad } m_{h\ell}(x^*(t_j)).f^-(z^*(t_j),p^*(z^*(t_j)))}$$

$$+ \frac{\sum_{k=1}^{N} \lambda_{ok}^- \text{ grad } V_k^+(x^*(t_j)).f^-(z^*(t_j),p^*(z^*(t_j)))}{\text{grad } m_{h\ell}(x^*(t_j)).f^-(z^*(t_j),p^*(z^*(t_j)))} \quad \text{and}$$

$$\sum_{k=1}^{N} \lambda_{ok}^+ a_k = \frac{\sum_{k=1}^{N} \lambda_{ok}^+ f_{ok}^+(z^*(t_j), p^*(z^*(t_j)))}{\text{grad } m_{h\ell}(x^*(t_j)).f^+(z^*(t_j),p^*(z^*(t_j)))}$$

$$+ \frac{\sum_{k=1}^{N} \lambda_{ok}^+ \text{ grad } V_k^-(x^*(t_j)).f^+(z^*(t_j), p^*(z^*(t_j)))}{\text{grad } m_{h\ell}(x^*(t_j)).f^+(z^*(t_j), p^*(z^*(t_j)))}$$

where superscripts + and - denote right-hand and left-hand limits, respectively, as t tends to t_j .

15. An Example of Pareto Equilibrium

We consider an undifferenciated duopoly over a short period. The quantities q_1 and q_2 produced in an unit of time by firms 1 and 2 respectively depend therefore only on y_1 and y_2, the labor per unit of time employed by these firms. Consequently $q_k = m_k y_k$, $k = 1, 2$, where m_1 and m_2 are positive constants related to firms' production techniques.

We suppose that :

1) the product may not be stocked and all production is sold immediately,

2) the unit price of the product is a given constant p,

3) the government practices a strict policy of full employment and the firms employ highly specialized labor, this specialized activity being peculiar to the two firms alone. If L is the number of laborers possessing the qualifications required by these firms, one has therefore : $y_1 + y_2 = L$. Moreover, the analysis being restricted to short periods, L is a positive constant,

4) for each firm the price of production consists of a constant cost and a cost arising from the salaries of the workers. We denote by c_1 and c_2 the constant per unit cost for firm 1 and firm 2, and we assume : $c_1 < p$, $c_2 < p$, $m_1(p - c_1) > m_2(p - c_2)$ and $m_1 < m_2$,

5) the hourly wage u_1 and u_2 offered by each firm is non negative and such that the instantaneous profit of the firm is also non negative,

6) labor is perfectly mobile between two firms and moves from one to the other at a rate proportional to the difference between the wages offered by the firms. We have therefore: $\dot{y}_1 = b(u_1 - u_2) = -\dot{y}_2$. Let $x_1 \triangleq y_1$, in view of Assumptions 3 and 5 the state equations of the model are :

$$\dot{x}_1 = b(u_1 - u_2), \quad 0 < x_1 < L$$
$$\dot{x}_2 = 1 ,$$

7) the target set θ is defined by :

$$\theta \triangleq \{ x : x \triangleq (x_1, x_2) , 0 < x_1 < L, x_2 = T \}$$

where T is fixed,

8) firm 1 wants to maximise his total profit while firm 2 maximises his final market share .

Let $\gamma_1 \triangleq (p-c_1)$ and $\gamma_2 \triangleq (p-c_2)$; from Assumptions 2, 3 and 4 we deduce that the instantaneous profits P_1 and P_2 of the firms are:

$$P_1 = (\gamma_1 m_1 - u_1)x_1$$
$$P_2 = (\gamma_2 m_2 - u_2)(L-x_1)$$

so that by invoking Assumption 5 we have :

$$0 \leqslant u_1 \leqslant m_1\gamma_1 \quad \text{and} \quad 0 \leqslant u_2 \leqslant m_2\gamma_2$$

If $\Pi^*(C)$ is a path generated by a Pareto optimal strategy pair $p^* \triangleq (p_1^*, p_2^*)$ and π^* its projection on the x_1,x_2-plane, given by $x^* : t \mapsto x^*(t)$, $t \in [t_o, T]$, since the constraints are independant of the state, the adjoint equations along $\Pi^*(C)$ are :

$$\dot{\lambda}_{o1} = \dot{\lambda}_{o2} = \dot{\lambda}_2 = 0$$
$$\dot{\lambda}_1 = -\lambda_{o1}(m_1\gamma_1 - u_1)$$

From condition d) of Theorem 1 we obtain

$$\tilde{\lambda}_1(T) = - \frac{\lambda_{o2}m_1m_2L}{((m_1-m_2)x_1^*(T)+m_2L)^2}$$

where

$$\tilde{\psi} \triangleq (\tilde{\lambda}_{o1},\tilde{\lambda}_{o2},\tilde{\lambda}_1,\tilde{\lambda}_2) : t \mapsto \tilde{\psi}(t),\ t \in [t_o,T]$$

is the solution of adjoint equations satisfying Theorem 1 .

Three cases need be considered, namely :

1) $\tilde{\lambda}_1(t) < 0$, $t \in [t_o,T]$; then by using condition a) of Theorem 1 we have :

$$p_1^*(x^*(t)) = 0 \text{ and } p_2^*(x^*(t)) = m_2\gamma_2$$
$$\text{for all } t \in [t_o,T],$$

2) $b\tilde{\lambda}_1(t) - \lambda_{o1}x_1^*(t) < 0$, $t \in [t_o,T]$, and there exists $t_1 \in [t_o,T]$ such that $\tilde{\lambda}_1(t_1) = 0$. From paragraph 13 we deduce that :

$$p_1^*(x^*(t)) = 0, \; t \in [t_o, T], \text{ and}$$

$$p_2^*(x^*(t)) = m_2\gamma_2 \text{ for all } t \in [t_o, T], \text{such that } \tilde{\lambda}_1(t) < 0$$
$$= 0 \text{ for all } t \in [t_o, T], \text{ such that } \tilde{\lambda}_1(t) > 0$$

3) there exist $t_1 \in [t_o, T]$ and $t_2 \in [t_o, T]$ such that : $\tilde{\lambda}_1(t_1) = 0$ and $b\tilde{\lambda}_1(t_2) - \lambda_{o1}x_1^*(t_2) = 0$.

Furthermore from adjoint equations, transversality condition and condition a) of Theorem 1 one deduces that there exists a neighborhood of θ where $p_1^*(x) = 0$ and $p_2^*(x) = m_2\gamma_2$. In this neighborhood the equations of the projections on the x_1, x_2-plane of the trajectories are :

$$x_1^*(t) = bm_2\gamma_2(T-t) + x_1^*(T)$$

$$x_2^*(t) = t - T + x_2^*(T)$$

and we have : $\tilde{\lambda}_1(t) = \lambda_{o1}m_1\gamma_1(T-t) + \tilde{\lambda}_1(T)$

For these trajectories we obtain :

$$T - t_1 = - \frac{\tilde{\lambda}_1(T)}{\lambda_{o1}m_1\gamma_1}$$

$$T - t_2 = \frac{\lambda_{o1}x_1^*(T) - b\tilde{\lambda}_1(T)}{b\lambda_{o1}(m_1\gamma_1 - m_2\gamma_2)}$$

As $0 < x_1^*(T) < L$, it follows from the transversality condition that : $T-t_2 > T-t_1$, for all $x_1^*(T) \in (0,L)$. Let $\bar{t}$ be such that : $bm_2\gamma_2(T-\bar{t}) = L-x_1^*(T)$, then case 2) or 3) occurs only if $T-t_1 < T-\bar{t}$, that is if :

$$\frac{\lambda_{o2}bL\gamma_1}{\lambda_{o1}\gamma_2\left(x_1^*(T)\dfrac{m_1-m_2}{m_2} + L\right)^2} < L - x_1^*(T)$$

As $m_2 > m_1$, this relation may be written : $y^3-3a^2y-2a^3-A > 0$, where

$$y \triangleq \frac{m_1-m_2}{m_2}\, x_1^*(T)+L(1-\frac{m_1}{3m_2}),\quad a \triangleq 3(\frac{m_1 L}{3m_2}),\quad A \triangleq \frac{\lambda_{o2} bL\gamma_2(m_2-m_1)}{\lambda_{o1}\gamma_1 m_2}$$

For $\lambda_{o1} \neq 0$, A is defined and non negative so that the equation : $y^3 - 3a^2y - 2a^3 - A = 0$, has a unique real root denoted by α. Let β be such that : $3(m_2-m_1)\beta = L(3m_2-m_1) - 3m_2\alpha$, then $\beta < L$ and

$$\beta \leqslant 0 \quad \text{if} \quad \lambda_{o2} b\gamma_2 \geqslant \lambda_{o1} L^2 \gamma_1$$

$$\beta > 0 \quad \text{if} \quad \lambda_{o2} b\gamma_2 < \lambda_{o1} L^2 \gamma_1$$

Therefore if $\lambda_{o2} b\gamma_2 \geqslant \lambda_{o1} L^2\gamma_1$ then for all $x_1^*(T) \in (0,T)$ we have: $T-t_1 \geqslant T-\bar{t}$, and case 1) occurs ; if $\lambda_{o2} b\gamma_2 < \lambda_{o1} L^2\gamma_1$ then $0 < \beta < L$ so that :

for $\beta \leqslant x_1^*(T) < L$ we have $T-t_1 \geqslant T-\bar{t}$ and case 1) occurs, and

for $0 < x_1^*(T) < \beta$ case 2) or 3) occurs .

Consider the case : $\lambda_{o2} b\gamma_2 \geqslant \lambda_{o1} L^2\gamma_1$. The corresponding trajectories do not meet any discontinuity manifold, the equations of their projections on the x_1,x_2-plane are :

$$x_1^*(t) = bm_2\gamma_2(T-t) + x^*(T)$$
$$x_2^*(t) = t - T + x_2^*(T)$$

Moreover there exists no optimal path emanating from a point in $E^2 \times D$, where $D \triangleq \{x : x \triangleq (x_1,x_2),\ x_1+bm_2\gamma_2x_2 < bm_2\gamma_2T,\ 0<x_1<L\}$ The projections of these paths on the x_1,x_2-plane are represented in figure 1 .

In case $\lambda_{o2} b\gamma_2 < \lambda_{o1} L^2\gamma_1$ then for $\beta \leqslant x_1^*(T) < L$, case 1) occurs, and the equations of the projections on the x_1,x_2-plane of the corresponding trajectories are the same as above.
For $0 < x_1^*(T) < \beta$, then for $t_1 < t < T$, the equations of the projections on the x_1,x_2-plane of the corresponding trajectories are :

$$x_1^*(t) = bm_2\gamma_2(T-t) + x_1^*(T)$$
$$x_2^*(t) = t - T + x_2^*(T)$$

and $\tilde{\lambda}_1(\cdot)$ takes the value zero on the manifold $E^2 \times M_1$, where

$$M_1 \triangleq \{x : x \triangleq (x_1,x_2),\ x_1-bm_2\gamma_2(T-x_2) = \frac{m_2}{m_2-m_1}\left(L-\left(\frac{\lambda_{o2}m_1L}{m_2(T-x_2)}\right)^{1/2}\right)$$
$$\frac{\lambda_{o2}b\gamma_2}{\lambda_{o1}L\gamma_1} < x_1 < L\}$$

Furthermore for $x \triangleq (x_1,x_2) \in M_1$ we have :

$$V_1^*(x) = m_1\gamma_1(T-x_2)\left(x_1 - \frac{bm_2\gamma_2(T-x_2)}{2}\right)$$

$$V_2^*(x) = \frac{m_2L - m_2x_1 + bm_2^2\gamma_2(T-x_2)}{m_2L + (m_1-m_2)(x_1-bm_2\gamma_2(T-x_2))}$$

Let $\tilde{\psi}^+(t_1) \triangleq \lim\limits_{\substack{t\to t_1\\ t>t_1}} \tilde{\psi}(t)$, $\tilde{\psi}^-(t_1) \triangleq \lim\limits_{\substack{t\to t_1\\ t<t_1}} \tilde{\psi}(t)$, where

$\tilde{\psi} \triangleq (\tilde{\lambda}_{o1}, \tilde{\lambda}_{o2}, \tilde{\lambda}_1, \tilde{\lambda}_2) : t \mapsto \tilde{\psi}(t),\ t \in [t_o, T]$, is the solution of adjoint equations satisfying Theorem 1, the projection on the x_1,x_2-plane of the corresponding Pareto optimal path be given by

$x^* : t \mapsto x^*(t),\ t \in [t_o,T]$, and $u_k^{*-}(t_1) = \lim\limits_{\substack{t\to t_1\\ t<t_1}} p_k^*(x^*(t))$,

$k = 1, 2.$

For $x^*(t_1) \triangleq (x_1, x_2) \in M_1$ we deduce from the transversality condition and adjoint equations that :

$$\tilde{\lambda}^{+}_{ok}(t_1) = \tilde{\lambda}_{ok}(t) \triangleq \lambda_{ok} \ , \ t_1 \leqslant t \leqslant T \qquad k=1,2$$

$$\tilde{\lambda}^{-}_{ok}(t_1) = \tilde{\lambda}_{ok}(t) \triangleq \lambda'_{ok} \ , \ t_1-\varepsilon \leqslant t \leqslant t_1, \ k=1,2, \quad \varepsilon > 0$$

$$\tilde{\lambda}^{+}_{1}(t_1) = 0$$

$$\tilde{\lambda}^{+}_{2}(t_1) = \tilde{\lambda}_{2}(T)$$

and from paragraph 14 that :

a) $$\lambda_{o1}\gamma_1 m_1 x_1 + \tilde{\lambda}^{+}_{2}(t_1) = 0$$

b) $$\lambda'_{o1}(m_1\gamma_1 - u^{*-}_1(t_1)x_1 + b\tilde{\lambda}^{-}_1(t_1)(u^{*-}_1(t_1) - u^{*-}_2(t_1)) + \tilde{\lambda}^{-}_2(t_1) = 0$$

c) $$\tilde{\lambda}^{-}_1(t_1) = \lambda'_{o1}m_1\gamma_1(T-x_2) - \frac{\lambda'_{o2}(T-x_2)}{\lambda_{o2}} + \lambda'_{o1}a_1 + \lambda'_{o2}a_2$$

$$\tilde{\lambda}^{-}_2(t_1) = bm_2\gamma_2(T-x_2)(\lambda'_{o1}m_1\gamma_1 - \frac{\lambda'_{o2}}{\lambda_{o2}})$$

$$+ \lambda'_{o1}a_1 + \lambda'_{o2}a_2 - \lambda'_{o1}m_1\gamma_1 x_1 \ ,$$

$$\lambda'_{o1}a_1 + \lambda'_{o2}a_2 = -\frac{A}{B} \ , \text{ where}$$

$$A = \lambda'_{o1}x_1(m_1\gamma_1 - u^{*-}_1(t_1)) - \lambda'_{o1}m_1\gamma_1 x_1 + bm_2\gamma_2(T-x_2)(\lambda'_{o1}m_1\gamma_1 - \frac{\lambda'_{o2}}{\lambda_{o2}})$$

$$+ b(u^{*-}_1(t_1) - u^{*-}_2(t_1))(T-x_2)(m_1\gamma_1\lambda'_{o1} - \frac{\lambda'_{o2}}{\lambda_{o2}}) \ ,$$

$$B = b(u^{*-}_1(t_1) - u^{*-}_2(t_1)) + \frac{3bm_2\gamma_2}{2} + (\frac{m_2 L}{m_2 - m_1} - x_1)\frac{1}{2(T-x_2)} \ ,$$

d) $$u^{*-}_1(t_1) = m_1\gamma_1 \text{ if } b\tilde{\lambda}^{-}_1(t_1) - \lambda'_{o1}x_1 \geqslant 0$$

$$= 0 \quad \text{if } b\tilde{\lambda}^{-}_1(t_1) - \lambda'_{o1}x_1 < 0$$

$$u^{*-}_2(t_1) = m_2\gamma_2 \text{ if } \tilde{\lambda}^{-}_1(t_1) < 0$$

$$= 0 \quad \text{if } \tilde{\lambda}^{-}_1(t_1) \geqslant 0$$

Moreover for $t = T$ we have :

$\lambda_{o1} m_1 \gamma_1 x_1^*(T) - \tilde{\lambda}_1(T) b m_2 \gamma_2 + \tilde{\lambda}_2(T) = 0$, and since

$\tilde{\lambda}_1(T) = -(T-x_2)$, $x \triangleq (x_1, x_2) \in M_1$, we have :

$\tilde{\lambda}_2(T) = b m_2 \gamma_2 (T-x_2)(\lambda_{o1} m_1 \gamma_1 - 1) - \lambda_{o1} m_1 \gamma_1 x_1$, $x \triangleq (x_1, x_2) \in M_1$,

so that we obtain from relation a) : $\lambda_{o1} = \dfrac{1}{m_1 \gamma_1}$.

In view of the expression of M_1, we have : $x_1 > 0$, $x \triangleq (x_1, x_2) \in M_1$, so that the value pair $(m_1\gamma_1, m_2\gamma_2)$ is incompatible with relations d) and consequently $(u_1^{*-}(t_1), u_2^{*-}(t_1)) \in \{(0,0), (m_1\gamma_1, 0), (0, m_2\gamma_2)\}$.

By using relations a) and b) relations c) may be written :

- for the value pair $(0,0)$: $\lambda'_{o1} a_1 + \lambda'_{o2} a_2 = 0$, $\dfrac{\lambda'_{o1}}{\lambda_{o1}} = \dfrac{\lambda'_{o2}}{\lambda_{o2}}$,

$$\tilde{\lambda}_1^-(t_1) = \tilde{\lambda}_1^+(t_1) = 0, \; \tilde{\lambda}_2^-(t_1) = \frac{\lambda'_{o1}}{\lambda_{o1}} \tilde{\lambda}_2^+(t_1) = \frac{\lambda'_{o2}}{\lambda_{o2}} \tilde{\lambda}_2^+(t) \; ;$$

- for the value pair $(m_1\gamma_1, 0)$: $\lambda'_{o1} a_1 + \lambda'_{o2} a_2 = 0$,

$$\frac{\lambda'_{o1} x_1}{\lambda_{o1}} = b m_1 \gamma_1 + m_2 \gamma_2 (T-x_2)\left(\frac{\lambda'_{o1}}{\lambda_{o1}} - \frac{\lambda'_{o2}}{\lambda_{o2}}\right) ,$$

$$\tilde{\lambda}_1^-(t_1) = (T-x_2)\left(\frac{\lambda'_{o1}}{\lambda_{o1}} - \frac{\lambda'_{o2}}{\lambda_{o2}}\right),$$

$$\tilde{\lambda}_2^-(t_1) = -\, b m_1 \gamma_1 (T-x_2)\left(\frac{\lambda'_{o1}}{\lambda_{o1}} - \frac{\lambda'_{o2}}{\lambda_{o2}}\right)$$

$$b\tilde{\lambda}_1^-(t_1) - \lambda_{o1} x_1 = -\lambda'_{o1} b m_2 \gamma_2 (T-x_2)\left(\frac{\lambda'_{o1}}{\lambda_{o1}} - \frac{\lambda'_{o2}}{\lambda_{o2}}\right)$$

- for the value pair $(0, m_2\gamma_2)$, M_1 is not a discontinuity manifold, therefore :

$\tilde{\lambda}_1^-(t_1) = \tilde{\lambda}_1^+(t_1)$, $\tilde{\lambda}_2^-(t_1) = \tilde{\lambda}_2^+(t_1)$, $a_1 = a_2 = 0$ and

$\lambda'_{ok} = \lambda_{ok}$, $k = 1, 2$. It is clear that the only value pair compatible with relations a) to d) is (0,0) , consequently $u_k^{*-}(t_1) = 0$, $k = 1, 2$, and case 3) occurs .

For $t_2 \leqslant t \leqslant t_1$ the equations of the projections on the x_1,x_2-plane of the trajectories are :

$$x_1^*(t) = x_1^*(t_1)$$

$$x_2^*(t) = t - t_1 + x_2^*(t_1)$$

with $t_1 - t_2 = \dfrac{x_1^*(t_1)}{bm_1\gamma_1}$, and $b\tilde{\lambda}_1(\cdot) - \lambda'_{o1}x_1^*(\cdot)$ takes the value zero on the manifold $E^2 \times M_2$, where

$$M_2 \triangleq \{x : x \triangleq (x_1,x_2),\ x_1(1 + \frac{bm_2\gamma_2}{m_1\gamma_1}) + bm_2\gamma_2(T-x_2) = \frac{m_2L}{m_2-m_1} +$$

$$+ \frac{1}{m_1-m_2}\left(\frac{\lambda_{o2}m_1^2m_2L\gamma_1}{m_1\gamma_1(T-x_2)-x_1}\right)^{1/2} ,\ \frac{\lambda_{o2}\gamma_2 b}{\lambda_{o1}\gamma_1 L} < x_1 < L\ \}$$

By similar arguments one can see that M_2 is a transition surface and that $u_1^{*-}(t_2) = m_1\gamma_1$ and $u_2^{*-}(t_2) = 0$, so that the equations of the projections on the x_1,x_2-plane of the trajectories are for $t \leqslant t_2$:

$$x_1^*(t) = bm_1\gamma_1(t-t_2) + x_1^*(t_2)$$

$$x_2^*(t) = t - t_2 + x_2^*(t_2)$$

Finally there exists no optimal path emanating from a point in the union of the sets :

$$D \triangleq \{x : x = (x_1,x_2),\ x_1 - bm_1\gamma_1(x_2 - T + \frac{L}{bm_1\gamma_1} + \frac{L-\beta}{bm_2\gamma_2}) \geq L$$

$$0 < x_1 < L \}$$

$$D' \triangleq \{x : x = (x_1,x_2),\ x_1 - bm_1\gamma_1(x_2 - T + \frac{\lambda_{o2}}{L}(\frac{m_1}{m_2} + \frac{b\gamma_2}{\gamma_1}) \leq \frac{\lambda_{o2}\gamma_2 b}{\lambda_{o1}\gamma_1 L},$$

$$0 < x_1 \leq \frac{\lambda_{o2}\gamma_2 b}{\lambda_{o1}\gamma_1} \}$$

$$D'' \triangleq \{x : x = (x_1,x_2),\ T - \frac{\lambda_{o2}}{L}(\frac{m_1}{m_2} + \frac{b\gamma_2}{\gamma_1}) \leq x_2 \leq T - \frac{\lambda_{o2}m_1}{m_2 L},$$

$$0 < x_1 \leq \frac{\lambda_{o2}\gamma_2 b}{\lambda_{o1}\gamma_1 L} \}$$

$$D''' \triangleq \{x : x = (x_1,x_2),\ x_1 + bm_2\gamma_2(x_2 - T) \leq 0,$$

$$0 < x_1 \leq \frac{\lambda_{o2}\gamma_2 b}{\lambda_{o1}\gamma_1 L} \}$$

The projections of paths on the x_1,x_2-plane and the transition surfaces M_1 and M_2 are shown in figure 2 .

16. Appendix

Let $z^*(t_f) + \eta \in C(\Sigma(C),\ z^*(t_f))$, then there exists $\eta^\nu \triangleq (\eta^\nu_{o1},\ldots\eta^\nu_{oN},\ldots\ \eta^\nu_1,\ldots\ \eta^\nu_n)$ and a strictly positive number ε^ν, $\nu \in \mathbf{N}$, such that $\eta^\nu \to \eta$ and $\varepsilon^\nu \to 0$ as $\nu \to \infty$, and $z^*(t_f) + \varepsilon^\nu\eta^\nu \in \Sigma(C)$, $\nu = 1,2,..\ell...$

Let $\chi^\nu \triangleq (\eta^\nu_{o1} + \varepsilon^\nu,\ldots\eta^\nu_{oN} + \varepsilon^\nu, \eta^\nu_1,\ldots\ \eta^\nu_n)$, then

$z^*(t_f) + \varepsilon^\nu\chi^\nu \in A/\Sigma(C)$, $\nu = 1,2,..\ell...$, and since $\chi^\nu \to \eta$ as $\nu \to \infty$, we can conclude that $z^*(t_f) + \eta \in C(A/\Sigma(C),\ z^*(t_f))$.

Hence

$$C(\Sigma(C),\ z^*(t_f)) \subset C(A/\Sigma(C),\ z^*(t_f))$$

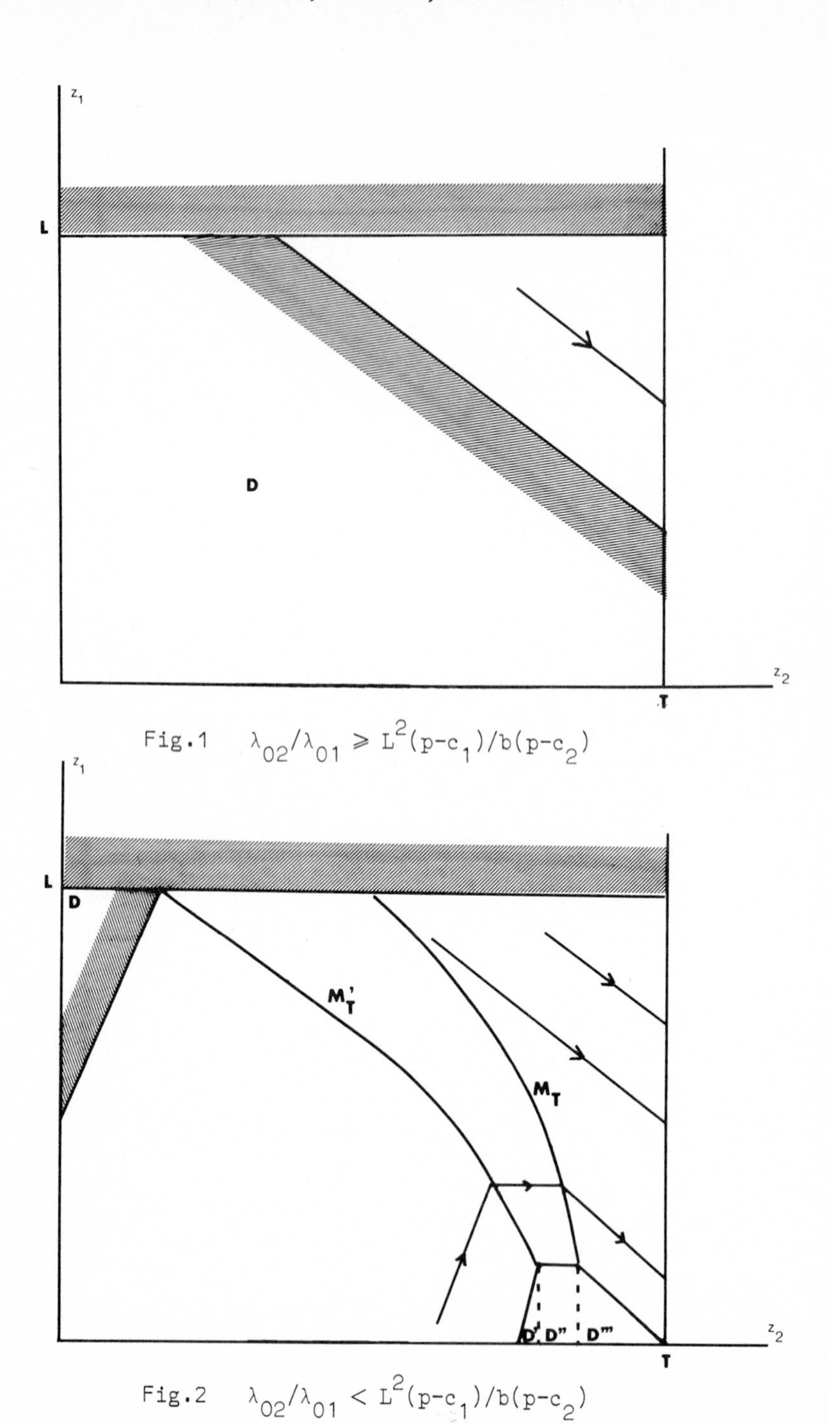

Fig.1 $\lambda_{02}/\lambda_{01} \geq L^2(p-c_1)/b(p-c_2)$

Fig.2 $\lambda_{02}/\lambda_{01} < L^2(p-c_1)/b(p-c_2)$

References

(1) A. BLAQUIERE, Sur la géométrie des surfaces de Pareto d'un jeu différentiel à N-joueurs, CRAS, Paris, t.271, p.744-746, 12 Octobre 1970, Série A.

(2) A. BLAQUIERE, L. JURICEK, and K.E. WIESE, Sur la géométrie des surfaces de Pareto d'un jeu différentiel à N-joueurs ; théorème du maximum, CRAS, Paris, t.271, p.1030-1032, 16 Novembre 1970, Série A.

(3) A. BLAQUIERE, F. GERARD, and G. LEITMANN, "Quantitative and Qualitative Games", Academic Press, New York, 1969.

(4) A. BLAQUIERE and G. LEITMANN, "Jeux quantitatifs", Mémorial des Sciences Mathématiques, Fasc. 168, Gauthier-Villars, Paris 1969

(5) A. HAURIE, Jeux Quantitatifs à M-Joueurs, Thèse de Doctorat de 3e Cycle, Université de Paris, Paris, June 1970

(6) Y.C. HO and G. LEITMANN, "Proceedings of the First International Conference on the Theory and Applications of Differential Games", University of Massachusetts, Amherst, Massachusetts, October 1969 .

(7) G. LEITMANN and T.L. VINCENT, "Control Space Properties of Cooperative Games", *J. Optimization Theory Appl.*, 6, n°2, August 1970.

(8) N.O. DA CUNHA and E. POLAK, Constrained minimization under vector-valued criteria in finite-dimensional spaces, *J. Math. Anal. Appl.*, 19, n°1, July 1967, 103-124 .

(9) N.O. DA CUNHA and E. POLAK, Constrained Minimization under Vector-valued Criteria in Linear Topological Spaces, in "Mathematical Theory of Control", Academic Press, New York, 1967 .

(10) L. ZADEH, "Optimality and Nonscalar Valued Performance Criteria", IEEE *Trans. Automatic Control*, ACS, 1963 .

(11) A. BLAQUIERE, L. JURICEK, and K.E. WIESE, Geometry of Pareto Equilibria and a Maximum Principle in N-Person Differential Games, *J. Math. Anal. Appl.*, 38, n°1, April 1972 .

(12) Y.C. HO, The First International Conference on the Theory and Applications of Differential Games, Final Report to Air Force Office of Scientific Research, AFOSR-69-1768, January 1970 .

(13) L. JURICEK, Jeux différentiels à N-joueurs coopératifs et non coopératifs, Thèse de Doctorat d'Etat, Université de Paris, 1972.

(14) R.D. LUCE and H. RAIFFA, "Games and Decisions", John Wiley & Sons, New York, 1967.

(15) A. RAPOPORT, "Two-Person Game Theory", Ann Arbor, The University of Michigan Press, 1966 .

GAMES WITH COALITIONS

L. JURICEK[†]

A characteristic of many player games is the possibility of cooperation among the players. Often this cooperation will take the form of coalitions, as von Neumann and Morgenstern were the first to point out.

In this chapter, which generalizes the notion of Pareto equilibrium, coalitions are discussed for the case of differential games.

We shall assume throughout that the players have perfect information. The game is defined as in the chapter on Pareto equilibria and one may employ the definitions and notations given there.

1. Definitions and Notations

Recall that following the introduction of the chapter on Pareto equilibria, the rules of the game specify the allowable coalitions at each point x of G. We shall assume that :

1 - the coalition constraint L is independent of x, so that card $L \triangleq I$ is finite[††]

2 - each player belongs to at least one coalition.

---

† Laboratoire d'Automatique Théorique, Université de Paris 7, France.

†† card L denotes the number of elements in the set L

Notations :

1) the coalitions will be indexed by i, $i = 1,\ldots,I$

2) for $C_i \in L$, $i = 1,\ldots,I$

$\alpha_i \triangleq \text{card } C_i$

β_i denotes the subset of $\{1,\ldots,N\}$ such that $k \in \beta_i \Leftrightarrow J_k \in C_i$

$\overline{C}_i$ denotes the set of players excluded from the coalition C_i

p_{C_i} is a strategy α_i-tuple selected by C_i in the set of functions $S_{C_i} \triangleq \prod_{k\in\beta_i} S_k$; the sets $S_1,\ldots,S_N$ being defined as in the paragraph 2 of the chapter on Pareto equilibria.

3) $E^i \triangleq E^{\alpha_i} \times E^n$, $i = 1,\ldots,I$, and $y^i \triangleq ((x_o)^i,x)$ denotes an element of E^i, where $(x_o)^i \triangleq (x_{ok})_{k\in\beta_i}$

4) $P^i(S)$ denotes the projection on E^i, $i = 1,\ldots,I$, of a set S, $S \subset E^{N+n}$.

Definition 1. Let $x \in G$ and $C_i \in L$. A strategy α_i-tuple p_{C_i} is called playable for the coalition C_i at point x, if there exists $p_{\overline{C}_i}$, $p_{\overline{C}_i} \in S_{\overline{C}_i} \triangleq S - S_{C_i}$, such that the strategy N-tuple $p \triangleq (p_{C_i}, p_{\overline{C}_i})$ is playable at point x. $J_{C_i}(x)$ denotes the set of strategy α_i-tuples playable at point x for the coalition C_i , $i = 1,\ldots,I$.

Definition 2. $\forall\, x \in G$, $\forall\, C_i \in L$, $\forall\, p_{C_i} \in J_{C_i}(x)$ and $\forall\, k \in \beta_i$, the gain guaranteed by the coalition C_i to player J_k is the value denoted $a_k(x,p_{C_i})$, of the function $a_k : (x,p_{C_i}) \mapsto R$, such that $a_k(x,p_{C_i}) \leqslant V_k(x,\theta,p_{C_i},p_{\overline{C}_i})$ for each $p_{\overline{C}_i} \in S_{\overline{C}_i}$ such that

$(p_{\mathcal{C}_i}, p_{\bar{\mathcal{C}}_i})$ is playable at point x.

Similarly $\forall x \in G$, $\forall \mathcal{C}_i \in L$, $\forall p_{\mathcal{C}_i} \in J_{\mathcal{C}_i}(x)$, the payoff guaranteed to $\mathcal{C}_i$ is the value of the function $a_{\mathcal{C}_i} : (x, p_{\mathcal{C}_i}) \mapsto R^{\alpha_i}$, where $a_{\mathcal{C}_i} \triangleq (a_k)_{k \in \beta_i}$

Definition 3. Let $x \in G$, p and $\bar{p}$ be playable strategy N-tuples at point x. We shall say that p dominates $\bar{p}$ if there exists $\mathcal{C}_i \in L$ such that :

i) $a_k(x, p_{\mathcal{C}_i}) \geqslant V_k(x, \theta, \bar{p})$, $k \in \beta_i$

ii) there exists $k_o \in \beta_i$ such that $a_{k_o}(x, p_{\mathcal{C}_i}) > V_{k_o}(x, \theta, \bar{p})$ where $p \triangleq (p_{\mathcal{C}_i}, p_{\bar{\mathcal{C}}_i})$

If $V_{\mathcal{C}_i}(x, \theta, \bar{p})$ denotes the α_i-tuple $(V_k(x, \theta, \bar{p}))_{k \in \beta_i}$, the relations i) and ii) will be written : $a_{\mathcal{C}_i}(x, p_{\mathcal{C}_i}) \succsim V_{\mathcal{C}_i}(x, \theta, \bar{p})$, and we shall say that, at point x, the coalition $\mathcal{C}_i$ blocks the strategy N-tuple $\bar{p}$.

Definition 4. A strategy N-tuple p^* is called core optimal on $X \subset G$ if :

i) p^* is playable at point x, for all $x \in X$

ii) $V_k(x, \theta, p^*) \triangleq V_k^*(x)$, $k = 1, \ldots, N$, is defined for all $x \in X$

iii) $\forall x \in X$, $\forall p$ playable at x, $\forall \mathcal{C}_i \in L$

$$a_{\mathcal{C}_i}(x, p_{\mathcal{C}_i}) \succsim V^*_{\mathcal{C}_i}(x) \Rightarrow a_{\mathcal{C}_i}(x, p_{\mathcal{C}_i}) = V^*_{\mathcal{C}_i}(x)$$

where $p \triangleq (p_{\mathcal{C}_i}, p_{\bar{\mathcal{C}}_i})$.

Definition 5. Let $\mathcal{C}_i \in L$. A strategy N-tuple $\bar{p}$ is called superparetian for the coalition $\mathcal{C}_i$ on $x \subset G$ if :

i) $\overline{p}$ is playable at point x, for all $x \in X$

ii) $V_k(x,\theta,\overline{p}) \triangleq \overline{V}_k(x)$, $k \in \beta_i$, is defined for all $x \in X$

iii) $\forall x \in X$, $\forall p$ playable at x

$$V_{C_i}(x,\theta,p) \succsim \overline{V}_{C_i}(x) \Rightarrow V_{C_i}(x,\theta,p) = \overline{V}_{C_i}(x)$$

Definition 6. A strategy N-tuple $\hat{p}$ is called a strong equilibrium on $X \subset G$ if :

i) $\hat{p}$ is playable at point x, for all $x \in X$

ii) $V_k(x,\theta,\hat{p}) \triangleq \hat{V}_k(x)$, $k = 1,\ldots,N$, is defined for all $x \in X$

iii) $\forall x \in X$, $\forall C_i \in L$, $\forall p_{C_i}$ such that $(p_{C_i}, \hat{p}_{\overline{C}_i})$ is playable at point x,

$$V_{C_i}(x,\theta,p_{C_i},\hat{p}_{\overline{C}_i}) \geqslant \hat{V}_{C_i}(x) \Rightarrow V_{C_i}(x,\theta,p_{C_i},\hat{p}_{\overline{C}_i}) = \hat{V}_{C_i}(x)$$

2. Preliminary Results

Lemma 1. *Let* $X \subset G$ *and* $C_i \in L$. *If* $\overline{p}$ *is a superparetian strategy N-tuple for* C_i *then* $\overline{p}$ *is not blocked by* C_i *on* X.

Demonstration : suppose that $\overline{p}$ is blocked by C_i on X. Then in view of Definition 3, there exist $x \in X$ and $p_{C_i} \in J_{C_i}(x)$ such that :

$$a_{C_i}(x,p_{C_i}) \succsim \overline{V}_{C_i}(x) \triangleq V_{C_i}(x,\theta,\overline{p})$$

From Definitions 1 and 2 one deduces that there exists $p_{\overline{C}_i} \in S_{\overline{C}_i}$ such that $p \triangleq (p_{C_i}, p_{\overline{C}_i})$ is playable at point x and $V_{C_i}(x,\theta,p) \geqslant a_{C_i}(x,p_{C_i})$. Consequently there exist $x \in X$ and $p \triangleq (p_{C_i}, p_{\overline{C}_i})$ playable at x such that :

$$V_{\mathcal{C}_i}(x,\theta,p) > a_{\mathcal{C}_i}(x,p_{\mathcal{C}_i}) \nless \overline{V}_{\mathcal{C}_i}(x)$$

which contradicts Definition 5.

Lemma 2. *If* $\overset{*}{p}$ *is a strategy* N*-tuple such that :*

i) $\overset{*}{p}$ *is playable at point* x*, for all* $x \in X$

ii) $V_k(x,\theta,\overset{*}{p})$, $k = 1,\ldots,N$, *is defined for all* $x \in X$

iii) $\forall x \in X$, $\forall C_i \in L$, $\overset{*}{p}$ *is superparetian for* C_i *at point* x
then $\overset{*}{p}$ *is core optimal on* X.

Demonstration : in view of the condition iii) of Lemma 2, the condition iii) of Definition 5 can be written : $\forall x \in X$, $\forall p$ playable at x, $\forall C_i \in L$

$$V_{\mathcal{C}_i}(x,\theta,p) \nless V_{\mathcal{C}_i}(x,\theta,p^*) \Rightarrow V_{\mathcal{C}_i}(x,\theta,p) = V_{\mathcal{C}_i}(x,\theta,p^*)$$

and the proposition follows from Definitions 2, 3 and 4.

3. Study of Certain Core Surfaces

The study is limited to strategy N-tuples $\overset{*}{p}$ satisfying Lemma 2.

3.1 - Core surfaces

Definition 7. If $\overset{*}{p}$ is a strategy N-tuple satisfying Lemma 2 and $C \in R^N$, the subset $\overset{*}{\Sigma}(C)$ of E^{N+n} defined by :

$$\overset{*}{\Sigma}(C) \triangleq \{z : z \triangleq (x_o,x),\ x \in X \cup \theta,\ x_{ok} + \overset{*}{V}_k(x) = C_k,\ \ k=1,\ldots,N\}$$

where $\overset{*}{V}_k(x) \triangleq V_k(x,\theta,\overset{*}{p})$, $k=1,\ldots,N$, and $C \triangleq (C_1,\ldots,C_N)$, is called a core surface.

We define also a subset $A/\overset{*}{\Sigma}(C)$ of E^{N+n} by :

$$A/\overset{*}{\Sigma}(C) \triangleq \{z : z \triangleq (x_o,x),\ x \in X \cup \theta,\ x_{ok} + \overset{*}{V}_k(x) > C_k,\ k=1,\ldots,N\}$$

Definition 8. Similarly if $\overset{*}{p}$ is a strategy N-tuple satisfying Lemma 2 and $C \in R^N$, we define

1) in E^i, i=1,...,I, a surface $\overset{*}{\Sigma}{}^i(C)$ and a set $A/\overset{*}{\Sigma}{}^i(C)$ by :

$$\overset{*}{\Sigma}{}^i(C) \triangleq \{y^i : y^i \triangleq ((x_o)^i, x),\ x \in X \cup \theta,\ x_{ok}+\overset{*}{V}_k(x) = C_k,\ k \in \beta_i\}$$

$$A/\overset{*}{\Sigma}{}^i(C) \triangleq \{y^i : x \in X \cup \theta,\ x_{ok}+\overset{*}{V}_k(x) > C_k,\ k \in \beta_i\}$$

2) in E_k, k=1,...,N, a surface $\overset{*}{\Sigma}_k(C)$ and a set $A/\overset{*}{\Sigma}_k(C)$ by :

$$\overset{*}{\Sigma}_k(C) \triangleq \{y_k : y_k \triangleq (x_{ok}, x),\ x \in X \cup \theta,\ x_{ok}+\overset{*}{V}_k(x) = C_k\}$$

$$A/\overset{*}{\Sigma}_k(C) \triangleq \{y_k : x \in X \cup \theta,\ x_{ok}+\overset{*}{V}_k(x) > C_k\}$$

As the values of parameters $(C_1,\ldots,C_N)$ are varied one obtains an N-parameter family of surfaces $\{\overset{*}{\Sigma}(C)\}$, I α_i-parameter families of surfaces $\{\overset{*}{\Sigma}{}^i(C)\}$, i=1,...,I, and N one-parameter families of surfaces $\{\overset{*}{\Sigma}_k(C)\}$, k=1,...,N. Some global properties of a core surface $\overset{*}{\Sigma}(C)$ are contained in :

Lemma 3. *Let* $\overset{*}{\Pi}(C)$ *be a terminating path generated by* $\overset{*}{p}$*, given by* $\overset{*}{z} : t \mapsto \overset{*}{z}(t),\ t \in [t_o, t_f]$, $\overset{*}{z}(t_f) = z^f \triangleq (C, x^f) \in \Theta$.
Then, $\overset{*}{\Pi}(C) \subset \overset{*}{\Sigma}(C)$.

Lemma 4. *Let* $z \in A/\overset{*}{\Sigma}(C)$ *and* $\overset{*}{\Sigma}(\overline{C})$ *be the core surface containing* z.
Then, $\overset{*}{\Sigma}(\overline{C}) \subset A/\overset{*}{\Sigma}(C)$

Lemma 5. *If* $z \in \overset{*}{\Sigma}(C)$ *then for every path* $\Pi(C')$ *emanating from* z *we have :*

$$\Pi(C') \cap A/\overset{*}{\Sigma}(C) = \phi$$

$$P^i(\Pi(C')) \cap A/\overset{*}{\Sigma}{}^i(C) = \phi \quad , \ i=1,\ldots,I$$

Lemma 6. *If* $\overset{*}{\Sigma}(C)$ *is a core surface then* $\overset{*}{\Sigma}{}^i(C)$, i=1,...,I, *is a Pareto surface in* E^i.

3.2 - Some local properties of a core surface.

Let us define

1) a subset $\hat{Y}$ of $\prod_{i=1}^{I} \mathcal{E}^i$ by :

$$\hat{Y} \triangleq \{(\mu^1,\ldots,\mu^I) : \mu^i \triangleq ((\xi_o)^i,\xi),\ (\xi_o)^i \in R^i,\ i=1,\ldots,I,\ \xi \in E^n\}$$

2) a mapping $\hat{P}: E^{N+n} \to \hat{Y}$ such that $\hat{P}(\eta) \triangleq (P^1(\eta),\ldots,P^I(\eta))$, $\eta \in E^{N+n}$.

3) the topology of

i) E^{N+n} as the topology induced by the distance d :

$$d(z,z') \triangleq \max\{(|x_{ok}-x'_{ok}|)_{k=1,\ldots,N},(|x_j-x'_j|)_{j=1,\ldots,n}\}$$

$$z \triangleq (x_o,x) \in E^{N+n},\ z' \triangleq (x'_o,x') \in E^{N+n}$$

ii) $\mathcal{E}^i$, i=1,...,I, as the topology induced by the distance d^i :

$$d^i(y^i,y'^i) \triangleq \max\{(|x_{ok}-x'_{ok}|)_{k\in\beta_i},(|x_j-x'_j|)_{j=1,\ldots,n}\}$$

$$y^i \triangleq (x_{ok})_{k\in\beta_i},\ x) \in \mathcal{E}^i,\ y'^i \triangleq ((x'_{ok})_{k\in\beta_i},\ x') \in \mathcal{E}^i$$

iii) $\prod_{i=1}^{I} \mathcal{E}^i$ as the product topology.

The topology of $\hat{Y}$ is the topology induced on $\hat{Y}$ by the topology of $\prod_{i=1}^{I} \mathcal{E}^i$

One can prove easily the following results :

Lemma 7. $\hat{P} : E^{N+n} \to \hat{Y}$ *is an isometry.*

Lemma 8. $\hat{P}(A/\overset{*}{\Sigma}(C)) = \hat{Y} \cap \prod_{i=1}^{I} A/\overset{*}{\Sigma}{}^i(C)$

Lemma 9. $\hat{P}(V_\delta(z+\eta)) = \hat{Y} \cap \prod_{i=1}^{I} V_\delta(P^i(z+\eta))$, $\delta > 0$, *where*

$$V_\delta(z+\eta) \triangleq \{z + \alpha\eta' : 0 < \alpha < \delta,\ z+\eta' \in B_\delta(z+\eta)\}$$

Let $\overset{*}{\Sigma}(C)$ be a terminating path generated by $\overset{*}{p}$, represented by $\overset{*}{z}$: $t \mapsto \overset{*}{z}(t)$, $t \in [t_o, t_f]$, $\overset{*}{z}(t_o) \in E^N \times X$, $\overset{*}{z}(t_f) = z^f \triangleq (C, x^f) \in \Theta$ and $\overset{*}{\pi}$ its projection on $X \cup \theta$.

Assumption 1. From now on we shall assume that $\overset{*}{p}$ agrees on $\Delta \cap X$, where Δ is a neighborhood of $\overset{*}{\pi}$ in G, with a function of class C^1 on Δ, and that $\overset{*}{\pi}$ does not meet θ tangentially.

Lemma 10. $\hat{P}(C(A/\overset{*}{\Sigma}(C), z)) = \hat{Y} \cap \prod_{i=1}^{I} C(A/\overset{*}{\Sigma}^i(C), P^i(z))$,
$z = \overset{*}{z}(t)$, $t \in [t_o, t_f)$.

Demonstration : as $z+\eta \in C(A/\overset{*}{\Sigma}(C), z) \iff V_\delta > 0$,
$V_\delta(z+\eta) \cap A/\overset{*}{\Sigma}(C) \neq \phi$, the proposition follows from Lemmas 7 to 9.

Lemma 11. *The contingent of* $A/\overset{*}{\Sigma}(C)$ *at the point* $z = \overset{*}{z}(t)$,
$t \in [t_o, t_f)$, *is a closed convex cone in* E^{N+n}.

The proposition is a direct consequence of Lemmas 6, 7 and 10 and of Lemma 4 of the chapter on Pareto equilibria.

Lemma 12. $z+\eta \in \overset{\circ}{C}(A/\overset{*}{\Sigma}(C), z) \neq \phi$, $z = \overset{*}{z}(t), t \in [t_o, t_f) \Rightarrow \exists\ \delta_o > 0$
$V_{\delta_o}(z+\eta) \subset A/\overset{*}{\Sigma}(C)$.

One may easily deduce this lemma from Lemmas 6 to 10, and from Lemma 5 of the chapter on Pareto equilibria.

3.3 - *Separability of contingents*

Definition 9. Let $S \subset E^m$, $s \in \overline{S}$ and $C(S,s)$. The set $I(S,s)$ defined by :

$$I(S,s) \triangleq \{s + \nu : s + \nu \in C(S,s),\ s - \nu \in C(S,s)\}$$

is called the symmetrical subset of $C(S,s)$.

From Definition 9, Lemmas 7 and 10, one deduces directly :

Lemma 13. $\hat{P}(I(A/\overset{*}{\Sigma}(C),z)) = \hat{Y} \cap \prod_{i=1}^{I} I(A/\overset{*}{\Sigma}{}^{i}(C),P^{i}(z))$,

$z = \overset{*}{z}(t),\ t \in [t_o,t_f)$.

From Definition 9 and Assumption 1 one can also deduce:

Lemma 14. $C(\overset{*}{\Pi}(C),z) \subset I(A/\overset{*}{\Sigma}(C),z),\ z = \overset{*}{z}(t),\ t \in [t_o,t_f)$.

Lemma 15. $C(\overset{*}{\Sigma}(C) \cap \Theta,\overset{*}{z}(t_f)) \subset I(A/\overset{*}{\Sigma}(C),\overset{*}{z}(t_f))$

Lemma 16. $\dim I(A/\overset{*}{\Sigma}(C),z) = n,\ z = \overset{*}{z}(t),\ t \in [t_o,t_f)$.

Demonstration : following Lemma 6 and the lemma 7 of the chapter on Pareto equilibria, $\dim I(A/\overset{*}{\Sigma}{}^{i}(C),P^{i}(z)) = n$, $i=1,\ldots,I$. Let $F \triangleq (\eta_\ell)_{\ell=1,\ldots,n}$ be the family of vectors in E^{N+n} such that :

i) $\eta_\ell \triangleq (\xi_o^\ell,\xi^\ell),\ \xi^\ell = (\delta_j^\ell)_{j=1,\ldots,n}\quad \ell = 1,\ldots,n$

ii) $P^{i}(z+F) \subset I(A/\overset{*}{\Sigma}{}^{i}(C),P^{i}(z)),\ i=1,\ldots,I$

Clearly $P^{i}(z+F)$ is a basis of $I(A/\overset{*}{\Sigma}{}^{i}(C),P^{i}(z))$, $i=1,\ldots,I$, and F a set of n independant non null vectors in E^{N+n}. The lemma follows directly from Lemmas 7 and 13.

Corollary 1. *Let* $z \triangleq (x_o,x) = \overset{*}{z}(t),\ t \in [t_o,t_f)$ and $F' \triangleq (N_h)_{h=1,\ldots,N}$ *be the family of vectors in* E^{N+n} *such that* $N_h \triangleq ((\delta_k^h)_{k=1,\ldots,N},\ \mathrm{grad}\ \overset{*}{V}_h(x)),\ h=1,\ldots,N$. *Then* $z+F'$ *constitutes a basis of the orthogonal complement of* $I(A/\overset{*}{\Sigma}(C),z)$ *in* E^{N+n}.

In view of Lemma 7 (chapter on Pareto equilibria) and of the hypothesis that $\bigcup_{i=1}^{I} \beta_i = \{1,\ldots,N\}$, the proposition is a direct consequence of Lemmas 13 and 16.

Lemma 17. *Let* $C^*(A/\overset{*}{\Sigma}(C),z) \triangleq \{z + n : n\cdot\eta \geqslant 0,\ z+\eta \in C(A/\overset{*}{\Sigma}(C),z)\}$ $z = \overset{*}{z}(t),\ t \in [t_o,t_f)$. *If* $z + n \in C^*(A/\overset{*}{\Sigma}(C),z)$ *then*

$$n = \sum_{h=1}^{N} a_h N_h\ ,\quad a_h \geqslant 0,\quad h=1,\ldots,N$$

Demonstration : for $z + \eta \in C(A/\overset{*}{\Sigma}(C),z)$, $z \triangleq (x_o,x)$, $\eta \triangleq (\xi_o,\xi)$, one has from the definition of F' : $N_h\cdot\eta = \xi_{oh} + \operatorname{grad} \overset{*}{V}_h(x)\cdot\xi$, $h=1,\ldots,N$. In view of Lemma 2 of the chapter on Pareto equilibria and of Lemma 10, $\xi_{oh} + \operatorname{grad} \overset{*}{V}_h(x)\cdot\xi \geqslant 0$, $h=1,\ldots,N$, so that $z + F' \subset C^*(A/\overset{*}{\Sigma}(C),z)$. $C^*(A/\overset{*}{\Sigma}(C),z)$ being a closed convex cone in E^{N+n} in virtue of Lemma 11, the proposition is established.

Corollary 2. *If* $I(z,\ N+n-1)$ *is a supporting hyperplane of* $C(A/\overset{*}{\Sigma}(C),z)$, $z = \overset{*}{z}(t)$, $t \in [t_o,t_f)$ *and* n *the normal vector to* $I(z,\ N+n-1)$ *such that* $z + n \in D(C(A/\overset{*}{\Sigma}(C),z),\ I(z,\ N+n-1))$, *then*

1) $n = \sum_{h=1}^{N} a_h N_h\ ,\ a_h \geqslant 0,\quad h=1,\ldots,N$

2) $n^i = \sum_{h\in\beta_i} a_h P^i(N_h)$ *is a normal vector to a supporting hyperplane of* $C(A/\overset{*i}{\Sigma^i}(C),P^i(z))$, $i=1,\ldots,I$.

The proposition follows from Lemmas 10 and 17.

3.4 - *Variational equations and their adjoints*

In addition to equations (10) (11) and (14) (15) of the chapter on paretian equilibria, we shall consider in E^i, $i=1,\ldots,I$, the variational equations :

$$\frac{d\xi_{ok}}{dt} = \sum_{\alpha=1}^{n} \left(\frac{\partial f_{ok}}{\partial x_\alpha} + \sum_{\beta=1}^{N} \sum_{\gamma=1}^{r_\beta} \frac{\partial f_{ok}}{\partial u_{\beta_\gamma}} \frac{\partial \overset{*}{p}_{\beta_\gamma}}{\partial x_\alpha} \right) \xi_\alpha\ ,\quad k \in \beta_i$$

$$\frac{d\xi_\nu}{dt} = \sum_{\alpha=1}^{n} \left(\frac{\partial f_\nu}{\partial x_\alpha} + \sum_{\beta=1}^{N} \sum_{\gamma=1}^{r_\beta} \frac{\partial f_\nu}{\partial u_{\beta_\gamma}} \frac{\partial p^*_{\beta_\gamma}}{\partial x_\alpha}\right) \xi_\alpha \ , \quad \nu = 1,\ldots,n$$

and the adjoint equations :

$$\frac{d\lambda_{ok}}{dt} = 0, \ k \in \beta_i$$

$$\frac{d\lambda_\nu}{dt} = - \sum_{k\in\beta_i} \left(\frac{\partial f_{ok}}{\partial x_\nu} + \sum_{\beta=1}^{N} \sum_{\gamma=1}^{r_\beta} \frac{\partial f_{ok}}{\partial u_{\beta_\gamma}} \frac{\partial p^*_{\beta_\gamma}}{\partial x_\nu}\right) \lambda_{ok}$$

$$- \sum_{\alpha=1}^{n} \left(\frac{\partial f_\alpha}{\partial x_\nu} + \sum_{\beta=1}^{N} \sum_{\gamma=1}^{r_\beta} \frac{\partial f_\alpha}{\partial u_{\beta_\gamma}} \frac{\partial p^*_{\beta_\gamma}}{\partial x_\nu}\right) \lambda_\alpha \ , \quad \nu=1,\ldots,n$$

where partial derivatives are evaluated for $z = \overset{*}{z}(t)$, $u_k = p^*_k(\overset{*}{z}(t))$, $k=1,\ldots,N$, $t \in [t_o,t_f)$.

The fundamental matrix of the variational equations in E^i denoted $A^i(t,t')$, $(t,t') \in [t_o,t_f] \times [t_o,t_f]$, $i=1,\ldots,I$, has the same properties as $A(t,t')$ (Section 10 of the chapter on Pareto equilibria).

3.5 - *Transformation of a contingent*

Lemma 18. $A(t,t')C(A/\overset{*}{\Sigma}(C),\overset{*}{z}(t')) \subset C(A/\overset{*}{\Sigma}(C),\overset{*}{z}(t''))$, $t' \in [t_o,t_f]$, $t'' \in [t_o,t_f)$. Using Lemmas 3 and 4 the proof is the same as in Section 11 of the chapter on Pareto equilibria.

Corollary 3. $A(t',t'')C(A/\overset{*}{\Sigma}(C),\overset{*}{z}(t')) = C(A/\overset{*}{\Sigma}(C),\overset{*}{z}(t''))$, $t' \in [t_o,t_f)$, $t'' \in [t_o,t_f)$.

The proposition is a direct consequence of Lemma 18 and of the fact that $\overset{*}{z}(t')$ and $\overset{*}{z}(t'')$ are interior points of $\overset{*}{\Sigma}(C)$.

From Lemmas 11 and 18 and from the properties of $A(t,t')$ one deduces the following :

Corollary 4. $C(A/\overset{*}{\Sigma}(C),\overset{*}{z}(t_f)) \subset A(t,t_f)C(A/\overset{*}{\Sigma}(C),\overset{*}{z}(t))$, $t \in [t_o,t_f)$

Corollary 5. $\Gamma(A/\overset{*}{\Sigma}(C),\overset{*}{z}(t_f)) \triangleq A(t,t_f)C(A/\overset{*}{\Sigma}(C),\overset{*}{z}(t))$, $t \in [t_o,t_f)$ *is a closed convex cone in* E^{N+n} *independent of* t .

Corollary 6. $I(A/\overset{*}{\Sigma}(C),\overset{*}{z}(t_f)) \subset A(t,t_f)I(A/\overset{*}{\Sigma}(C),\overset{*}{z}(t))$, $t \in [t_o,t_f)$.

3.6 - Separability of local cones

Assumption 2. From now on we shall assume that $\forall z \triangleq (x_o,x) \in \overset{*}{\Pi}(C)$ $\forall u \triangleq (u_1,\ldots,u_N) \in K(z) \triangleq K(x)$,
$\exists\ \alpha(z,u,\cdot) \triangleq (\alpha_1(x,u,\cdot),\ldots,\alpha_N(x,u,\cdot)) \in S$ of class C^1 on some neighborhood of z in $E^N \times G$ such that $\alpha(z,u,z) = u$.

In addition to the sets $R(t)$, $L(t)$, $K(\overset{*}{z}(t))$ and $V(\overset{*}{z}(t))$ defined in Section 12 of the chapter on Pareto equilibria, we shall consider for $i=1,\ldots,I$, the following sets :

$$R^i(t) \triangleq \{v^i : v^i \triangleq A^i(\tau,t)F^i(\overset{*}{z}(\tau),u),\ u \in K(\overset{*}{z}(\tau)),\ \tau \in [t_o,t_f)\},\ t \in [t_o,t_f\,]$$

where $F^i \triangleq ((f_{ok})_{k\in\beta_i},f)$, $i=1,\ldots,I$,

$$L^i(t) \triangleq \text{the convex cone generated by } R^i(t),\ t \in [t_o,t_f\,],$$

$$K^i(P^i(\overset{*}{z}(t))) \triangleq P^i(\overset{*}{z}(t)) + L^i(t),\ t \in [t_o,t_f\,],$$

$$V^i(P^i(\overset{*}{z}(t))) \triangleq \{P^i(\overset{*}{z}(t)) + \gamma F^i(\overset{*}{z}(t),u) : \gamma \geqslant 0,\ u \in K(\overset{*}{z}(t))\}\ ,\ t \in [t_o,t_f)$$

From these definitions it follows that :

1) $R^i(t) = P^i(R(t))$, $L^i(t) = P^i(L(t))$, $K^i(P^i(\overset{*}{z}(t))) = P^i(K(\overset{*}{z}(t)))$, $t \in [t_o,t_f\,]$, $i=1,\ldots,I$,

2) $K(\overset{*}{z}(t))$ and $K^i(P^i(\overset{*}{z}(t))$ are closed convex cones, $t \in [t_o,t_f]$,

3) $A(t,t_f)V(\overset{*}{z}(t)) \subset K(\overset{*}{z}(t_f))$ and

$A^i(t,t_f)V^i(P^i(\overset{*}{z}(t))) \subset K^i(P^i(\overset{*}{z}(t_f)))$, $t \in [t_o,t_f)$, $i=1,..,I$.

Lemma 19. $K(\overset{*}{z}(t)) \cap \overset{\circ}{C}(A/\overset{*}{\Sigma}(C),z(t)) = \phi$, $t \in [t_o,t_f)$.

Demonstration : as $L(t_o) = \phi$, the proposition follows for $t = t_o$ from the definition of $K(\overset{*}{z}(t))$ and from Lemma 11. Suppose that there exists t, $t \in (t_o,t_f)$, such that

$\overset{*}{z}(t) + \rho \in K(\overset{*}{z}(t)) \cap \overset{\circ}{C}(A/\overset{*}{\Sigma}(C),\overset{*}{z}(t)) \neq \phi$. According to the definition of $K(\overset{*}{z}(t))$ there exist a family $(\gamma_h)_{h=1,\ldots,s}$, $\gamma_h \geqslant 0$ $h=1,\ldots,s$, and a family $(v_h)_{h=1,\ldots,s}$, $v_h \in R(t)$, $h=1,\ldots,s$, such that $\rho = \sum_{h=1}^{s} \gamma_h v_h$, where $v_h \triangleq A(\tau_h,t)F(\overset{*}{z}(\tau_h),u^h)$,

$u^h \in K(\overset{*}{z}(\tau_h))$, $h=1,\ldots,s$, the indexes being selected so that $t_o \leqslant \tau_1 \leqslant \tau_2 \leqslant \ldots \leqslant \tau_s < t < t_f$.

Let $\gamma_o \triangleq 0$, $\varepsilon > 0$,

$I(\tau_h) \triangleq \{z : \tau_h+(\gamma_o+\ldots+\gamma_{h-1})\varepsilon \leqslant x_n < \tau_h+(\gamma_o+\ldots+\gamma_h)\varepsilon\}$, ,$h=1,\ldots s$

and the strategy N-tuple $\widetilde{p} \in S$ such that

$$\widetilde{p}(z) = \alpha(\overset{*}{z}(\tau_h),u^h,z) \quad \text{for} \quad z \in I(\tau_h),\ h=1,\ldots,s$$

$$= \overset{*}{p}(z) \quad \text{for} \quad z \notin \bigcup_{h=1}^{s} I(\tau_h)$$

where $\alpha(\overset{*}{z}(\tau_h),u^h,\cdot) \in S$ and satisfies Assomption 2 for $h=1,\ldots,s$.

From Assumptions 1 and 2 and from the fact that $\overset{*}{z}(t)$ is an interior point of $\overset{*}{\Sigma}(C)$, it follows that, for ε sufficiently small, there exists a path $\Pi(C')$ generated by $\widetilde{p}$, given by $\widetilde{z} : t \mapsto \widetilde{z}(t)$, $t \in [t_o,\ t + (\gamma_o+\ldots+\gamma_s)\varepsilon]$ such that $\widetilde{z}(t_o) = \overset{*}{z}(t_o)$. Then

$\widetilde{z}(t + (\gamma_o+\ldots+\gamma_s)\varepsilon) = \overset{*}{z}(t) + \varepsilon\rho + o(\varepsilon)$.

As $\overset{*}{z}(t) + \rho \in \overset{\circ}{C}(A/\overset{*}{\Sigma}(C), \overset{*}{z}(t))$, it follows from Lemma 12 that, for ε sufficiently small, we have $\overset{*}{z}(t) + \varepsilon(\rho + \frac{o(\varepsilon)}{\varepsilon}) \in A/\overset{*}{\Sigma}(C)$ so that $\Pi(C') \cap A/\overset{*}{\Sigma}(C) \neq \phi$ which contradicts Lemma 5.

Corollary 7. $K^i(P^i(\overset{*}{z}(t)) \cap \overset{\circ}{C}(A/\overset{*}{\Sigma}{}^i(C), P^i(\overset{*}{z}(t))) = \phi$, $t \in [t_o, t_f)$, $i=1,\ldots,I$.

Corollary 7 is a direct consequence of Lemmas 10 and 19 and of the fact that $K^i(P^i(\overset{*}{z}(t)) = P^i(K(\overset{*}{z}(t)))$, $t \in [t_o, t_f]$, $i=1,\ldots I$

Lemma 20. $K(\overset{*}{z}(t_f)) \cap \overset{\circ}{\Gamma}(A/\overset{*}{\Sigma}(C), \overset{*}{z}(t_f)) = \phi$

By invoking Lemma 19 the demonstration is the same as in Lemma 14 of the chapter on Pareto equilibria.

Corollary 8. $K^i(P^i(\overset{*}{z}(t_f))) \cap \overset{\circ}{\Gamma}(A/\overset{*}{\Sigma}{}^i(C), P^i(\overset{*}{z}(t_f))) = \phi$, $i=1,\ldots,I$ *where*

$\Gamma(A/\overset{*}{\Sigma}{}^i(C), P^i(\overset{*}{z}(t_f))) \triangleq A^i(t,t_f)C(A/\overset{*}{\Sigma}{}^i(C), P^i(\overset{*}{z}(t)))$,

$t \in [t_o, t_f)$, $i=1,\ldots,I$.

This corollary follows from Lemmas 10 and 20 and from the fact that: $A^i(t_f,t)P^i(\overset{*}{z}(t_f) + \eta) = P^i(\overset{*}{z}(t) + A(t_f,t)\eta)$, $\eta \in E^{N+n}$, $t \in [t_o, t_f]$, $i=1,\ldots,I$.

From the definition of $K(\overset{*}{z}(t_f))$ and from Lemmas 11 and 20, one deduces :

Lemma 21. $K(\overset{*}{z}(t_f))$ *and* $\Gamma(A/\overset{*}{\Sigma}(C), \overset{*}{z}(t_f))$ *are separated in* E^{N+n} .

Let $I(\overset{*}{z}(t_f), N+n-1)$ be an $(N+n-1)$-dimensional hyperplane separating $K(\overset{*}{z}(t_f))$ and $\Gamma(A/\overset{*}{\Sigma}(C), \overset{*}{z}(t_f))$ in E^{N+n} .

From the definition of $I(\overset{*}{z}(t_f), N+n-1)$, from Corollaries 4 and 5, and the fact that $A(t,t_f)V(\overset{*}{z}(t)) \subset K(\overset{*}{z}(t_f))$, $t \in [t_o,t_f)$, one readily deduces :

Corollary 9. *$C(A/\overset{*}{\Sigma}(C),\overset{*}{z}(t_f))$ and $A(t,t_f)V(\overset{*}{z}(t))$, $t \in [t_o,t_f)$, are separated in E^{N+n}* by *$I(\overset{*}{z}(t_f), N+n-1)$*

Corollary 10. *$V(\overset{*}{z}(t))$ and $C(A/\overset{*}{\Sigma}(C),\overset{*}{z}(t))$, $t \in [t_o,t_f)$, are separated in E^{N+n} by $I(\overset{*}{z}(t), N+n-1) \triangleq A(t_f,t)I(\overset{*}{z}(t_f), N+n-1)$.*

By similar arguments we have :

Lemma 22. *$K^i(P^i(\overset{*}{z}(t_f)))$ and $\Gamma(A/\overset{*}{\Sigma}{}^i(C), P^i(\overset{*}{z}(t_f)))$ are separated in E^i, $i=1,\dots,I$.*

For $i=1,\dots,I$, let $I(P^i(\overset{*}{z}(t_f)),\alpha_i+n-1)$ be an (α_i+n-1)-dimensional hyperplane separating $K^i(P^i(\overset{*}{z}(t_f)))$ and $\Gamma(A/\overset{*}{\Sigma}{}^i(C),P^i(\overset{*}{z}(t_f)))$ in E^i.

Corollary 11. *$C(A/\overset{*}{\Sigma}{}^i(C),P^i(\overset{*}{z}(t_f)))$ and $A^i(t,t_f)V^i(P^i(\overset{*}{z}(t)))$, $t \in [t_o,t_f)$, are separated in E^i by $I(P^i(\overset{*}{z}(t_f)),\alpha_i+n-1)$, $i=1,..I$.*

Corollary 12. *$V^i(P^i(\overset{*}{z}(t)))$ and $C(A/\overset{*}{\Sigma}{}^i(C),P^i(\overset{*}{z}(t)))$, $t \in [t_o,t_f)$, are separated in E^i by*

$$I(P^i(\overset{*}{z}(t)),\alpha_i+n-1) \triangleq A^i(t_f,t)I(P^i(\overset{*}{z}(t_f)),\alpha_i+n-1),\ i=1,\dots,I.$$

Let $I(\overset{*}{z}(t), N+n-1)$ be a separating hyperplane of $V(\overset{*}{z}(t))$ and $C(A/\overset{*}{\Sigma}(C),\overset{*}{z}(t))$, $t \in [t_o,t_f)$, and n the normal vector to this hyperplane such that $\overset{*}{z}(t)+n \in D(C(A/\overset{*}{\Sigma}(C),\overset{*}{z}(t)),I(\overset{*}{z}(t), N+n-1))$. Following Corollary 2 we have :

$$n = \sum_{h=1}^{N} a_h N_h \quad , a_h \geqslant 0,\ h=1,\dots,N,$$

where $N_h \triangleq ((\delta^h_k)_{k=1,\dots,N}, \text{grad}\, \overset{*}{V}_h(\overset{*}{x}(t)))$, $h=1,\dots,N$, and $\overset{*}{x}(t)$ is the projection on E^n of $\overset{*}{z}(t)$.

Lemma 23. *Let* $\eta^i = \sum_{h\in\beta_i} a_h P^i(N_h)$ *and*

$I(P^i(\overset{*}{z}(t)),\alpha_i+n-1) \triangleq \{P^i(\overset{*}{z}(t)) + \eta^i : \eta^i \in E^i, n^i\cdot\eta^i = 0\}$

$i=1,\dots,I$. *If* $\{J_k\} \subset L$, $k=1,\dots,N$, *then* $V^i(P^i(\overset{*}{z}(t)))$ *and* $C(A/\overset{*}{\Sigma}{}^i(C),P^i(\overset{*}{z}(t)))$ *are separated in* E^i *by* $I(P^i(\overset{*}{z}(t)),\alpha_i+n-1)$, $i=1,\dots,I$.

Demonstration : in view of Corollary 2 we have :
$n^i\cdot\eta^i \geqslant 0$, $P^i(\overset{*}{z}(t)) + \eta^i \in C(A/\overset{*}{\Sigma}{}^i(C),P^i(\overset{*}{z}(t)))$. Since $\{J_k\} \subset L$, $k=1,\dots,N$, there exist $i_1,\dots,i_N$, $i_k \in \{1,\dots,I\}$, $k=1,\dots,N$, such that $\overset{*}{\Sigma}{}^i(C) = \overset{*}{\Sigma}_k(C)$, $i = i_k$, $k=1,\dots,N$, so that in virtue of Assumption 1 :

$$C(A/\overset{*}{\Sigma}{}^i(C),P^i(\overset{*}{z}(t))) = \{y_k + \zeta_k : y_k \triangleq (x_{ok},x),\ \zeta_k \triangleq (\xi_{ok},\xi),\ \xi_{ok} + \text{grad}\, \overset{*}{V}_k(x)\cdot\zeta \geqslant 0\}$$

$y_k = P^i(\overset{*}{z}(t))$, $i = i_k$, $k=1,\dots,N$.

From Corollary 12 we have :
$\overset{\circ}{C}(A/\overset{*}{\Sigma}{}^i(C),P^i(\overset{*}{z}(t))) \cap V^i(P^i(\overset{*}{z}(t))) = \phi$, $i = i_k$, $k=1,\dots,N$, so that $n^i\cdot\eta^i \leqslant 0$, $P^i(\overset{*}{z}(t)) + \eta^i \in V^i(P^i(\overset{*}{z}(t)))$, $i = i_k$, $k=1,\dots,N$. Therefore $n^i\cdot\eta^i \leqslant 0$, $P^i(\overset{*}{z}(t)) + \eta^i \in V^i(P^i(\overset{*}{z}(t)))$, $i=1,\dots,I$, which concludes the proof.

3.7 - *A maximum principle*

From Corollaries 2, 10, 12 and Lemmas 14, 15, 23 one can deduce easily :

Theorem 1. *If* $\overset{*}{\Pi}(C)$ *is a path generated by a strategy* N*-tuple* $\overset{*}{p}$ *satisfying Lemma 2 given by* $\overset{*}{z} : t \mapsto \overset{*}{z}(t)$, $t \in [t_o, t_f]$, *and if*

Assumptions 1 and 2 are satisfied, then there exist a non-zero continuous solution $\tilde{\psi} \triangleq (\tilde{\lambda}_{o1},\dots,\tilde{\lambda}_{oN},\tilde{\lambda}_1,\dots,\tilde{\lambda}_n) : t \to \tilde{\psi}(t)$, $t \in [t_o,t_f]$, *of adjoint equations* (14) (15) *given in the chapter on Pareto equilibria, and a non-zero continuous solution* $\hat{\psi}^i \triangleq ((\hat{\lambda}^i_{ok})_{k\in\beta_i}, \hat{\lambda}^i_1,\dots,\hat{\lambda}^i_n) : t \to \hat{\psi}^i(t)$, $t \in [t_o,t_f]$, *of adjoint equations in paragraph* 3.4, $i=1,\dots,I$, *such that :*

a) $\max \tilde{\psi}(t)\cdot F(\overset{*}{z}(t),u) = \tilde{\psi}(t)\cdot F(\overset{*}{z}(t),\overset{*}{p}(\overset{*}{z}(t)))$, $t \in [t_o,t_f)$,
$u \in K(\overset{*}{z}(t))$

b) $\max \hat{\psi}^i(t)\cdot F^i(\overset{*}{z}(t),u) = \hat{\psi}^i(t)\cdot F^i(\overset{*}{z}(t),\overset{*}{p}(\overset{*}{z}(t)))$, $t \in [t_o,t_f)$
$i=1,\dots,I$, $u \in K(\overset{*}{z}(t))$

c) $\tilde{\psi}(t)\cdot F(\overset{*}{z}(t),\overset{*}{p}(\overset{*}{z}(t))) = \hat{\psi}^i(t)\cdot F^i(\overset{*}{z}(t),\overset{*}{p}(\overset{*}{z}(t))) = 0$,
$t \in [t_o,t_f)$, $i=1,\dots,I$

d) $\tilde{\lambda}_{o1}(t),\dots,\tilde{\lambda}_{oN}(t)$ and $\hat{\lambda}^i_{ok}(t)$, $k \in \beta_i$, $i=1,\dots,I$, *are non negative constants,* $t \in [t_o,t_f)$; *furthermore if* $\{J_k\} \subset L$, $k=1,\dots,N$, $\hat{\lambda}^i_{ok}(t) = \tilde{\lambda}_{ok}(t)$, $k \in \beta_i$, $i=1,\dots,I$,

e) $(\tilde{\lambda}_1(t_f),\dots,\tilde{\lambda}_n(t_f))$ and $(\hat{\lambda}^i_1(t_f),\dots,\hat{\lambda}^i_n(t_f))$, $i=1,\dots,I$,
are normal to θ

By using Assumption 2 one can prove easily that conditions a),b), c) and d) of Theorem 1 hold on $[t_o,t_f]$ by continuity.

4. Study of Strong Equilibria Surfaces

4.1 - *Strong equilibria surfaces*

Definition 10. If $\hat{p}$ is a strategy N-tuple satisfying Definition 6 and $C \in R^N$, the subset $\hat{\Sigma}(C)$ of E^{N+n} defined by

$$\hat{\Sigma}(C) \triangleq \{z : z \triangleq (x_o,x),\ x \in X \cup \theta,\ x_{ok} + \hat{V}_k(x) = C_k,\ k=1,\dots,N\}$$

where $\hat{V}_k(x) \triangleq V_k(x,\theta,\hat{p})$, $k=1,\ldots,N$, and $C \triangleq (C_1,\ldots,C_N)$, is called a strong equilibrium surface.

We define also a subset $A/\hat{\Sigma}(C)$ of E^{N+n} by :

$$A/\hat{\Sigma}(C) \triangleq \{z : z \triangleq (x_o,x),\ x \in X \cup \theta,\ x_{ok} + \hat{V}_k(x) > C_k,\ k=1,..,N\}$$

Definition 11. Similarly if $\hat{p}$ is a strong equilibrium strategy N-tuple and $C \in R^N$, we define in E^i, $i=1,\ldots,I$, a surface $\hat{\Sigma}^i(C)$ and a set $A/\hat{\Sigma}^i(C)$ by :

$$\hat{\Sigma}^i(C) \triangleq \{y^i : y^i \triangleq ((x_o)^i,x),\ x \in X \cup \theta,\ x_{ok}+\hat{V}_k(x)=C_k,\ k \in \beta_i\}$$

$$A/\hat{\Sigma}^i(C) \triangleq \{y^i : x \in X \cup \theta,\ x_{ok} + \hat{V}_k(x) > C_k,\ k \in \beta_i\} .$$

As the values of parameters $(C_1,\ldots,C_N)$ are varied, one obtains an N-parameter family of surfaces $\{\hat{\Sigma}(C)\}$, and I α_i-parameter families of surfaces $\{\hat{\Sigma}^i(C)\}$.

Some global properties of a strong equilibrium surface $\hat{\Sigma}(C)$ are contained in :

Lemma 24. *Let* $\hat{\Pi}(C)$ *be a terminating path generated by* $\hat{p}$, *given by* $\hat{z} : t \mapsto \hat{z}(t)$, $t \in [t_o,t_f]$, $\hat{z}(t_f) = z^f \triangleq (C,x^f) \in \Theta$.
Then $\hat{\Pi}(C) \subset \hat{\Sigma}(C)$.

Lemma 25. *If* $\hat{\Sigma}(C)$ *is a strong equilibrium surface then* $\hat{\Sigma}^i(C)$, $i=1,\ldots,I$, *is a Pareto surface in* E^i.

This proposition is a direct consequence of Definition 6 and of the definition of a Pareto optimal strategy N-tuple.

By invoking this lemma, all the results concerning the contingents and their transforms established in the chapter on Pareto equilibria remain valid, if we use the variational and adjoint equations defined in 3.4 .

4.2 - Separability of local cones

Assumption 2 is supposed satisfied, and we consider for $i=1,\dots,I$, the following sets :

$$\hat{R}^i(t) \triangleq \{v^i : v^i = A(\tau,t)F^i(\hat{z}(\tau),u_{\mathcal{C}_i},\hat{p}_{\bar{\mathcal{C}}_i}(\hat{z}(\tau))),$$
$$(u_{\mathcal{C}_i},\hat{p}_{\bar{\mathcal{C}}_i})(\hat{z}(\tau))) \in K(\hat{z}(\tau)),\tau \in [t_o,t),t \in [t_o,t_f]$$

$\hat{L}^i(t) \triangleq$ the convex cone generated by $\hat{R}^i(t)$, $t \in [t_o,t_f]$,

$\hat{\mathcal{K}}^i(P^i(\hat{z}(t))) \triangleq P^i(\hat{z}(t)) + \hat{L}^i(t)$, $t \in [t_o,t_f]$,

$$\hat{\mathcal{V}}^i(P^i(\hat{z}(t))) \triangleq \{P^i(\hat{z}(t)) + \gamma F^i(\hat{z}(t),u_{\mathcal{C}_i},\hat{p}_{\bar{\mathcal{C}}_i}(\hat{z}(t))) : \gamma \geqslant 0,$$
$$(u_{\mathcal{C}_i},\hat{p}_{\bar{\mathcal{C}}_i}(\hat{z}(t))) \in K(\hat{z}(t))\} \text{ , } t \in [t_o,t_f).$$

From these definitions it follows that :

1) $\hat{\mathcal{K}}^i(P^i(\hat{z}(t)))$, $t \in [t_o,t_f]$, is a convex cone in $\mathcal{E}^i$, $i=1,\dots,I$.

2) $A^i(t,t_f)\hat{\mathcal{V}}^i(P^i(\hat{z}(t))) \subset \hat{\mathcal{K}}^i(P^i(\hat{z}(t_f)))$, $t \in [t_o,t_f)$, $i=1,\dots,I$.

In view of Lemma 25 it follows from Lemmas 13 and 14 of the chapter on Pareto equilibria that :

3) $\mathcal{K}^i(P^i(\hat{z}(t))) \cap \overset{\circ}{C}(A/\hat{\Sigma}^i(C),P^i(\hat{z}(t))) = \phi$, $t \in [t_o,t_f)$, $i=1,..,I$.

4) $\mathcal{K}^i(P^i(\hat{z}(t_f))) \cap \overset{\circ}{\Gamma}(A/\hat{\Sigma}^i(C),P^i(\hat{z}(t_f))) = \phi$, $i=1,\dots,I$.

Consequently $\mathcal{K}^i(P^i(\hat{z}(t_f)))$ and $\Gamma(A/\hat{\Sigma}^i(C),P^i(\hat{z}(t_f)))$ are separated in $\mathcal{E}^i$ by $\mathcal{I}(P^i(\hat{z}(t_f)),\alpha_i+n-1)$, $i=1,\dots,I$, so that $\mathcal{V}^i(P^i(\hat{z}(t)))$ and $C(A/\hat{\Sigma}^i(C),P^i(\hat{z}(t)))$ are separated by $\mathcal{I}(P^i(\hat{z}(t),\alpha_i+n-1) \triangleq A^i(t_f,t)\mathcal{I}(P^i(\hat{z}(t_f)),\alpha_i+n-1)$, $t \in [t_o,t_f)$, $i=1,\dots,I$.

4.3 - A maximum principle

From the theorem in the chapter on Pareto equilibria and from paragraph 4.2 one readily deduces :

Theorem 2. *If* $\hat{\Pi}(C)$ *is a path generated by a strong equilibrium strategy* N*-tuple* $\hat{p}$, *given by* $\hat{z} : t \mapsto \hat{z}(t)$, $t \in [t_o, t_f]$, *and if Assumptions* 1 *and* 2 *are satisfied, then there exist a non-zero continuous solution* $\hat{\psi}^i \triangleq ((\hat{\lambda}^i_{ok})_{k \in \beta_i}, \hat{\lambda}^i_1, \ldots, \hat{\lambda}^i_n) : t \mapsto \hat{\psi}^i(t)$, $t \in [t_o, t_f]$, $i=1,\ldots,I$, *of adjoint equations in paragraph* 3.4 *such that* :

a) $\max \hat{\psi}^i(t) \cdot F^i(\hat{z}(t), u_{c_i}, \hat{p}_{\bar{c}_i}(\hat{z}(t))) = \hat{\psi}^i(t) \cdot F^i(\hat{z}(t), \hat{p}(\hat{z}(t)))$,

$(u_{c_i}, \hat{p}_{\bar{c}_i}(\hat{z}(t))) \in K(\hat{z}(t)))$, $\qquad t \in [t_o, t_f)$, $i=1,..,I$.

b) $\hat{\psi}^i(t) \cdot F^i(\hat{z}(t), \hat{p}(\hat{z}(t))) = 0$, $t \in [t_o, t_f)$, $i=1,\ldots,I$

c) $\hat{\lambda}^i_{ok}(t)$, $k \in \beta_i$, $i=1,\ldots,I$, *are non negative constants*, $t \in [t_o, t_f)$

d) $(\hat{\lambda}^i_1(t_f), \ldots, \hat{\lambda}^i_n(t_f))$, $i=1,\ldots,I$, *are normal to* θ.

By using Assumption 2 one can easily prove that conditions a), b) and c) of Theorem 2 hold on $[t_o, t_f]$ by continuity.

5. An Example of Core Equilibrium

We consider an economy producing a single homogeneous output, the agents of this economy being the two firms producing the output, the consumers and the government. The analysis is restricted to short periods so that the quantities q_1 and q_2 produced in an unit of time by firm 1 and 2 respectively depend only on y_1 and y_2, the labor per unit of time employed by these firms. Therefore : $q_1 = m_1 y_1$ and $q_2 = m_2 y_2$, where m_1 and m_2 are positive constants related to the firms' production techniques. We suppose that :

1) $m_1 = m_2 \triangleq m$,

2) the product may not be stocked and all production is immediately consummed or invested,

3) the firms do not make any investment,

4) the labor is perfectly mobile between two firms and moves from one to the other at a rate proportional to the difference between the hourly wages u_1 and u_2 offered by the firms,

5) $0 \leqslant u_1 \leqslant \bar{u}$ and $0 \leqslant u_2 \leqslant \bar{u}$, $\bar{u}$ being a positive constant,

6) each firm produces as long as its instantaneous profit is positive.

The government practices a strict policy of full employment and controls foreing trade. We suppose that :

7) foreign trade consists only of imports and exports of the commodity produced by this economy. We agree to consider as positive a quantity Z imported and as negative a quantity exported,

8) the size of the foreign trade is proportional to national output so that : $Z = u(q_1 + q_2)$, $-1 \leqslant u \leqslant 1$.

The consumers are supposed to be the wage earners. They fix their consumption expenditure C following a "keynesian" law : $C = cY_d$, where c is their consumption ratio and Y_d their income after taxes, $Y_d > 0$.

Furthermore we suppose that :

9) the consumption goods market is balanced by variations of the monetary price p of these goods, this balance being represented by :

$\dot{p} = cY_d - pQ$, where Q is the quantity offered per unit of time.

10) the players are :

$J_1 \triangleq$ the consumers

$J_2 \triangleq$ firm 1

$J_3 \triangleq$ firm 2

$J_4 \triangleq$ the government

11) the target is a final fixed time T

12) the coalition constraint $\mathcal{L}$ is independant of the state and such that :

$\mathcal{L} \triangleq \{(J_1,J_2,J_3,J_4),\ (J_1,J_3,J_4),\ (J_2,J_3,J_4)\}$

13) J_1 wants to maximise his total consumption expenditure,
J_2 maximises his sale receipts,
J_3 maximises his final market share,
J_4 minimises total rise in consumption price.

If L denotes the number of laborers in this economy, L is a constant and $y_1+y_2 = L$. Let $x_1 \triangleq y_1$ and $x_2 \triangleq p$, then according to Assumptions 1 to 13, the state equations of the model are :

$$\dot{x}_1 = b(u_1-u_2)$$

$$\dot{x}_2 = cx_1(u_1-u_2) + cu_2L - (u+1)mLx_2$$

$$\dot{x}_3 = 1$$

$$\dot{x}_{o1} = cx_1(u_1-u_2) + cu_2L$$

$$\dot{x}_{o2} = mx_1x_2$$

$$x_{o3}(T) = 1 - \frac{x_1(T)}{L}$$

$$\dot{x}_{o4} = (u+1)mLx_2 - cx_1(u_1-u_2) - cu_2L$$

The constraints are : $0 \leqslant u_1 \leqslant \bar{u}$, $0 \leqslant u_2 \leqslant \bar{u}$, $0 \leqslant c \leqslant 1$ and $x_1(u_1-u_2) + u_2L > 0$.

The game is defined on $E^4 \times G$, where

$G \triangleq \{x : x = (x_1,x_2,x_3),\ x_3 \leqslant T,\ 0 < x_1 < L,\ x_2 > \bar{x}_2 \triangleq m\bar{u}\ \}$

If $\overset{*}{\Pi}(C)$ is a path generated by a strategy 4-tuple $\overset{*}{p}$ satisfying Lemma 2 and $\overset{*}{\pi}$ its projection on G, given by $\overset{*}{x} : t \mapsto \overset{*}{x}(t)$, $t \in [t_o,T]$, since the constraints are independant of the state, the adjoint equations along $\overset{*}{\Pi}(C)$ are :

1) for the coalition $C_1 \triangleq (J_1,J_2,J_3,J_4)$

$$\dot{\lambda}_{ok} = 0, \quad k=1,2,3,4$$

$$\dot{\lambda}_1 = - c(\lambda_{o1}^{(1)} - \lambda_{o4}^{(1)} + \lambda_2)(u_1-u_2) - \lambda_{o2}^{(1)} mx_2$$

$$\dot{\lambda}_2 = - \lambda_{o2}^{(1)} mx_1 - (u+1)mL(\lambda_{o4}^{(1)} - \lambda_2)$$

$$\dot{\lambda}_3 = 0$$

and the transversality conditions are :

$$\tilde{\lambda}_1(T) = - \lambda_{o3}^{(1)}/L, \qquad \tilde{\lambda}_2(T) = 0$$

where $\tilde{\Psi}_1 \triangleq (\tilde{\lambda}_{o1},\tilde{\lambda}_{o2},\tilde{\lambda}_{o3},\tilde{\lambda}_{o4},\tilde{\lambda}_1,\tilde{\lambda}_2,\tilde{\lambda}_3) : t \mapsto \tilde{\Psi}_1(t)$, $t \in [t_o,T]$, is the solution of these adjoint equations satisfying Theorem 1,and $\tilde{\lambda}_{ok}(t) = \lambda_{ok}^{(1)}$, $k=1,2,3,4$, $t \in [t_o,T]$,

2) for the coalition $C_2 \triangleq (J_1,J_3,J_4)$

$$\dot{\lambda}_{ok} = 0, \quad k=1,3,4$$

$$\dot{\lambda}_1 = - c(\lambda_{o1}^{(2)} - \lambda_{o4}^{(2)} + \lambda_2)(u_1-u_2)$$

$$\dot{\lambda}_2 = - mL(u+1)(\lambda_{o4}^{(2)} - \lambda_2)$$

$$\dot{\lambda}_3 = 0$$

and the transversality conditions are :

$$\hat{\lambda}_1(T) = - \lambda_{o3}^{(2)}/L , \qquad \tilde{\lambda}_2(T) = 0$$

where $\hat{\Psi}_2 \triangleq (\hat{\lambda}_{o1},\hat{\lambda}_{o3},\hat{\lambda}_{o4},\hat{\lambda}_1,\hat{\lambda}_2,\hat{\lambda}_3) : t \mapsto \hat{\Psi}_2(t)$, $t \in [t_o,T]$, is the solution of these adjoint equations satisfying Theorem 1,and $\hat{\lambda}_{ok}(t) = \lambda_{ok}^{(2)}$, $k=1,3,4$, $t \in [t_o,T]$,

3) for the coalition $C_3 \triangleq (J_2, J_3, J_4)$

$$\dot{\lambda}_{ok} = 0, \quad k=2,3,4$$

$$\dot{\lambda}_1 = - c(\lambda_2 - \lambda_{o4}^{(3)})(u_1 - u_2) - \lambda_{o2}^{(3)} m x_2$$

$$\dot{\lambda}_2 = - \lambda_{o2}^{(3)} m x_1 - mL(u+1)(\lambda_{o4}^{(3)} - \lambda_2)$$

$$\dot{\lambda}_3 = 0$$

and the transversality conditions are :

$$\tilde{\lambda}_1(T) = - \lambda_{o3}^{(3)}/L, \qquad \tilde{\lambda}_2(T) = 0$$

where $\tilde{\psi}_3 \triangleq (\tilde{\lambda}_{o2}, \tilde{\lambda}_{o3}, \tilde{\lambda}_{o4}, \tilde{\lambda}_1, \tilde{\lambda}_2, \tilde{\lambda}_3) : t \mapsto \tilde{\psi}_3(t)$, $t \in [t_o, T]$, is the solution of these adjoint equations satisfying Theorem 1, and $\tilde{\lambda}_{ok}(t) = \lambda_{ok}^{(3)}$, $k=2,3,4$, $t \in [t_o, T]$,

In view of the transversality conditions, we deduce from Theorem 1 that :

- for $\lambda_{o4}^{(1)} \neq 0$ and $\lambda_{o1}^{(i)} < \lambda_{o4}^{(i)}$, $i=1,2,3$, we have :

$$\begin{aligned} \overset{*}{p}(\overset{*}{x}(T)) &\triangleq (\overset{*}{p}_1(\overset{*}{x}(T)), \overset{*}{p}_2(\overset{*}{x}(T)), \overset{*}{p}_3(\overset{*}{x}(T)), \overset{*}{p}_4(\overset{*}{x}(T))) \\ &= (0,0,\overline{u},1) \text{ if } \lambda_{o3}^{(i)} \neq 0, \ i=1,2,3 \\ &= (0,\overline{u},0,1) \text{ if } \lambda_{o3}^{(i)} = 0, \text{and } \lambda_{o2}^{(i)} \neq 0, \ i=1,2,3 \end{aligned}$$

- for $\lambda_{o4}^{(i)} = 0$ and $\lambda_{o2}^{(i)} \neq 0$, $i=1,2,3$, we have :

$$\begin{aligned} \overset{*}{p}(\overset{*}{x}(T)) &= (1,0,\overline{u},-1) \text{ if } \lambda_{o3}^{(i)} \neq 0, \ i=1,2,3 \text{ , and} \\ &\qquad - \frac{\lambda_{o3}^{(i)} b}{L} + \lambda_{o1}^{(i)} \overset{*}{x}_1(T) < 0, \ i=1,2 \\ &= (1,\overline{u},\overline{u},-1) \text{ if } \lambda_{o3}^{(i)} = 0, \ i=1,2,3 \end{aligned}$$

Consider the case : $\lambda_{o4}^{(i)} \neq 0$ and $\lambda_{o1}^{(i)} < \lambda_{o4}^{(i)}$, i=1,2,3 . It follows from Theorem 1, adjoint equations and transversality conditions, that there exists a neighborhood of the target in G where $\overset{*}{p}(x) = \overset{*}{p}(\overset{*}{x}(T))$.

If $\lambda_{o3}^{(i)} \neq 0$, i=1,2,3 , then in this neighborhood $\overset{*}{p}(x) = (0,0,\bar{u},1)$, and the equations of the projections on the x_1,x_2-plane of the corresponding paths are :

$$\overset{*}{x}_1(t) = b\bar{u}(T-t) + \overset{*}{x}_1(T)$$

$$\overset{*}{x}_2(t) = \overset{*}{x}_2(T)e^{2mL(T-t)}$$

Moreover from adjoint equations we obtain :

$$\tilde{\lambda}_1(t) = \frac{\lambda_{o2}^{(1)}\overset{*}{x}_2(T)}{2L}\left(e^{2mL(T-t)}-1\right) - \frac{\lambda_{o3}^{(1)}}{L}$$

$$\tilde{\lambda}_2(t) = \frac{\lambda_{o2}^{(1)}b\bar{u}}{2L}(T-t) + \left(\lambda_{o4}^{(1)} + \frac{\lambda_{o2}^{(1)}\overset{*}{x}_1(T)}{2L} - \frac{\lambda_{o2}^{(1)}mb\bar{u}}{(2mL)^2}\right)\left(1-e^{2mL(t-T)}\right)$$

$$\hat{\lambda}_1(t) = -\lambda_{o3}^{(2)}/L$$

$$\hat{\lambda}_2(t) = \lambda_{o4}^{(2)}\left(1-e^{2mL(T-t)}\right)$$

$$\tilde{\lambda}_1(t) = \frac{\lambda_{o2}^{(3)}\overset{*}{x}_2(T)}{2L}\left(e^{2mL(T-t)}-1\right) - \frac{\lambda_{o3}^{(3)}}{L}$$

$$\tilde{\lambda}_2(t) = \frac{\lambda_{o2}^{(3)}b\bar{u}}{2L}(T-t) + \left(\lambda_{o4}^{(3)} + \frac{\lambda_{o2}^{(3)}\overset{*}{x}_1(T)}{2L} - \frac{\lambda_{o2}^{(3)}mb\bar{u}}{(2mL)^2}\right)\left(1-e^{2mL(t-T)}\right)$$

Hence the above-mentioned solution satisfies Theorem 1 for all $t \leqslant T$ such that :

$$\lambda_{o1}^{(1)} - \lambda_{o4}^{(1)} + \tilde{\lambda}_2(t) \leqslant 0$$

$$\lambda_{o1}^{(2)} - \lambda_{o4}^{(2)} + \hat{\lambda}_2(t) \leqslant 0$$

$$\overset{\sim}{\lambda}_2(t) - \lambda_{o4}^{(3)} \leqslant 0$$

$$\tilde{\lambda}_1(t) \leqslant 0$$

$$\overset{\sim}{\lambda}_1(t) \leqslant 0$$

$$\overset{*}{x}_1(t) < L$$

If $\lambda_{o3}^{(i)} = 0$ and $\lambda_{o2}^{(i)} \neq 0$, $i=1,2,3$, then in a neighborhood of the target in G, $\overset{*}{p}(x) = (0,\bar{u},0,1)$ and the equations of the projections on the x_1,x_2-plane of the corresponding paths are :

$$\overset{*}{x}_1(t) = b\bar{u}(t-T) + \overset{*}{x}_1(T)$$

$$\overset{*}{x}_2(t) = \overset{*}{x}_2(T)\, e^{2mL(T-t)}$$

From the adjoint equations we obtain :

$$\tilde{\lambda}_1(t) = \frac{\lambda_{o2}^{(1)}}{2L}\left(e^{2mL(T-t)} - 1\right)$$

$$\tilde{\lambda}_2(t) = -\frac{\lambda_{o2}^{(1)} mb\bar{u}}{2L}(T-t) + \left(\lambda_{o4}^{(1)} + \frac{\lambda_{o2}^{(1)}\overset{*}{x}_1(T)}{2L} + \frac{\lambda_{o2}^{(1)} mb\bar{u}}{(2mL)^2}\right)\left(1-e^{2mL(t-T)}\right)$$

$$\hat{\lambda}_1(t) = 0$$

$$\hat{\lambda}_2(t) = \lambda_{o4}^{(2)}\left(1- e^{2mL(t-T)}\right)$$

$$\overset{\sim}{\lambda}_1(t) = \frac{\lambda_{o2}^{(3)}}{2L}\left(e^{2mL(T-t)} - 1\right)$$

$$\overset{\sim}{\lambda}_2(t) = -\frac{\lambda_{o2}^{(3)} mb\bar{u}}{2L}(T-t) + \left(\lambda_{o4}^{(3)} + \frac{\lambda_{o2}^{(3)}\overset{*}{x}_1(T)}{2L} + \frac{\lambda_{o2}^{(3)} mb\bar{u}}{(2mL)^2}\right)\left(1-e^{2mL(t-T)}\right)$$

Hence the above mentioned solution satisfies Theorem 1 for all $t \leqslant T$ such that :

$$\lambda_{o1}^{(1)} - \lambda_{o4}^{(1)} + \tilde{\lambda}_2(t) \leqslant 0$$

$$\lambda_{o1}^{(2)} - \lambda_{o4}^{(2)} + \hat{\lambda}_2(t) \leqslant 0$$

$$\tilde{\lambda}_2(t) - \lambda_{o4}^{(3)} \leqslant 0$$

$$\overset{*}{x}_1(t) > 0$$

In the case : $\lambda_{o4}^{(i)} = 0$ and $\lambda_{o2}^{(i)} \neq 0$, i=1,2,3, by similar arguments one can prove the existence of a neighborhood of the target in G where $\overset{*}{p}(x) = \overset{*}{p}(\overset{*}{x}(T))$.

If $\lambda_{o3}^{(i)} \neq 0$, i=1,2,3 , and $-\dfrac{\lambda_{o3}^{(i)}b}{L} + \lambda_{o1}^{(i)}\overset{*}{x}_1(T) < 0$, i=1,2 , then in this neighborhood $\overset{*}{p}(x) = (1,0,\bar{u},-1)$ and the equations of the projections on the x_1,x_2-plane of the corresponding paths are :

$$\overset{*}{x}_1(t) = b\bar{u}(T-t) + \overset{*}{x}_1(T)$$

$$\overset{*}{x}_2(t) = b\bar{u}^2 \frac{(T-t)^2}{2} + \bar{u}(L-\overset{*}{x}_1(T))(t-T) + \overset{*}{x}_2(T)$$

Moreover from adjoint equations we obtain :

$$\tilde{\lambda}_1(t) = -\lambda_{o2}^{(1)}m\bar{u}L\frac{(T-t)^2}{2} + (\lambda_{o1}^{(1)}\bar{u} - \lambda_{o2}^{(1)}m\overset{*}{x}_2(T))(t-T) - \frac{\lambda_{o3}^{(1)}}{L}$$

$$\tilde{\lambda}_2(t) = \lambda_{o2}^{(1)}m\overset{*}{x}_1(T)(T-t) + \lambda_{o2}^{(1)}mb\bar{u}\frac{(T-t)^2}{2}$$

$$\hat{\lambda}_1(t) = \lambda_{o1}^{(2)}\bar{u}(t-T) - \frac{\lambda_{o3}^{(2)}}{L}$$

$$\hat{\lambda}_2(t) = 0$$

$$\tilde{\lambda}_1(t) = -\lambda_{o2}^{(3)}m\bar{u}L\frac{(T-t)^2}{2} - \lambda_{o2}^{(3)}m\overset{*}{x}_2(T)(t-T) - \frac{\lambda_{o3}^{(3)}}{L}$$

$$\tilde{\lambda}_2(t) = \lambda_{o2}^{(3)}m\overset{*}{x}_1(T)(T-t) + \lambda_{o2}^{(3)}mb\bar{u}\frac{(T-t)^2}{2}$$

This solution satisfies Theorem 1 for all $t \leqslant T$ such that :

$$b\tilde{\lambda}_1(t) + (\lambda_{o1}^{(1)} + \tilde{\lambda}_2(t))\overset{*}{x}_1(t) \leqslant 0$$

$$b\hat{\lambda}_1(t) + \lambda_{o1}^{(2)}\overset{*}{x}_1(t) \leqslant 0$$

$$b\tilde{\tilde{\lambda}}_1(t) + \tilde{\tilde{\lambda}}_2(t)\overset{*}{x}_1(t) \leqslant 0$$

$$\overset{*}{x}_1(t) < L$$

If $\lambda_{o3}^{(i)} = 0$, $i=1,2,3$, then in a neighborhood of the target in G we have $\overset{*}{p}(x) = (1,\bar{u},\bar{u},-1)$ and the equations of the projections on the x_1,x_2-plane of the corresponding paths are :

$$\overset{*}{x}_1(t) = \overset{*}{x}_1(T)$$

$$\overset{*}{x}_2(t) = \bar{u}L(t-T) + \overset{*}{x}_2(T)$$

From adjoint equations we obtain :

$$\tilde{\lambda}_1(t) = -\lambda_{o2}^{(1)}\ m\bar{u}L\ \frac{(T-t)^2}{2} + \lambda_{o2}^{(1)}\ m\overset{*}{x}_2(T)(T-t)$$

$$\tilde{\lambda}_2(t) = \lambda_{o2}^{(1)}\ m\overset{*}{x}_1(T)(T-t)$$

$$\hat{\lambda}_1(t) = \hat{\lambda}_2(t) = 0$$

$$\tilde{\tilde{\lambda}}_1(t) = -\lambda_{o2}^{(3)}\ m\bar{u}L\ \frac{(T-t)^2}{2} + \lambda_{o2}^{(3)}\ m\overset{*}{x}_2(T)(T-t)$$

$$\tilde{\tilde{\lambda}}_2(t) = \lambda_{o2}^{(3)}\ m\overset{*}{x}_1(T)(T-t)$$

so that this solution satisfies Theorem 1 for all $t \leqslant T$ such that

$$b\tilde{\lambda}_1(t) + \tilde{\lambda}_2(t)\overset{*}{x}_1(T) \geqslant 0$$

$$-b\tilde{\lambda}_1(t) + \tilde{\lambda}_2(t)(L-\overset{*}{x}_1(T)) \geqslant 0$$

$$b\tilde{\tilde{\lambda}}_1(t) + \tilde{\tilde{\lambda}}_2(t)\overset{*}{x}_1(T) \geqslant 0$$

$$-b\tilde{\tilde{\lambda}}_1(t) + \tilde{\tilde{\lambda}}_2(t)(L-\overset{*}{x}_1(T)) > 0$$

$$\overset{*}{x}_2(t) > \bar{x}_2 \triangleq m\bar{u}$$

6. An Example of Strong Equilibrium

We study the same model as in paragraph 5, but we replace Assumption 12 by the following : the coalition constraint L is independant of the state and such that : $L \triangleq \{(J_1,J_4),(J_2,J_3)\}$

The state equations are the same as previously, and if $\hat{\Pi}(C)$ is a path generated by a strong equilibrium strategy 4-tuple $\hat{p}$ and $\hat{\pi}$ its projection on G, given by $\hat{x} : t \mapsto \hat{x}(t)$, $t \in [t_o,T]$, since the constraints are independant of the state, the adjoint equations along $\hat{\Pi}(C)$ are :

1) for the coalition $C_1 \triangleq (J_1,J_2)$

$$
\begin{aligned}
\dot{\lambda}_{o1} &= \dot{\lambda}_{o4} = 0 \\
\dot{\lambda}_1 &= -\, c(\lambda_{o1} + \lambda_2 - \lambda_{o4})(u_1-u_2) \\
\dot{\lambda}_2 &= mL(u+1)(\lambda_2 - \lambda_{o4}) \\
\dot{\lambda}_3 &= 0
\end{aligned}
$$

and the transversality conditions are :

$$
\hat{\lambda}_1(T) = \hat{\lambda}_2(T) = 0
$$

where $\hat{\psi}_1 \triangleq (\hat{\lambda}_{o1},\hat{\lambda}_{o4},\hat{\lambda}_1,\hat{\lambda}_2,\hat{\lambda}_3): t \mapsto \hat{\psi}_1(t)$, $t \in [t_o,T]$, is the solution of these adjoint equations satisfying Theorem 2, and $\hat{\lambda}_{o1}(t) = \lambda_{o1}$, $\hat{\lambda}_{o2}(t) = \lambda_{o2}$, $t \in [t_o,T]$,

2) for the coalition $C_2 \triangleq (J_2,J_3)$

$$
\begin{aligned}
\dot{\mu}_{o2} &= \dot{\mu}_{o3} = 0 \\
\dot{\mu}_1 &= -\, \mu_{o2}mx_2 - \mu_2 c(u_1-u_2) \\
\dot{\mu}_2 &= -\, \mu_{o2}mx_1 + \mu_2 mL(u+1) \\
\dot{\mu}_3 &= 0
\end{aligned}
$$

and the transversality conditions are :

$$\hat{\mu}_1(T) = -\mu_{o3}/L, \quad \hat{\mu}_2(T) = 0$$

where $\hat{\psi}_2 \triangleq (\hat{\mu}_{o2},\hat{\mu}_{o3},\hat{\mu}_1,\hat{\mu}_2,\hat{\mu}_3) : t \mapsto \hat{\psi}_2(t)$, $t \in [t_o,T]$, is the solution of these adjoint equations satisfying Theorem 2, and $\hat{\mu}_{o2}(t) = \mu_{o2}$, $\hat{\mu}_{o3}(t) = \mu_{o3}$, $t \in [t_o,T]$.

From condition a) of Theorem 2, from transversality conditions and adjoint equations its follows that :

$$\hat{p}_1(\hat{x}(T)) \triangleq \hat{c}(T) = 1 \text{ if } \lambda_{o1}-\lambda_{o4} > 0$$
$$= 0 \text{ if } \lambda_{o1}-\lambda_{o4} < 0$$

$$\hat{p}_2(\hat{x}(T)) \triangleq \hat{u}_1(T) = 0 \text{ if } \mu_{o3} \neq 0$$
$$= \overline{u} \text{ if } \mu_{o3} = 0$$

$$\hat{p}_3(\hat{x}(T)) = \hat{u}_2(T) = \overline{u} \text{ if } \mu_{o3} \neq 0 \text{ or if } \mu_{o3} = 0,\ \hat{c}(T)L^2-4b\hat{x}_2(T) > 0$$
$$\text{and } \frac{L-A^{1/2}}{2} < \hat{x}_1(T) < \frac{L+A^{1/2}}{2}$$
$$= 0 \text{ if } \hat{c}(T)L^2-4b\hat{x}_2(T) < 0 \text{ or } \hat{c}(T)L^2-4b\hat{x}_2(T) > 0$$
$$\text{and } 0 < \hat{x}_1(T) < \frac{L-A^{1/2}}{2} \text{ or } \frac{L+A^{1/2}}{2} < \hat{x}_1(T) < L$$
$$\text{where } A \triangleq L^2 - \frac{4b\hat{x}_2(T)}{\hat{c}(T)}$$

$$\hat{p}_4(\hat{x}(T)) \triangleq \hat{u}(T) = 1 \text{ if } \lambda_{o4} \neq 0$$

Consider the case : $\lambda_{o1}-\lambda_{o4} < 0$. From adjoint equations and transversality conditions, we deduce that there exists a neighborhood of the target in G where $\hat{p}(x) = \hat{p}(\hat{x}(T))$.

If $\mu_{o3} \neq 0$, then in this neighborhood $\hat{p}(x) = (0,0,\overline{u},1)$ and the equations of the projections on the x_1,x_2-plane of the corresponding paths are :

$$\hat{x}_1(t) = b\overline{u}(T-t) + \hat{x}_1(T)$$

$$\hat{x}_2(t) = \hat{x}_2(T)\ e^{2mL(T-t)}$$

and from adjoint equations we obtain :

$$\hat{\lambda}_1(t) = 0$$

$$\hat{\lambda}_2(t) = \lambda_{o4}(1-e^{2mL(t-T)})$$

$$\hat{\mu}_1(t) = \frac{\mu_{o2}\hat{x}_2(T)}{2L}\ (e^{2mL(T-t)}-1) - \frac{\mu_{o3}}{L}$$

$$\hat{\mu}_2(t) = (\frac{\mu_{o2}\hat{x}_1(T)}{2L} - \frac{\mu_{o2}mb\overline{u}}{(2mL)^2})(1-e^{2mL(t-T)}) + \frac{\mu_{o2}b\overline{u}}{2L}(T-t)$$

Hence the solution satisfies Theorem 2 for all $t \leqslant T$ such that :

$$\lambda_{o1} + \hat{\lambda}_2(t) - \lambda_{o4} \leqslant 0$$

$$\hat{\mu}_1(t) \leqslant 0$$

$$\hat{x}_1(t) < L$$

If $\mu_{o3} = 0$, then in a neighborhood of the target in G we have $\hat{p}(x) = (0,\overline{u},0,1)$ and the equations of the projections on the x_1,x_2-plane of the corresponding paths are :

$$\hat{x}_1(t) = b\overline{u}(t-T) + \hat{x}_1(T)$$

$$\hat{x}_2(t) = \hat{x}_2(T)e^{2mL(T-t)}$$

Moreover from adjoint equations we have :

$$\hat{\lambda}_1(t) = 0$$

$$\hat{\lambda}_2(t) = \lambda_{o4}(1-e^{2mL(t-T)})$$

$$\hat{\mu}_1(t) = \frac{\mu_{o2}\hat{x}_2(T)}{2L}\ (e^{2mL(T-t)}-1)$$

$$\hat{\mu}_2(t) = \frac{\mu_{o2}}{2L}(\hat{x}_1(T) + \frac{b\overline{u}}{2mL})(1-e^{2mL(t-T)}) + \frac{\mu_{o2}b\overline{u}}{2L}\ (t-T)$$

so that the above-mentioned solution satisfies Theorem 2 for all $t \leqslant T$ such that :

$$\lambda_{o1} + \hat{\lambda}_2(t) - \lambda_{o4} \leqslant 0$$
$$\hat{x}_1(t) > 0$$

In the case : $\lambda_{o1} - \lambda_{o4} > 0$ and $\lambda_{o4} \neq 0$, as previously there exists a neighborhood of the target in G where $\hat{p}(x) = \hat{p}(\hat{x}(T))$.

If $\mu_{o3} \neq 0$, then in this neighborhood $\hat{p}(x) = (1,0,\bar{u},1)$ and the equations of the projections on the x_1,x_2-plane of the corresponding paths are :

$$\hat{x}_1(t) = - b\bar{u}(t-T) + \hat{x}_1(T)$$

$$\hat{x}_2(t) = \frac{\bar{u}}{2mL}(L-\hat{x}_1(T)- \frac{b\bar{u}}{2mL})(1-e^{2mL(T-t)})+ \hat{x}_2(T)e^{2mL(T-t)} + \frac{b\bar{u}^2}{2mL}(t-T)$$

and from adjoint equations we obtain :

$$\hat{\lambda}_1(t) = \lambda_{o1}\bar{u}(t-T) + \frac{\lambda_{o4}\bar{u}}{2mL}(1-e^{2mL(t-T)})$$

$$\hat{\lambda}_2(t) = \lambda_{o4}(1-e^{2mL(t-T)})$$

$$\hat{\mu}_1(t) = - \frac{\mu_{o3}}{L} + \frac{\mu_{o2}\bar{u}}{2L}(2\hat{x}_1(T)-L)(t-T) - \frac{\mu_{o2}b\bar{u}^2}{2L}(T-t)^2 + \frac{\mu_{o2}}{2L}(\hat{x}_2(T) - \frac{\bar{u}}{2m})(e^{2mL(T-t)}-1) + \frac{\mu_{o2}\bar{u}}{2mL}(\hat{x}_1(T) - \frac{b\bar{u}}{2mL})(1-\cosh 2mL(T-t))$$

$$\hat{\mu}_2(t) = \frac{\mu_{o2}}{2L}(\hat{x}_1(T) - \frac{b\bar{u}}{2mL})(1-e^{2mL(t-T)}) + \frac{\mu_{o2}b\bar{u}}{2L}(T-t)$$

It follows from the above equations that $\hat{\lambda}_2(t)$ and $\hat{\mu}_2(t)$ are non negative for $t \leqslant T$, so that the solution satisfies Theorem 2 for all $t \leqslant T$ such that :

$$b\hat{\mu}_1(t) + \hat{x}_1(t)\hat{\mu}_2(t) \leqslant 0$$
$$\hat{x}_1(t) < L$$

If $\mu_{o3} = 0$, then we deduce from Theorem 2, adjoint equations and transversality conditions that $\mu_{o2} \neq 0$. Then

1) if $L^2 - 4b\hat{x}_2(T) > 0$ and $\frac{L-A^{1/2}}{2} < \hat{x}_1(T) < \frac{L+A^{1/2}}{2}$ we have in a neighborhood of the target in G, $\hat{p}(x) = (1,\bar{u},\bar{u},1)$, therefore the equations of the projections on the x_1,x_2-plane of the corresponding paths are :

$$\hat{x}_1(t) = \hat{x}_1(T)$$

$$\hat{x}_2(t) = \frac{\bar{u}}{2m} + (\hat{x}_2(T) - \frac{\bar{u}}{2m})\ e^{2mL(T-t)}$$

and from adjoint equations we obtain :

$$\hat{\lambda}_1(t) = 0$$

$$\hat{\lambda}_2(t) = \lambda_{o4}(1-e^{2mL(t-T)})$$

$$\hat{\mu}_1(t) = \frac{\mu_{o2}\bar{u}}{2}(T-t) + \frac{\mu_{o2}}{2L}(\hat{x}_2(T) - \frac{\bar{u}}{2m})(e^{2mL(T-t)}-1)$$

$$\hat{\mu}_2(t) = \frac{\mu_{o2}\hat{x}_1(T)}{2L}(1-e^{2mL(t-T)})$$

As for $t \leqslant T$, $\hat{\lambda}_2(t)$, $\hat{\mu}_1(t)$ and $\hat{\mu}_2(t)$ are non negative, the above mentioned solution satisfies Theorem 2 for all $t \leqslant T$ such that :

$$-b\hat{\mu}_1(t) + \hat{\mu}_2(t)(L-\hat{x}_1(T)) \geqslant 0$$

2) if $L^2 - 4b\hat{x}_2(T) < 0$ or $L^2 - 4b\hat{x}_2(T) > 0$ and $0 < \hat{x}_1(T) < \frac{L-A^{1/2}}{2}$ or $\frac{L+A^{1/2}}{2} < \hat{x}_1(T) < L$, then in a neighborhood of the target in G we have $\hat{p}(x) = (1,\bar{u},0,1)$, and the equations of the projections on the x_1,x_2-plane of the corresponding paths are :

$$\hat{x}_1(t) = b\bar{u}(t-T) + \hat{x}_1(T)$$

$$\hat{x}_2(t) = \hat{x}_2(T)e^{2mL(T-t)} + \frac{b\bar{u}^2}{2mL}(t-T) + \frac{\bar{u}}{2mL}(\hat{x}_1(T) - \frac{b\bar{u}}{2mL})(e^{2mL(T-t)}-1)$$

and from adjoint equations we have :

$$\hat{\lambda}_1(t) = \lambda_{o1}\bar{u}(T-t) - \frac{\lambda_{o4}\bar{u}}{2mL}(1-e^{2mL(t-T)})$$

$$\hat{\lambda}_2(t) = \lambda_{o4}(1-e^{2mL(t-T)})$$

$$\hat{\mu}_1(t) = \frac{\mu_{o2}b\bar{u}^2}{2mL^2}(T-t) - \frac{\mu_{o2}b\bar{u}^2}{2L}(T-t)^2 + \frac{\mu_{o2}\hat{x}_2(T)}{2L}(e^{2mL(T-t)}-1)$$

$$- \frac{\mu_{o2}b\bar{u}}{4m^2L^3}\sinh 2mL(T-t) - \frac{\mu_{o2}\hat{x}_1(T)\bar{u}}{2mL^2}(1-\cosh 2mL(T-t))$$

$$\hat{\mu}_2(t) = \frac{\mu_{o2}\hat{x}_1(T)}{2L}(1-e^{2mL(t-T)})$$

This solution satisfies Theorem 2 for all $t \leqslant T$ such that :

$$b\hat{\mu}_1(t) + \hat{\mu}_2(t)\hat{x}_1(t) \geqslant 0$$

$$- b\hat{\mu}_1(t) + \hat{\mu}_2(t)(L-\hat{x}_1(t)) \leqslant 0$$

$$\hat{x}_1(t) > 0$$

References

(1) L. JURICEK, Thèse de Doctorat d'Etat, Jeux différentiels à N-joueurs coopératifs et non coopératifs, Paris 1972

(2) R. LUCE, H. RAIFFA, Games and Decisions, John Wiley 1957

(3) M. SHUBIK, Stratégies de marchés et structures oligopolistiques, Dunod 1964

(4) J. von NEUMANN, O. MORGENSTERN, Theory of Games and Economic Behavior, Princeton University Press 1953

SUFFICIENCY CONDITIONS FOR NASH EQUILIBRIA IN N-PERSON DIFFERENTIAL GAMES[†]

H. STALFORD[††] and G. LEITMANN[†††]

1. Introduction and Problem Statement

In this chapter we discuss some aspects of Nash equilibrium solutions for N-person differential games. That is, we consider games in which the evolution in time of the state of the game is governed by ordinary differential equations. Each player influences this evolution of the state through his choice of an admissible strategy. Each player's cost is a function of the strategies of all players and of a corresponding evolution of the state. The players desire the transfer of the state to a given target set in state space while minimizing their own costs. Each player possesses the same information, namely, the governing differential equations, the target set, the cost functions of all players and the current state. In the absence of any knowledge about the strategies of his opponents, each player assumes that all of his opponents are rational and that each plays "optimally", that is, so as to minimize his own cost. Such a "mood of play" is charac-

---

† Based on research supported in part by NSF and ONR.

†† Naval Research Laboratory, Washington, D.C.

††† University of California, Berkeley .

terized by the Nash equilibrium condition ; see (1). Thus, provided all other players utilize equilibrium strategies, a player cannot gain by departing from his equilibrium strategy.

Necessary conditions for Nash equilibria in N-person differential games may be found in the literature ; e.g., see (2)-(4). Here we give sufficient conditions. After defining the problem and giving some preliminary lemmas, we present two sufficiency theorems, one of the field type involving a function of the state, the other involving a function of time only. The use of these theorems is illustrated by some examples. In addition, a relation between sufficiency conditions and game surfaces, (4), is established.

Consider a dynamical system with state equation

$$\dot{x} = f(x, v^1, v^2, \ldots, v^N) \tag{1}$$

where state $x \in E^n$, control $v^i \in E^{m_i}$, $i \in \{1,2,\ldots,N\}$, and state velocity function f is Borel measurable on $E^n \times E^{m_1} \times E^{m_2} \times \ldots \times E^{m_N}$ †.

The playing space (state space) X is a Lebesgue measurable subset of E^n ††. The initial state x^o and target set Θ are given sets contained in X. To include the possibility that the system is non-autonomous or that the target set depends on t, we take $x_n \equiv t$ with $x_n^o = t_o$.

Let P^i, $i \in \{1,2,\ldots,N\}$, denote the spaces of all Borel measurable function from X into E^{m_i} ; space P^i constitutes the space of *strategies* $p^i : X \to E^{m_i}$ for player i. A strategy p^i is *admissible* if, and only if, $p^i \in P^i$ and

$$p^i(x) \in U^i(x) \qquad \forall x \in X \tag{2}$$

---

† E^p denotes p-dimensional Euclidean space.

†† X has the induced topology from E^n

for given functions

$$(3) \qquad U^i : X \to \text{set of all non-empty subsets of } E^{m_i} .$$

For a given $x \in X$, the set $U^i(x)$ is the set of control values available to player i at state x. Thus, state constraints are introduced by specifying X and control constraints by specifying the U^i.

Given an admissible strategy N-tuple $p = (p^1, p^2, .., p^N)$ and initial state x^o, an absolutely continuous function $\varphi : [t_o, t_f] \to X$ is called a *trajectory* if, and only if,

$$(4) \qquad \varphi(t) = x^o + \int_{t_o}^{t} f(\varphi(\tau), p(\varphi(\tau)))d\tau$$

for all $t \in [t_o, t_f]$. The pair $(p ; \varphi)$ is said to be a *play* commencing at time t_o from state x^o. A play *terminates* at time t_f if, and only if, there exists a $t_f \in [t_o, \infty)$ such that $\varphi(t_f) \in \Theta$ and $\varphi(t) \notin \Theta$ for $t < t_f$. Then the trajectory φ is said to be a *terminating trajectory.*

A strategy N-tuple is *playable at* x^o if, and only if, it is admissible and generates at least one terminating trajectory. Let $I(x^o)$ denote the collection of all strategy N-tuples that are playable at x^o ; we assume that $I(x^o)$ is non-empty. Let $\Phi(x^o ; p)$ denote the set of all trajectories generated by $p \in I(x^o)$, and let $I = \cap \{I(x^o) : x^o \in X\}$. That is, I is the set of all strategy N-tuples which are playable on the entire playing place X ; we assume that I is non-empty.

The i-th player's *cost,* J^i, is a function of strategy N-tuple $p \in I(x^o)$ and trajectory $\varphi \in \Phi(x^o ; p)$, and is given by

$$(5) \qquad J^i(p ; \varphi) = \int_{t_o}^{t_f} f_o^i(\varphi(\tau), p(\varphi(\tau)))d\tau$$

where the f_o^i are given real valued bounded Borel measurable functions with domain $E^n \times E^{m_1} \times .. \times E^{m_N}$.

Let $p^* = (p^{1*}, p^{2*}, \ldots, p^{N*}) \in I(x^o)$ and $\varphi^* \in \Phi(x^o; p^*)$. Strategy N-tuple p^* is *optimal at* x^o if, and only if, it satisfies Nash equilibrium condition, [1],

$$(6) \qquad J^i(p^*;\varphi^*) \leq J^i(p^{1*}, p^{2*}, \ldots, p^i, \ldots, p^{N*}; \varphi^i)$$

$$i \in \{1,2,\ldots,N\}$$

for all $(p^{1*}, p^{2*}, \ldots, p^i, \ldots, p^{N*}) \in I(x^o)$

and $\varphi^i \in \Phi(x^o \; ; \; p^{1*}, p^{2*}, \ldots, p^i, \ldots, p^{N*})$, and

$$(7)^{\dagger} \qquad J^i(p^*;\varphi^*) = J^i(p^*;\varphi^{**}), \qquad i \in \{1,2,\ldots,N\}$$

for all $\varphi^{**} \in \Phi(x^o; p^*)$.

A strategy N-tuple $p^* \in I$ is *optimal on* X if, and only if, it is optimal at all x^o contained in X.

2. A Sufficiency Theorem of the Field Type

In preparation for the sufficiency theorem, we give some definitions and prove some preliminary lemmas.

Definition 1. A *denumerable decomposition* D of a set $X \subset E^n$ is a denumerable collection of pairwise disjoint subsets whose union is X. This is usually written as $D = \{X_j : j \in J\}$, where J is a denumerable index set of the pairwise disjoint subsets.

In the sequel it is not essential that the members of D be pairwise disjoint, since finer decompositions having pairwise disjoint members always exist.

Let B be a subset of E^n. A mapping $F : B \to E^1$ is said to be

---

† For two-person zero-sum games, condition (6) implies (7)

continuously differentiable if there is an open set W containing B such that F may be extended to a function which is continuously differentiable on W.

Definition 2. Let X be a subset of E^n and D a denumerable decomposition of X. A real-valued continuous function V on X is said to be *continuously differentiable with respect to* D if, for $j \in J$, $V|X_j : X_j \to E^1$ is continuously differentiable ; that is, there exists a collection $\{(W_j,V_j) : j \in J\}$ such that W_j is an open set containing X_j, $V_j : W_j \to E^1$ is continuously differentiable, and $V_j(x) = V(x)$ for $x \in X_j$. We say that the collection $\{(W_j,V_j) : j \in J\}$ is associated with V and D. This collection is not necessarily unique.

A subset of the real line is called a *null set* if it has Lebesgue measure zero. A countable union of null sets is again a null set ; see (5). If a certain proposition concerning the points of a subset T of the real line is true for every point with the exception at most of a set of points which form a null set, then it is customary to say that the proposition is true almost everywhere in T ; this is abbreviated to a.e. in T.

We now state the first of four preliminary lemmas.

Lemma 1. *Let* X *be a subset of* E^n *and* $D = \{X_j : j \in J\}$ *a denumerable decomposition of* X. *If* $\varphi : [t_o,t_f] \to X$ *is an absolutely continuous function and* $V : X \to E^1$ *is a function which is continuously differentiable with respect to* D, *then the composite function* $V \circ \varphi : [t_o,t_f] \to E^1$ *maps null sets into null sets* .

Proof. This proof does not require that V be continuous on X. Let K be a null set in $[t_o,t_f]$ and for each $j \in J$ define $K_j = \{t \in K : \varphi(t) \in X_j\}$. It suffices to show that $(V \circ \varphi)[K_j]$ is a null set for each $j \in J$, since $(V \circ \varphi)[K] = \cup\{(V \circ \varphi)[K_j] : j \in J\}$ and the union is denumerable.

Let $j \in J$. By hypothesis, there exists an open neighborhood W_j of X_j and a continuously differentiable function $V_j : W_j \to E^1$ such that $V_j(x) = V(x)$ for each $x \in X_j$.
Let $\theta_j = \{t \in (t_o, t_f) : \varphi(t) \in W_j\}$.
Note that $(V_j \circ \varphi)[K_j] = (V \circ \varphi)[K_j]$. The function φ being continuous implies that θ_j is open in (t_o, t_f). θ_j, being open, is the denumerable union of disjoint open intervals $\theta_j(i)$, $i \in I_j$, where I_j is the denumerable index set of these disjoint open intervals ; Cantor (6) proved that every open set of real numbers is a denumerable union of pairwise disjoint open intervals. Now it suffices to show that $(V_j \circ \varphi)[K_j \cap \theta_j(i)]$ is a null set for each $i \in I_j$.

Let $i \in I_j$ and suppose that $\theta_j(i) = (a,b)$. Let δ be a positive number such that $b - a - 2\delta > 0$.
Let $C_o = [a + \delta, b - \frac{1}{2}\delta]$, $C_n = [a + \delta/(n+2), a + \delta/n]$ for n an odd positive integer, and $C_n = [b - \delta/n, b - \delta/(n+2)]$ for n an even positive integer. Since $(a,b) = \cup\{C_n : n = 0, 1, 2, \ldots\}$, it suffices now only to show that $(V_j \circ \varphi)[K_j \cap C_n]$ is a null set for each $n \in \{0, 1, 2, \ldots\}$

Let $n \in \{0, 1, 2, \ldots\}$. Consider the mapping $V_j \circ \varphi : C_n \to E^1$. Since C_n is a compact interval, V_j is continuously differentiable and φ is absolutely continuous, then $V_j \circ \varphi$ is absolutely continuous on C_n ; see (5). Since an absolutely continuous function maps null sets into null sets, the set $(V_j \circ \varphi)[K_j \cap C_n]$ is a null set ; see (5). This completes the proof.

Since an absolutely continuous function is differentiable almost everywhere (e.g., (5)) and $V_j \circ \varphi : C_n \to E^1$ is absolutely continuous on C_n, it follows that $V_j \circ \varphi : \theta_j \to E^1$ is differentiable a.e. in θ_j . We shall make use of this result in the proof of the following lemma.

Lemma 2. *Let* X *be a subset of* E^n *and* D *a denumerable decomposition of* X. *Let* $\varphi : [t_o, t_f] \to X$ *be absolutely continuous and* $V : X \to E^1$ *be continuously differentiable with respect to* D. *Let* $T_j = \{t \in [t_o, t_f] : \varphi(t) \in X_j\}$. *If* $\{(W_j^1, V_j^1) : j \in J\}$ *and* $\{(W_j^2, V_j^2) : j \in J\}$ *are two collections associated with* V *and* D, *then for* $j \in J$

$$(8) \qquad (d/dt)(V_j^1 \circ \varphi)(t) = (d/dt)(V_j^2 \circ \varphi)(t) \qquad \text{a.e. in } T_j$$

Proof. As noted above, the derivatives $(d/dt)(V_j^1 \circ \varphi)(t)$ and $(d/dt)(V_j^2 \circ \varphi)(t)$ exist a.e. in $\theta_j^1 \cap \theta_j^2$, where $\theta_j^i = \{t \in (t_o, t_f) : \varphi(t) \in W_j^i\}$, $i \in \{1,2\}$. Let $A = \{ t \in T_j :$ $(d/dt)(V_j^i \circ \varphi)(t)$ exists, $i = 1,2\}$, $B = \{t \in A : t$ is a limit point of $A\}$, and $C = \{t \in T_j : t \notin B\}$. The set C is a null set since the set $\{t \in A : t$ is an isolated point in $A\}$ is denumerable and the set $\{t \in T_j : (d/dt)(V_j^1 \circ \varphi)(t)$ does not exist or $(d/dt)(V_j^2 \circ \varphi)(t)$ does not exist$\}$ is a null set.

If $t \in B$, then (8) holds, since : $(V_j^1 \circ \varphi)(\tau) = (V \circ \varphi)(\tau) = (V_j^2 \circ \varphi)(\tau)$ for all $\tau \in A$. Note that T_j is the union of B and the null set C. This completes the proof.

It follows from Lemma 2 that, in Theorem 1 below, there is no need to distinguish among the various existing collections $\{(W_j, V_j) : j \in J\}$ associated with V and D.

Lemma 3. *Let* $h : [a,b] \to E^1$ *be a continuous function which maps null sets into null sets such that* $h(a) > h(b)$ *and* $h(a) \geqslant h(\tau)$ *for all* $\tau \in [a,b]$. *For* $s \in [h(b), h(a)]$ *define* $r(s) = \max\{t \in [a,b] : h(t) = s\}$. *Let* G *be the image of* $[h(b), h(a)]$ *in* $[a,b]$ *by the mapping* r ; *that is,* $G = r([h(b), h(a)]) = \{t \in [a,b] : r(s) = t \text{ for some } s\}$. *Then* G *is a measurable set having positive measure.*

Proof . If G is a measurable set, then G has positive measure ; for, otherwise, h would map the null set G into the set $[h(b), h(a)]$ which has measure $h(a)-h(b) > 0$. We shall show that there exists a denumerable set F_1 such that $G \cup F_1$ is a closed subset of $[a,b]$, thus showing that G is a measurable set and, therefore, has positive measure.

Let $E = \{t \in (a,b) : \exists\ \tau \in (t,b) \ni h(t) < h(\tau)\}$. The set E is open by the rising-sun lemma in (5). Note that $G \cap E$ is empty.

Let $E_1 = \{t \in (a,b) : \exists\ \tau \in (t,b) \ni h(t) = h(\tau)$ and $s \in (t,\tau) \Rightarrow h(s) \leqslant h(t)\}$. Define $K : E_1 \to (a,b)$ such that $K(t) = \max\ \{\tau \in (t,b) : h(t) = h(\tau)$ and $s \in (t,\tau) \Rightarrow h(s) \leqslant h(t)\}$. Note that $(t, K(t))$ is a well defined open interval for all $t \in E_1$. Consequently, the set $E_2 = \cup\{(t, K(t)) : t \in E_1\}$ is an open set in (a,b). Note that $G \cap E_2$ is empty. The set E_2, being open, is the denumerable union of disjoint open intervals ; $E_2 = \cup\ \{E_2(i) : i \in I\}$, where I is the index set of these disjoint open intervals.

Let $F = \{t \in [a,b] : t$ is an endpoint of $E_2(i)$ for some $i \in I\}$. The set F is denumerable and the intersection $F \cap E_2$ is empty.

Since $E \cup E_2$ is open in $[a,b]$, the intersection $G \cap (E \cup E_2)$ is empty, and F is denumerable, it now suffices to show that $G \cup F \cup E \cup E_2 = [a,b]$. For this purpose let $t \in [a,b]$. It must be shown that t belongs to the union $G \cup F \cup E \cup E_2$. If $t \notin E$, then $h(t) \geqslant h(\tau)$ for all $\tau \in [t,b]$. If $h(t) > h(\tau)$ for all $\tau \in (t,b]$, then $t \in G$; this follows since, in this case , $t = r(h(t)) = \max\ \{\tau \in [a,b] : h(\tau) = h(t)\}$. Consequently, if $t \notin E \cup G$, then there exists a $\tau \in (t,b]$ such that $h(t) = h(\tau)$ and such that $s \in (t,\tau) \Rightarrow h(s) \leqslant h(t)$, which implies that $(t,\tau) \subset E_2$ and, therefore, $t \in E_2 \cup F$. Thus $G \cup F \cup E \cup E_2 = [a,b]$.

Since $G \subset [a,b] - E \cup E_2$ and $G \cup F \supset [a,b] - E \cup E_2$, there exists $F_1 \subset F$ such that $G \cup F_1 = [a,b] - E \cup E_2$. The set $G \cup F_1$ is closed, since $E \cup E_2$ is open. The set F_1 is denumerable since F is denumerable. This completes the proof of the lemma.

Lemma 4. *Let* X *be a subset of* E^n *and* $D = \{X_j : j \in J\}$ *be a denumerable decomposition of* X. *Let* $\varphi : [t_o,t_f] \to X$ *be absolutely continuous and* $h_o : [t_o,t_f] \to E^1$ *integrable. Let* $V : X \to E^1$ *be continuous and continuously differentiable with respect to* D. *Let* $\{(W_j, V_j) : j \in J\}$ *be a collection which is associated with* V *and* D. *Let* $T_j = \{t \in [t_o,t_f] : \varphi(t) \in X_j\}$ *for* $j \in J$. *Suppose that, for each* $j \in J$,

$$h_o(t) + (d/dt)(V_j \circ \varphi)(t) \geqslant 0 \qquad \text{a.e. } \textit{in } T_j \tag{9}$$

Then, the function $g : [t_o,t_f] \to E^1$ *defined by*

$$g(t) = \int_{t_o}^{t} h_o(\tau)d\tau + (V \circ \varphi)(t) \tag{10}$$

for $t \in [t_o,t_f]$ *is monotone nondecreasing, continuous, maps null sets into null sets, and hence is absolutely continuous.*

Proof. Let $\psi(t) = \int_{t_1}^{t} h_o(\tau)d\tau$ for $t \in [t_o,t_f]$. The function ψ is absolutely continuous (e.g., (5)-(7)). Therefore, g is continuous. Lemma 1 implies that $V \circ \varphi$ maps null sets into null sets. Therefore, g maps null sets into null sets since both ψ and $V \circ \varphi$ do, and $g = \psi + V \circ \varphi$.

It is shown in (5) that a monotone nondecreasing function on the real line is absolutely continuous if, and only if, it is continuous and maps null sets into null sets. It suffices, therefore, to show that g is monotone nondecreasing. This is accomplished by assuming that g is not monotone nondecreasing and obtaining a contradiction.

Suppose that g is not monotone nondecreasing. Then there exist α and $b \in [t_o, t_f]$ with $\alpha < b$ and $g(\alpha) > g(b)$.

Let $h(t) = g(t) + \varepsilon t$, where $0 < \varepsilon < [g(\alpha) - g(b)]/(b-\alpha)$. The function h is continuous, maps null sets into null sets, and the relation $h(\alpha) > h(b)$ is satisfied.
Let $a = \max\{t \in [\alpha, b) : h(t) = h(\alpha)\}$. Note that $h(\alpha) > h(\tau)$ for all $\tau \in (a, b]$.

For $s \in [h(b), h(a)]$ define $r(s) = \max\{t \in [t_o, t_f] : h(t) = s\}$. The function r is strictly decreasing and continuous from the left. Let G be the image of $[h(b), h(a)]$ in $[a, b]$ by the mapping r. G is a measurable set having positive measure ; this follows from Lemma 3.

For $j \in J$, let $\theta_j = \{t \in [t_o, t_f] : \varphi(t) \in W_j\}$ and let

$$h_j(t) = \psi(t) + (V_j \circ \varphi)(t) + \varepsilon t \quad \text{for } t \in \theta_j. \tag{11}$$

If $t \in T_j$, then $h_j(t) = h(t)$.
Let $S_j = \{t \in T_j : (d/dt)(V_j \circ \varphi)(t)$ exists, $d\psi(t)/dt$ exists, $d\psi(t)/dt = h_o(t)$ and condition (9) is satisfied$\}$.
Let $K_j = \{t \in T_j : (d/dt)(V_j \circ \varphi)(t)$ does not exist or $d\psi(t)/dt$ does not exist or $d\psi(t)/dt \neq h_o(t)$ or condition (9) is not satisfied$\}$. The set K_j is a null set and $t \in S_j$ implies $(d/dt)\, h_j(t) > 0$.

Let $K = \cup\{K_j : j \in J\}$. The set K is a null set.
Let $G_1 = G - K$. The set G_1 is measurable and has positive measure. Since $[a, b] \subset [t_o, t_f] = \cup\{(S_j \cup K_j) : j \in J\}$ and G is uncountable, it follows that there exists $j_o \in J$ such that $G_1 \cap S_{j_o}$ is uncountable. Let $A = \{t \in G_1 \cap S_{j_o} : t$ is a limit point in $G_1 \cap S_{j_o}\}$ and let $B = \{t \in G_1 \cap S_{j_o} : t$ is an isolated point in $G_1 \cap S_{j_o}\}$. Since $A \cup B = G_1 \cap S_{j_o}$ and B is denumerable, it follows that A is uncountable.

Let $t \in A$. There is a convergent sequence (t_n) in $G_1 \cap S_{j_o}$ converging to t such that either $t_n < t$ for all n or $t_n > t$ for all n . In either case, we have $(d/dt)h_{j_o}(t) \leqslant 0$ since $t_n < t$ implies $h_{j_o}(t_n) > h_{j_o}(t)$ and $t_n > t$ implies $h_{j_o}(t_n) < h_{j_o}(t)$. Recall that $h_{j_o}(\tau) = h(\tau)$ for all $\tau \in G_1 \cap S_{j_o}$. Note here that $t_n \in G$ implies that there exists s_n contained in $[h(b), h(a)]$ such that $r(s_n) = t_n$; therefore, $s_n = h(t_n) > h(\tau)$ for all $\tau \in (t_n, b]$, and hence $t_n < t$ implies $h(t_n) > h(t)$. Also, note here that $t_n = r(s_n) = \max \{\tau \in [a,b] : h(\tau) = s_n\}$. Therefore, we have our contradiction since $(d/dt)h_{j_o}(t) > 0$ for t contained in S_{j_o} . This contradiction proves that g defined by Equation (10) is monotone nondecreasing ; that is, g is monotone increasing. This completes the proof.

For each $i \in \{1,2,...,N\}$ suppose there exist a denumerable decomposition D^i of X and a continuous function $V^i : X \to E^1$ which is continuously differentiable with respect to D^i . It can be easily shown that there exists a denumerable decomposition D of X such that V^i is continuously differentiable with respect to D for all $i \in \{1,2,...,N\}$. Therefore, it is unnecessary for the players to have distinct denumerable decompositions. For this reason, only one denumerable decomposition is utilized in the statement of Theorem 1 below .

Theorem 1. *Let* $p^* = (p^{1*},...,p^{N*})$ *be contained in* $I(x^o)$. *For the optimality of* p^* *at* x^o, *it is sufficient that there exists a denumerable decomposition* D *of* X *and for each* $i \in \{1,...,N\}$ *there exists a continuous function* $V^{i*} : X \to E^1$ *which is continuously differentiable with respect to* D *such that*

(i) $$\int_{t_o}^{t_f^*} f_o^i(\varphi^*(\tau), p^*(\varphi^*(\tau)))d\tau = V^{i*}(x^o)$$

for all $\varphi^* \in \Phi(x^o ; p^*)$
where t_f^* *is the terminating time for* φ^* ;

(ii)† $f_o^i(x,p^{1*}(x),\ldots,v^i,\ldots p^{N*}(x)) +$

$$+ \text{grad } V_j^{i*}(x) \cdot f(x,p^{1*}(x),\ldots,v^i,\ldots,p^{N*}(x)) \geqq 0$$

for all $x \in X_j$, $v^i \in U^i(x)$, $j \in J$;

(iii) $V^{i*}(x) = 0$ *for all* $x \in \Theta$;

where $\{(W_j, V_j^{i*}) : j \in J\}$ *is a collection associated with* V^{i*} *and* $D = \{X_j : j \in J\}$ *for each* $i \in \{1,2,\ldots,N\}$.

Proof. Choose any $i \in \{1,\ldots,N\}$. Let $p = (p^{1*},\ldots,p^i,\ldots,p^{N*})$ be contained in $I(x^o)$ and let φ be contained in $\Phi(x^o ; p)$. Let t_f be the terminating time for the trajectory φ. In view of (i), it suffices to show that

$$V^{i*}(x^o) \leqslant \int_{t_o}^{t_f} f_o^i(\varphi(\tau), p(\varphi(\tau)))d\tau \tag{12}$$

For t contained in $[t_o,t_f]$ define

$$g^i(t) = \int_{t_o}^{t} f_o^i(\varphi(\tau), p(\varphi(\tau)))d\tau + (V^{i*} \circ \varphi)(t) \tag{13}$$

Let $h_o^i(t) = f_o^i(\varphi(t), p(\varphi(t)))$ for all $t \in [t_o,t_f]$. The function h_o^i is an integrable function since f_o^i is a bounded Borel measurable function. It follows from (ii) that condition (9) of

---

† Dot notation is used to denote inner product ; e.g., for $a = (a_1,a_2,\ldots,a_n)$ and $b = (b_1,b_2,\ldots b_n)$, $a \cdot b = \sum_{i=1}^{n} a_i b_i$.

Lemma 4 is satisfied. Therefore, by applying Lemma 4, we have that g^i is a monotone increasing continuous function. Thus, since $g^i(t_o) = V^{i*}(x^o) = (V^{i*} \circ \varphi)(t_o)$, it follows that $g^i(t) \geq V^{i*}(x^o)$. But this together with (13) gives

$$V^{i*}(x^o) \leq \int_{t_o}^{t} f_o^i(\varphi(\tau), p(\varphi(\tau)))d\tau + (V^{i*} \circ \varphi)(t) \tag{14}$$

The inequality (12) follows from (14) and (iii). This completes the proof.

For two-person zero-sum games, $N = 2$ and $f_o^1 = -f_o^2 = f_o$. For such games we have the following corollary.

Corollary 1. *Let* $p^* = (p^{1*}, p^{2*})$ *be contained in* $I(x^o)$. *For the optimality of* p^* *at* x^o, *it is sufficient that there exist a denumerable decomposition* D *of* X *and a continuous function* $V^*: X \to E^1$ *which is continuously differentiable with respect to* D *such that*

(a) $$f_o(x, p^{1*}(x), v^2) + \text{grad } V_j^*(x) \cdot f(x, p^{1*}(x), v^2) \geq 0$$

for all $x \in X_j$, $v^2 \in U^2(x)$, $j \in J$;

(b) $$f_o(x, v^1, p^{2*}(x)) + \text{grad } V_j^*(x) \cdot f(x, v^1, p^{2*}(x)) \leq 0$$

for all $x \in X_j$, $v^1 \in U^1(x)$, $j \in J$;

(c) $V^*(x) = 0$ *for all* $x \in \Theta$.

Proof. Let (p^{1*}, p^2) and (p^1, p^{2*}) be contained in $I(x^o)$. Consider φ_1, φ^*, φ_2 contained in $\Phi(x^o ; p^{1*}, p^2)$, $\Phi(x^o ; p^{1*}, p^{2*})$ and $\Phi(x^o ; p^1, p^{2*})$, respectively. Let t_f^1 , t_f^* and t_f^2 be the terminating times for trajectories φ_1 , φ^* and φ_2 , respectively.

Finally, let

$$(15)\quad g_1(t) = \int_{t_o}^{t} f_o(\varphi_1(\tau),p^{1*}(\varphi_1(\tau)),p^2(\varphi_1(\tau)))d\tau + (V^*\circ\varphi_1)(t)$$

for all t in $[t_o,t_f^1]$,

$$(16)\quad g^*(t) = \int_{t_o}^{t} f_o(\varphi^*(\tau),p^{1*}(\varphi^*(\tau)),p^{2*}(\varphi^*(\tau)))d\tau + (V^*\circ\varphi^*)(t)$$

for all t in $[t_o,t_f^*]$, and

$$(17)\quad g_2(t) = \int_{t_o}^{t} f_o(\varphi_2(\tau),p^1(\varphi_2(\tau)),p^{2*}(\varphi_2(\tau)))d\tau + (V^*\circ\varphi_2)(t)$$

for all t in $[t_o,t_f^2]$.

In view of (a) and (b), it follows from Lemma 4 that g^*, $-g^*$, g_2 and $-g_1$ are monotone nondecreasing on the respective intervals. But $g_1(t_o) = g^*(t_o) = g_2(t_o)$, so that

$$(18)\quad g_1(t_f^1) \leqslant g^*(t_f^*) \leqslant g_2(t_f^2)$$

Finally, as a consequence of (c), (18) leads to

$$(19)\quad \int_{t_o}^{t_f^1} f_o(\varphi_1(\tau),p^{1*}(\varphi_1(\tau)),p^2(\varphi_2(\tau)))d\tau$$

$$\leqslant \int_{t_o}^{t_f^*} f_o(\varphi^*(\tau),p^{1*}(\varphi^*(\tau)),p^{2*}(\varphi^*(\tau)))d\tau$$

$$\leqslant \int_{t_o}^{t_f^2} f_o(\varphi_2(\tau),p^1(\varphi_2(\tau)),p^{2*}(\varphi_2(\tau)))d\tau$$

This concludes the proof.

Remark 1. In Theorem 1, if $p^* \in I$ and condition (i) is met for all $x^o \in X$, then p^* is optimal on X. Similarly, in Corollary 1, if $p^*=(p^{1*},p^{2*}) \in I$, then p^* is optimal on X .

3. Sufficiency and Game Surfaces

Lemma 4 is utilized now to point out a relationship between *game surfaces* and the sufficient conditions (i)-(iii) of Theorem 1. For this purpose the following definitions are needed.

Definition 3. Let p be contained in $I(x^o)$ and let φ be contained in $\Phi(x^o,p)$ where $\varphi : [t_o,t_f] \to X$. Let r be a real number and for $i \in \{1,\ldots,N\}$ let $\psi^i_r : [t_o,t_f] \to E^1 \times X$ be defined by $\psi^i_r(t) = (z^i(t), \varphi(t))$ where

$$(20) \qquad z^i(t) + \int_t^{t_f} f^i_o(\varphi(\tau),p(\varphi(\tau)))d\tau = r$$

for all t contained in $[t_o,t_f]$. The mapping ψ^i_r is referred to as an *augmented admissible path for player i*.

Definition 4. For $i \in \{1,2,\ldots,N\}$ let $V^{i*} : X \to E^1$ be continuous and let r be a real number. Let

$$(21) \qquad \Sigma^i_r = \{(z,x) \in E^1 \times X : z + V^{i*}(x) = r\}$$

The set Σ^i_r is referred to as a *game surface for player i*.

Definition 5. Let

$$(22) \qquad A/\Sigma^i_r = \{(z,x) \in E^1 \times X : z + V^{i*}(x) > r\}$$

$$(23) \qquad B/\Sigma^i_r = \{(z,x) \in E^1 \times X : z + V^{i*}(x) < r\}$$

A point $(z,x) \in A/\Sigma^i_r$ is called an A *(above) point relative to the game surface* Σ^i_r for player i and a point $(z,x) \in B/\Sigma^i_r$ is called a B *(below) point relative to the game surface* Σ^i_r for player i .

We now prove a theorem which is analogous to Theorem 1 in (8, page 13) and Lemmas 2.1 and 2.2 in (9, pages 17 f.) .

Theorem 2. *Let* $p^* = (p^{1*},\dots,p^{N*})$ *be contained in* I. *Let* D *be a denumerable decomposition of* X *and for each* $i \in \{1,2,\dots,N\}$ *let* $V^{i*} : X \to E^1$ *be continuously differentiable with respect to* D *such that* (i)-(iii) *of Theorem* 1 *are satisfied for each* $x_o \in X$. *Let* $i \in \{1,2,\dots,N\}$. *If players* $k \neq i$, $k \in \{1,2,\dots,N\}$, *play* p^{k*}, *then an augmented admissible path for player* i *(optimal or non-optimal), whose initial point belongs to a given game surface for player* i, *cannot have a* B *point relative to the given game surface for player* i. *Moreover, if all players agree to play* p^* *then all augmented admissible paths for player* i *belong to the given game surface for player* i.

Proof. Let $x^o \in X$ and let $i \in \{1,2,\dots,N\}$. Consider $p = (p^{1*},\dots,p^i,\dots,p^{N*}) \in I(x^o)$. Let $\varphi \in \Phi(x^o ; p)$ where $\varphi : [t_o,t_f] \to X$.

Consider the following game surface Σ^i_r for player i given by

$$\Sigma^i_r = \{(z,x) \in E^1 \times X : z + V^{i*}(x) = r\} \tag{24}$$

Let $\psi^i_{r_1} : [t_o,t_f] \to E^1 \times X$ be the augmented admissible path corresponding to φ such that

$$z^i_1(t) + \int_t^{t_f} f^i_o(\varphi(\tau), p(\varphi(\tau)))d\tau = r_1 \tag{25}$$

where

$$r_1 = r + \int_{t_o}^{t_f} f^i_o(\varphi(\tau), p(\varphi(\tau)))d\tau - V^{i*}(x_o) \tag{26}$$

and where $\psi^i_{r_1}(t) = (z^i_1(t), \varphi(t))$ for all $t \in [t_o,t_f]$. Note that the initial point of the augmented admissible path $\psi^i_{r_1}$ belongs to the given game surface Σ^i_r for player i .

Using Lemma 4 and (i)-(iii) of Theorem 1, we have the inequality (14). Using (14), (25) and (26), we obtain

$$r \leqslant z_1^i(t) + (V^{i*} \circ \varphi)(t) \tag{27}$$

The inequality (27) implies that $\psi_{r_1}^i(t) \in \Sigma_r^i \cup A/\Sigma_r^i$ for all $t \in [t_o, t_f]$.

Now, if $p \equiv p^*$ then the inequality (14) reduces to

$$V^{i*}(x^o) = \int_{t_o}^{t} f_o^i(\varphi^*(\tau), p^*(\varphi^*(\tau)))d\tau + (V^{i*} \circ \varphi^*)(t) \tag{28}$$

for all $\varphi^* \in \Phi(x^o; p^*)$; otherwise, (i) could not be met. For the case that $p \equiv p^*$, (25) and (26) reduce to

$$z_1^i(t) = r + \int_{t_o}^{t} f_o^i(\varphi^*(\tau), p^*(\varphi^*(\tau)))d\tau - V^{i*}(x^o). \tag{29}$$

From (28) and (29) it follows that

$$r = z_1^i(t) + (V^{i*} \circ \varphi^*)(t) \tag{30}$$

for all $t \in [t_o, t_f^*]$ where t_f^* is the terminal time for φ^*. Thus, if $p \equiv p^*$ then $\psi_r^i(t) \in \Sigma_r^i$ for all $t \in [t_o, t_f^*]$. This completes the proof.

4. Application of Sufficiency Conditions

In this section we present two problems, one simple and one rather complicated, to illustrate the utilization of the sufficiency conditions embodied in Theorem 1 and Corollary 1.

First we consider a two-person zero-sum game to illustrate Corollary 1.

Example 1. Consider the system with state equations

$$(31) \quad \begin{aligned} \dot{x}_1 &= x_2 \\ \dot{x}_2 &= a\, v^1 + v^2 \\ \dot{x}_3 &= 1 \end{aligned}$$

where a = constant $\geqslant 1$. The control constraints are

$$(32) \quad \begin{aligned} U^1(x) &= \{v^1 : |v^1| \leqslant 1\} \\ U^2(x) &= \{v^2 : |v^2| \leqslant 1\} \end{aligned}$$

for all $x \in E^3$.

The target set Θ is given by

$$(33) \quad \Theta = \{x : x_3 = x_3^f = t_f\}$$

The state space $X = \{x : x_3 \leqslant x_3^f\}$. The player's cost is

$$(34) \quad J(p;\varphi) = \int_{t_o}^{t_f} \varphi_1(t)\varphi_2(t)dt$$

Application of necessary conditions, e.g. (3) or (4), leads to the following candidate strategies :

$$(35) \quad \begin{aligned} p^{1*}(x) &\equiv -1 \\ p^{2*}(x) &\equiv \ \ 1 \end{aligned}$$

for all $x \in X_1$ where

$$X_1 = \{x : x_1+x_2(x_3^f-x_3)- \tfrac{1}{2}(a-1)(x_3^f-x_3)^2 \geqslant 0,\ x_3 \leqslant x_3^f\}$$

and

$$(36) \quad \begin{aligned} p^{1*}(x) &\equiv \ \ 1 \\ p^{2*}(x) &\equiv -1 \end{aligned}$$

for all $x \in X_2$ where

$$X_2 = \{x : x_1+x_2(x_3^f-x_3)+ \tfrac{1}{2}(a-1)(x_3^f-x_3)^2 \leqslant 0,\ x_3 \leqslant x_3^f\}$$

Finally, (p^{1*}, p^{2*}) is any admissible strategy pair for $x \in X_3$ where

$$X_3 = \{x : -\tfrac{1}{2}(a-1) < x_1(x_3^f - x_3)^{-2} + x_2(x_3^f - x_3)^{-1} < \tfrac{1}{2}(a-1),\ x_3 \leqslant x_3^f\}$$

Now let

$$V^{1*}(x) = \tfrac{1}{2}[x_1 + x_2(x_3^f - x_3) - \tfrac{1}{2}(a-1)(x_3^f - x_3)^2]^2 - \tfrac{1}{2}x_1^2$$

for $x \in X_1$

$$V^{2*}(x) = \tfrac{1}{2}[x_1 + x_2(x_3^f - x_3) + \tfrac{1}{2}(a-1)(x_3^f - x_3)^2]^2 - \tfrac{1}{2}x_1^2$$

for $x \in X_2$, and

$$V^{3*}(x) = -\tfrac{1}{2}x_1^2$$

for $x \in X_3$.

It is readily verified that the conditions of Corollary 1 are met for all $x^o \in X$, so that (p^{1*}, p^{2*}) is indeed optimal on X.

Before describing a special decomposition of the plane E^2, which is needed in Example 2, attention is called here to variants of a famous subset of the unit interval, Cantor's discontinuum set. Cantor (10), by removing middle thirds ad infinitum from the unit interval $[0,1]$, constructed a subset of $[0,1]$ which is uncountable, closed, nowhere-dense, perfect, of zero length, and contains the endpoints 0 and 1. Here, a set is called *perfect* if it is closed and contains no isolated points. Subsets of the real numbers that are perfect, nowhere-dense, and of positive length are called here variants of Cantor's discontinuum. A perfect subset of the real line necessarily has an uncountable number of points. A constructive proof of the existence of these variants is given to provide the reader with a deeper insight into Example 2.

Lemma 5. *Given a compact interval* $[a,b]$ *in* E^1 *with* $a < b$, *and given a positive number* δ *contained in* $(0, b-a)$, *there exists a perfect nowhere-dense subset* C *of* $[a,b]$ *such that* C *has length* δ

and such that C *contains the endpoints* a *and* b.

Proof. Define C_o^1 as the compact interval $[a,b]$. Remove at the midpoint of C_o^1 an open interval θ_1^1 of length $\mu/2$, where $\mu = b-a-\delta$. There remain two disjoint closed intervals each of length $(\ell-\mu)/2 + \mu/2^2$ where $\ell = b-a$. Denote them by C_1^1 and C_1^2.

Remove at the midpoint of C_1^i an open interval θ_2^i of length $\mu/2^3$, $i = 1,2$. There remain 2^2 closed disjoint intervals each of length $(\ell-\mu)/2^2 + \mu/2^4$. Denote them by C_2^j, $j = 1,\ldots,2^2$. Repeat this procedure ad infinitum.

Note that the nth step proceeds as follows. There are 2^{n-1} closed disjoint intervals C_{n-1}^i, $i = 1,2,\ldots,2^{n-1}$, which remain after the (n-1)th step, each of length $(\ell-\mu)/2^{n-1} + \mu/2^{2n-2}$. Remove at the midpoint of C_{n-1}^i an open interval θ_n^i of length $\mu/2^{2n-1}$, $i = 1,\ldots,2^{n-1}$. There remain 2^n closed disjoint intervals each of length $(\ell-\mu)/2^n + \mu/2^{2n}$. Denote them by C_n^j, $j = 1,\ldots,2^n$.

Let θ_n be the union of all open intervals removed during the nth step, $\theta_n = \cup\{\theta_n^i : i = 1,2,\ldots,2^{n-1}\}$. Note that θ_n has length $\mu/2^n$. Let θ be the union of the intervals removed at each step, $\theta = \cup\{\theta_n : n = 1,2,\ldots\}$. From the geometric series $\Sigma_{n=1}^{\infty}(\mu/2^n)$ it is easily seen that the length of θ is μ.

Let $C = [a,b] - \theta$. The set C is a closed set having length δ; that is, the length of C is equal to the length of $[a,b]$ minus the length of θ. The set C has no interior, since the length $(\ell-\mu)/2^n + \mu/2^{2n}$ of C_n^j, $j \in \{1,\ldots,2^n\}$, converges to zero as $n \to \infty$. Recall here that $C \subset \cup\{C_n^j : j = 1,2,\ldots,2^n\}$ for all n and that the C_n^j are pairwise disjoint for n fixed and $j \in \{1,2,\ldots,2^n\}$. Note that $C = \cap_{n=1}^{\infty} \cup_{j=1}^{2^n} C_n^j$. C is, therefore, nowhere-dense. It is now to be shown that C contains no isolated points; or, equivalently, it is to be shown that each point of

C is an accumulation point (limit point) of C. Suppose that $x \in C$. Then, for each $n \in \{1,2,\ldots\}$, it follows that $x \in \cup \{C_n^j : j = 1,2,\ldots,2^n\}$. Therefore, the exists j_x^n such that $x \in C_n^{j_x^n}$, where $1 \leqq j_x^n \leqq 2^n$. Let y_n be an endpoint of $C_n^{j_x^n}$ such that $y_n \neq x$. Then, for each $n \in \{1,2,\ldots\}$, it follows that y_n is contained in C and y_n converges to x as $n \to \infty$. Recall here that the length of C_n^j goes to zero as $n \to \infty$. This completes the proof.

We are now ready to describe a special denumerable decomposition of E^2 which is used in Example 2. This decomposition of E^2 is given the descriptive title "Cantor δ-pie decomposition of E^2", since every member of this decomposition except one looks like a piece of pie (i.e., open wedge) sitting in E^2; the exception is a perfect nowhere-dense subset of E^2 having infinite area in two dimensions and defined by utilizing a perfect nowhere-dense subset of the interval $[0, 2\pi]$ having length δ where δ is contained in $(\pi, 2\pi)$.

Let δ be contained in $(\pi, 2\pi)$. It follows from Lemma 4 that there exists a perfect nowhere-dense subset C of $[0, 2\pi]$ such that C has length δ and contains the endpoints 0 and 2π. It is supposed that C has been constructed as in the proof of Lemma 4. Let $\theta = [0, 2\pi] - C$. The set θ, being open, is the countable union of disjoint open intervals (α_i, β_i) for $i \in I$, where I is the countable index set of these disjoint open intervals whose union is θ. It follows from the construction of C that

$$\delta + \sum_{i \in I} (\beta_i - \alpha_i) = 2\pi \tag{37}$$

$$\beta_i - \alpha_i \leqq \pi - \delta/2 \quad \text{for all } i \in I \tag{38}$$

Let (r,ψ) denote polar coordinates of a point in E^2. Let

$$\hat{X}_o = \{(r \cos \psi, r \sin \psi) \in E^2 : 0 \leqq r < \infty, \psi \in C\} \tag{39}$$

and, for each $i \in I$, let

(40) $$\hat{X}_i = \{(r \cos \psi, r \sin \psi) \in E^2 : 0 < r < \infty, \alpha_i < \psi < \beta_i\}$$

The subset $\hat{X}_o$ is a perfect nowhere-dense subset of E^2 having infinite area. For $i \in I$, the subset $\hat{X}_i$ is an open wedge with angle $\beta_i - \alpha_i$; it resembles a piece of pie whose angle of cut is $\beta_i - \alpha_i$.

Let $\hat{D} = \{\hat{X}_o, \hat{X}_i : i \in I\}$. Note that the members of $\hat{D}$ are pairwise disjoint, $\hat{D}$ is denumerable and $\cup \hat{D} = E^2$. $\hat{D}$ is therefore, a denumerable decomposition of E^2 .

For each $i \in I$, let $\hat{A}_i$ be the subset of $\hat{X}_i$ having $(\alpha_i + \beta_i)/2 < \psi < \beta_i$, let $\hat{B}_i$ be the subset of $\hat{X}_i$ having $\alpha_i < \psi < (\alpha_i + \beta_i)/2$ and let $\hat{C}_i$ be the subset of $\hat{X}_i$ such that $\psi = (\alpha_i + \beta_i)/2$. Let $\hat{D}_2 = \{\hat{X}_o , \hat{A}_i , \hat{B}_i, \hat{C}_i : i \in I\}$. It can be easily verified that $\hat{D}_2$ is a denumerable decomposition of E^2 .

Now we consider a nonzero-sum game to illustrate the use of Theorem 1.

Example 2. Consider the system with state equations

$$\dot{x}_1 = - v^1 \cos v^2$$

(41) $$\dot{x}_2 = - v^1 \sin v^2$$

$$\dot{x}_3 = 1$$

and control constraints

(42) $$U^1(x) = \{v^1 \in E^1 : 0 \leqslant v^1 \leqslant 1\} \text{ for all } x \in E^3 ;$$

(43) $$U^2(x) = \{v^2 \in E^1 : \alpha_i \leqslant v^2 \leqslant \beta_i\} \text{ for all } x \in X_i, i \in I ;$$

(44) $$U^2(x) = \{v^2 \in E^1 : 0 \leqslant v^2 \leqslant 2\pi\} \text{ for all } x \in X_o ;$$

where $X_i = \hat{X}_i \times E^1$ and $X_o = \hat{X}_o \times E^1$.

The target set $\Theta = \{x : x_1 = x_2 = 0\}$. The state space X is all of E^3 .

The player's costs are given by

$$(45) \qquad J^1(p,\varphi) = \int_{t_o}^{t_f} dt$$

$$(46) \qquad J^2(p,\varphi) = \int_{t_o}^{t_f} f_o^2(\varphi(t))dt$$

where $f_o^2(x) = 1$ for all $x \in X_i$, $i \in I$, and where $f_o^2(x) = 0$ for all $x \in X_o$.

Let $p^* = (p^{1*}, p^{2*})$ with

$$(47) \qquad p^{1*}(x) \equiv 1 \qquad \text{for all } x \in E^3$$

$$(48) \qquad p^{2*}(x) = \begin{cases} \alpha_i & \text{for all } x \in A_i \cup C_i \text{ , } i \in I \\ \beta_i & \text{for all } x \in B_i \text{ , } i \in I \end{cases}$$

$$(49) \qquad p^{2*}(x) = \cos^{-1}(x_1/\sqrt{[(x_1)^2+(x_2)^2]})$$

$$= \sin^{-1}(x_2/\sqrt{[(x_1)^2+(x_2)^2]})$$

$$\text{for all } x \in X_o$$

where $A_i = \hat{A}_i \times E^1$, $B_i = \hat{B}_i \times E^1$ and $C_i = \hat{C}_i \times E^1$.

Define

$$(50) \qquad V_o^{1*}(x) = \sqrt{[(x_1)^2+(x_2)^2]}$$

$$\text{for all } x \in E^3$$

$$(51) \qquad V_i^{1*}(x) = [x_1(\sin \beta_i - \sin \alpha_i)$$

$$+ \; x_2(\cos \alpha_i - \cos \beta_i)] / \sin(\beta_i - \alpha_i)$$

$$\text{for all } x \in X_i, \; i \in I$$

$$(52) \qquad V_o^{2*}(x) = 0 \qquad \text{for all } x \in E^3$$

$$V^{2*}_{ai}(x) = (x_1 \sin \beta_i - x_2 \cos \beta_i)/\sin(\beta_i - \alpha_i)$$

for all $x \in A_i$, $i \in I$;

(53)
$$V^{2*}_{bi}(x) = (x_2 \cos \alpha_i - x_1 \sin \alpha_i)/\sin(\beta_i - \alpha_i)$$

for all $x \in B_i$, $i \in I$;

$$V^{2*}_{ci}(x) = [\sin((\beta_i - \alpha_i)/2)/\sin(\beta_i - \alpha_i)] \sqrt{[(x_1)^2 + (x_2)^2]}$$

for all $x \in X_i$, $i \in I$.

The collections $\{(V_o^{1*}, E^3), (V_i^{1*}, A_i), (V_i^{1*}, B_i), (V_i^{1*}, X_i) : i \in I\}$ and $\{(V_o^{2*}, E^3), (V_{ai}^{2*}, A_i), (V_{bi}^{2*}, B_i), (V_{ci}^{2*}, X_i) : i \in I\}$ are associated with V^{1*} and D, and with V^{2*} and D, respectively, where $D = \{ X_o , A_i , B_i , C_i : i \in I \}$ and

$$(54) \qquad V^{1*}(x) = \begin{cases} V_o^{1*}(x) & \text{for all } x \in X_o \\ V_i^{1*}(x) & \text{for all } x \in X_i , \ i \in I \end{cases}$$

$$(55) \qquad V^{2*}(x) = \begin{cases} V_o^{2*}(x) & \text{for all } x \in X_o \\ V_{ai}^{2*}(x) & \text{for all } x \in A_i , \ i \in I \\ V_{bi}^{2*}(x) & \text{for all } x \in B_i , \ i \in I \\ V_{ci}^{2*}(x) & \text{for all } x \in C_i , \ i \in I \end{cases}$$

Note that the continuous functions V^{1*} and V^{2*} are continuously differentiable with respect to D .

It is easily verified that $p^* \in I(x_o)$ for all $x_o \in E^3$. For x^o contained in X_o the trajectory $\varphi^* = (\varphi_1^*, \varphi_2^*, \varphi_3^*) \in \Phi(x^o ; p^*)$ is given by

$$(56)\qquad \varphi_1^*(t) = x_1^o - x_1^o(t-t_o)/\sqrt{[(x_1^o)^2 + (x_2^o)^2]}$$

$$(57)\qquad \varphi_2^*(t) = x_2^o - x_2^o(t-t_o)/\sqrt{[(x_1^o)^2 + (x_2^o)^2]}$$

for all t contained in $[t_o, t_o + \sqrt{[(x_1^o)^2 + (x_2^o)^2]}]$. For $x^o \in X_i$, $i \in I$, the trajectory φ^* is given by

$$(58)\qquad \varphi_1^*(t) = x_1^o - (t-t_o)\cos a_i$$

$$(59)\qquad \varphi_2^*(t) = x_2^o - (t-t_o)\sin a_i$$

for all t contained in $[t_o, t_o + t_1^*]$ and by

$$(60)\qquad \varphi_1^*(t) = (t_2^* + t_o - t)\cos b_i$$

$$(61)\qquad \varphi_2^*(t) = (t_2^* + t_o - t)\sin b_i$$

for all t contained in $[t_o + t_1^*, t_o + t_2^*]$ where

$$(62)\qquad t_1^* = (x_1^o \sin b_i - x_2^o \cos b_i)/\sin(b_i - a_i)$$

$$(63)\qquad t_2^* = [x_1^o(\sin \beta_i - \sin \alpha_i) + x_2^o(\cos \alpha_i - \cos \beta_i)]/\sin(\beta_i - \alpha_i)$$

where $a_i = \alpha_i$ and $b_i = \beta_i$ for all $x^o \in A_i \cup C_i$ and where $a_i = \beta_i$ and $b_i = \alpha_i$ for all $x^o \in B_i$.

Recall here that $0 \leqslant (\beta_i - \alpha_i) < \pi/2$. Of course, $\varphi_3^*(t) \equiv t$.

First it will be shown that (i) of Theorem 1 holds. Let $x^o \in X_o$. It follows from (56) and (57) that $(\varphi_1^*(t), \varphi_2^*(t)) = (0,0)$ at $t_o + \sqrt{[(x_1^o)^2 + (x_2^o)^2]} = t_f^*$. Using (45) we have $J^1(p^*; \varphi^*) = t^* - t_o$. Making use of (50) and (54) we see that (i) holds for player 1 at all $x^o \in X_o$.

Let $x^o \in X_i$ for any $i \in I$. It follows from (58)-(61) that $(\varphi_1^*(t), \varphi_2^*(t)) = (0,0)$ at $t = t_o + t_2^*$. Using (45) we have $J^1(p^*;\varphi^*) = t_2^*$. Making use of (51) and (54) we see that (i) holds for player i at all $x^o \in X_i$, $i \in I$.

Let $x^o \in X_o$. It follows from (41),(47),(49),(56) and (57) that if $(\varphi_1^*(t_o),\varphi_2^*(t_o),t_o) = x^o \in X_o$ then $(\varphi_1^*(t),\varphi_2^*(t),t) \in X_o$ for all t contained in $[t_o, t_o + \sqrt{[(x_1^o)^2 + (x_2^o)^2]}]$. Consequently since $f_2^o(x) = 0$ for all $x \in X_o$ it follows from (46) that $J^2(p^*;\varphi^*) = 0$. From (52) and (55) we see that (i) holds for player 2 at all $x^o \in X_o$.

Let $x^o \in X_i$ for any $i \in I$. It follows from (58) and (59) that $(\varphi_1^*(t),\varphi_2^*(t),t) \in X_i$ for all t contained in $[t_o, t_o + t_1^*)$, and from (60) and (61) that $(\varphi_1^*(t),\varphi_2^*(t),t) \in X_o$ for all $t \in [t_o + t_1^*, t_o + t_2^*]$. Thus, since $f_o^2(x) = 1$ for all $x \in X_i$, it follows that $J^2(p^*;\varphi^*) = t_1^*$ for all $x^o \in X_i$, $i \in I$. From (53) and (55) we see that (i) holds for player 2 at all $x^o \in X_i$, $i \in I$.

It will now be shown that (ii) of Theorem 1 holds. In view of (41),(45),(49) and (50) the left-hand side (LHS) of (ii) becomes

(64) $\quad$ LHS of (ii) $= 1 - v^1$ $\quad$ for all $x \in X_o$

As a consequence of (41),(45),(48) and (51)

(65) $\quad$ LHS of (ii) $= 1 - v^1$ $\quad$ for all $x \in X_i$, $i \in I$.

Since $0 \leqq v^1 \leqq 1$, condition (ii) of Theorem 1 is met for player 1.

Because of (41),(46),(47) and (52) the left-hand side of (ii) is identically equal to zero for all $x \in X_o$. Thus, (ii) is met for player 2 for all $x \in X_o$.

In view of (41),(46),(47) and (53)

$$\text{(66)} \qquad \text{LHS of (ii)} = \begin{cases} 1-\sin(\beta_i - v^2)/\sin(\beta_i - \alpha_i) & \text{for } x^o \in A_i, i \in I \\ 1-\sin(v^2-\alpha_i)/\sin(\beta_i-\alpha_i) & \text{for } x^o \in B_i, i \in I \\ 1-\sin((\beta_i-\alpha_i)\cos(v^2-(\beta_i-\alpha_i)/2)/\sin(\beta_i-\alpha_i) & \\ & \text{for } x^o \in C_i, i \in I \end{cases}$$

Since $\alpha_i < v^2 < \beta_i$ and since $\beta_i - \alpha_i < \pi/2$, it follows from (66) that (ii) is met for player 2 for all $x \in X_i$, $i \in I$.

Finally, as a consequence of (39),(50),(52),(54) and (55), (iii) of Theorem 1 holds. Consequently, all conditions of Theorem 1 hold. Therefore, p^* described by (47)-(49) is optimal over all of the state space. This concludes Example 2.

5. Nash Equilibrium Games with Fixed Playing Time

In this section we give a sufficiency theorem applicable to games with fixed playing time. Consequently, it is convenient to let $x = (y,t)$ and $x^o = (y^o, t_o)$ where y is an (n-1) dimensional vector. The interval $[t_o, t_f]$ is fixed with $t_f \in [t_o, \infty)$. The playing space X is denoted by $Y \times [t_o, t_f]$. The target set is such that $t = t_f$ if $(y,t) \in \Theta$.

For $i \in \{1,2,\ldots,N\}$ let $\lambda^i : [t_o, t_f] \to E^n$ denote absolutely continuous functions.

Theorem 3. *Let* $p^* = (p^{1*}, \ldots, p^{N*})$ *be contained in* $I(x^o)$. *For the optimality of* p^* *at* x^o, *it is sufficient that the following conditions hold :*

For $i \in \{1,2,\ldots,N\}$

$$\text{(I)} \qquad \int_{t_o}^{t_f} f_o^i(\varphi^*(\tau), p^*(\varphi^*(\tau)))d\tau = \text{constant}$$

$$\textit{for all } \varphi^* \in \Phi(x^o ; p^*)$$

There exists a $\varphi^* \in \Phi(x^o; p^*)$ *and there exist* $\lambda^i, i \in \{1,2,\ldots,N\}$ *such that for* $i \in \{1,2,\ldots,N\}$

(II) $$f_o^i(\varphi^*(t),p^*(\varphi^*(t))) + \lambda^i(t) \cdot f(\varphi^*(t),p^*(\varphi^*(t)))$$
$$- f_o^i(x,p^{1*}(x),\ldots,v^i,\ldots,p^{N*}(x))$$
$$- \lambda^i(t) \cdot f(x,p^{1*}(x),\ldots,v^i,\ldots,p^{N*}(x))$$
$$+ \dot{\lambda}^i(t) \cdot (\varphi^*(t) - x) \leqslant 0$$

a.e. *in* $[t_o,t_f]$, *for all* $y \in Y$ *and for all* $v^i \in U^i(x)$ *where* $x = (y,t)$.

(III) $\lambda^i(t_f) \cdot (\varphi^*(t_f) - x) \geqslant 0$ *for all* $x \in \Theta$.

Proof. Let $i \in \{1,2,\ldots,N\}$. Consider a strategy N-tuple $p = (p^{1*}, p^{2*},\ldots,p^i,\ldots,p^{N*}) \in I(x^o)$, $x^o = (y^o,t_o)$, and a trajectory $\varphi^i \in \Phi(x^o;p)$. By (II)

$$f_o^i(\varphi^*(t),p^*(\varphi^*(t))) - f_o^i(\varphi^i(t),p(\varphi^i(t))) + \frac{d}{dt}[\lambda^i(t) \cdot (\varphi^*(t) - \varphi^i(t))] \leqslant 0 \tag{67}$$

On integration of (67), and use of $\varphi^*(t_o) = \varphi^i(t_o) = x^o$ and condition (III), we have

$$\int_{t_o}^{t_f} f_o^i(\varphi^*(\tau),p^*(\varphi^*(\tau)))d\tau \leqslant \int_{t_o}^{t_f} f_o^i(\varphi^i(\tau),p(\varphi^i(\tau)))d\tau \tag{68}$$

Thus, Nash equilibrium condition (6) is met. Since (I) implies (7) the proof is complete.

For a two-person zero-sum game, $N = 2$ and $f_o^1 = - f_o^2 = f_o$. For such a game we have the following corollary.

Corollary 2. *Strategy pair* $p^* = (p^{1*},p^{2*}) \in I(x^o)$ *is optimal (saddle-point) at* $x^o = (y^o,t_o)$ *if there exist a* $\varphi^* \in \Phi(x^o;p^*)$ *and an absolutely continuous function* $\lambda : [t_o,t_f] \to E^n$ *such that*

$$(\alpha) \qquad f_o(\varphi^*(t),p^*(\varphi^*(t))) + \lambda(t) \cdot f(\varphi^*(t),p^*(\varphi^*(t)))$$

$$- f_o(x,v^1,p^{2*}(x))-\lambda(t)\cdot f(x,v^1,p^{2*}(x))+\dot{\lambda}(t)\cdot(\varphi^*(t)-x) \leqslant 0$$

$$(\beta) \qquad f_o(\varphi^*(t),p^*(\varphi^*(t))) + \lambda(t) \cdot f(\varphi^*(t),p^*(\varphi^*(t)))$$

$$- f_o(x,p^{1*}(x),v^2)-\lambda(t)\cdot f(x,p^{1*}(x),v^2)+\dot{\lambda}(t)\cdot(\varphi^*(t)-x) \geqslant 0$$

a.e. *in* $[t_o,t_f]$, *for all* $y \in Y$, *and for all* $v^1 \in U^1(x)$ *and* $v^2 \in U^2(x)$, *where* $x = (y,t)$.

$$(\gamma) \qquad \lambda(t_f) \cdot (\varphi^*(t_f) - x) = 0 \quad \textit{for all} \quad x \in \Theta$$

Remark 2. The solutions of the adjoint equations in the necessary conditions, e.g., Equation (12) of [3] or Equation (93) of [4], may be used in place of functions λ^i . Then, if $\varphi_i^*(t_f)$, $i \neq n$, are free, that is $\Theta = E^{n-1}\times\{t_f\}$, the necessary transversality condition assures satisfaction of conditions (III) and (γ), respectively.

Remark 3. If the solutions of the adjoint equations are used in place of the λ^i, it may be necessary to allow the λ^i to be piecewise absolutely continuous on $[t_o,t_f]$. In that event, condition (III) is modified as follows :

$$\text{(III)}' \quad \lambda^i(t_f) \cdot (\varphi^*(t_f) - x_f)$$

$$+ \sum_{k=1}^{r_i} [\lambda^i(t_k-0)-\lambda^i(t_k+0)]\cdot(\varphi^*(t_k)-x] \geqslant 0$$

for all $x_f \in \Theta$ and all $x \in X$, where the t_k, $k = 1,2,\ldots,r_i$, are the points of discontinuity of λ^i. Condition (γ) is similarly modified, except that the strict equality holds.

Remark 4. In Theorem 3, if $p^* \in I$ and conditions (I)-(III) hold for all $x^o \in X$, then p^* is optimal on X. An analogous observation applies to Corollary 2 .

6. Application of the Sufficiency Conditions of Section 5

We give an example to illustrate the use of the sufficiency conditions of Section 5. We treat a two-person zero-sum game so that Corollary 2 may be utilized. Recall that x is denoted by (y,t) for such games.

Example 3. Consider a game whose state equations are

$$\begin{aligned} \dot{x}_i &= \dot{y}_i = y_{i+1}, \quad i = 1,2,\dots,n-2 \\ \dot{x}_{n-1} &= \dot{y}_{n-1} = v^1 + v^2 \\ \dot{x}_n &= 1 \end{aligned} \tag{69}$$

with control constraints

$$U^j(y,t) = \{v^1 \in E^1 : -1 \leqq v^j \leqq 1\} \quad j = 1,2 \tag{70}$$

and costs

$$J^1(p^1,p^2;\varphi) = -J^2(p^1,p^2;\varphi) = \int_{t_o}^{t_f} \sum_{i=1}^{n-1} \varphi_i(\tau)d\tau \tag{71}$$

where t_f is fixed. The target set $\Theta = E^{n-1} \times \{t_f\}$.

Candidates for optimal strategies may be obtained from existing necessary conditions. The H-function is given by

$$\begin{aligned} H(\lambda(t),y,v^1,v^2) = & \sum_{i=1}^{n-1} y_i + \sum_{i=1}^{n-2} \lambda_i(t)y_{i+1} \\ & + \lambda_{n-1}(t)(v^1+v^2) + \lambda_n(t) \end{aligned} \tag{72}$$

where

$$\begin{aligned} \dot{\lambda}_1(t) &= -1 \\ \dot{\lambda}_{i+1}(t) &= -1 - \lambda_i(t), \quad i = 1,2,\dots,n-2 \\ \dot{\lambda}_n(t) &= 0 \end{aligned} \tag{73}$$

with $\lambda_i(t_f) = 0, \quad i = 1,2,\dots,n-1$. Consequently

(74) $$\lambda_i(t) = \sum_{k=1}^{i} \frac{(t_f - t)^k}{k!} \qquad i = 1,2,\ldots,n-1$$

$$\lambda_n(t) = \text{constant}$$

Note that $\lambda_{n-1}(t) > 0$ for $t \in [t_o, t_f)$. Utilizing the necessary conditions, e.g. (3) or (4),

(75) $$\begin{aligned} &\min_{v^1} H(\lambda(t), \varphi^*(t), v^1, p^{2*}(\varphi^*(t))) \\ &= \max_{v^2} H(\lambda(t), \varphi^*(t), p^{1*}(\varphi^*(t)), v^2) \\ &= H(\lambda(t), \varphi^*(t), p^{1*}(\varphi^*(t)), p^{2*}(\varphi^*(t))) = 0 \end{aligned}$$

we obtain the following candidates for optimal strategies :

(76) $$\begin{aligned} p^{1*}(y,t) &= -1 \text{ for all } (y,t) \in E^{n-1} \times [t_o, t_f] \\ p^{2*}(y,t) &= \;\;\,1 \text{ for all } (y,t) \in E^{n-1} \times [t_o, t_f] \end{aligned}$$

For $x^o = (y^o, t_o) \in E^n$, $t_o \leqslant t_f$, the corresponding trajectory is given by

(77) $$\varphi_i^*(t) = \sum_{k=0}^{n-i} y_{i+k}^o \frac{(t - t_o)^k}{k!} \qquad i = 1,2,\ldots,n-1$$

where $0! = 1$.

Upon use of (74) and (76), conditions (α) and (β) of Corollary 2 reduce to

(78) $$-\lambda_{n-1}(t)(v^1+1) \leqslant 0 \quad \text{for} \quad -1 \leqslant v^1 \leqslant 1$$

(79) $$-\lambda_{n-1}(t)(v^2-1) \geqslant 0 \quad \text{for} \quad -1 \leqslant v^2 \leqslant 1$$

Condition (γ) of Corollary 2 follows from $\lambda_i(t_f) = 0$, $i = 1,2,\ldots,n-1$. Thus, $p^* = (p^{1*}, p^{2*})$ is optimal for all $(y^o, t_o) \in E^n$, $t_o \leqslant t_f$.

References

(1) J. NASH, Non-Cooperative Games, *Annals of Mathematics*, Vol.54,N°2, 1951

(2) J.H. CASE, Toward a Theory of Many Player Differential Games, *SIAM Journal on Control*, Vol.7, N°2, 1969

(3) A.W. STARR and Y.C. HO, Nonzero-Sum Differential Games, *Journal of Optimization Theory and Applications*, Vol.3, N°3, 1969

(4) G. LEITMANN, Differential Games, in *Differential Games : Theory and Applications*, (eds. Ciletti, M.D. and Starr, A.W.), ASME, New York, 1970

(5) E. ASPLUND and L. BUNGART, *A First Course in Integration*, Chapter 7, Holt-Rinehart-Winston, New York, 1966

(6) G. CANTOR, Über unendliche, lineare Punktmannichfaltigkeiten, 3, *Mathematische Annalen*, Vol.20, 1882

(7) P. HALMOS, *Measure Theory*, Chapter 5, van Nostrand Co., New York, 1950

(8) G. LEITMANN, *An Introduction to Optimal Control*, McGraw Hill Book Co., New York, 1966

(9) A. BLAQUIERE, F. GERARD, and G. LEITMANN, *Quantitative and Qualitative Games*, Academic Press, New York, 1969

(10) G. CANTOR, Über unendliche, lineare Punktmannichfaltigkeiten, 5, *Mathematische Annalen*, Vol.21, 1883

DIFFERENTIAL TRADING GAMES

James CASE[†]

Elsewhere in this volume, the general many-player differential game is discussed. Here we restrict our attention to a special class of such games, wherin a certain structure is evident. For it is our belief that different differential games demand different solution concepts. So we wish to establish first that a certain solution concept is appropriate for the games we consider, then that such solutions may sometimes be calculated is a straight forward manner.

All scalars discussed herein are real, and all vectors are elements of the real n-dimensional Euclidean space E_n. The symbol 0 denotes either the null vector of that space or the number zero, as context demands. We say that the vector $V = (V_1,\dots,V_N)$ is greater than $v = (v_1,\dots,v_n)$ if $V_i > v_i$ for each $i = 1,\dots,n$, and we write $V > v$ in this case. The inner product of the vectors V and v is written $\langle V,v \rangle$. Finally, if φ is a real function of $x = (x_1,\dots,x_n)$, we denote by φ_{x_i} the resultant of φ with the operator $\partial/\partial x_i$, and by φ_x the vector function $(\varphi_{x_1},\dots,\varphi_{x_n})$.

† Mathematical Sciences, The Johns Hopkins University, Baltimore, Maryland 21218, USA.

1. An Example

We begin with an example. In the seventeenth century, England was unable to produce sufficient naval stores (tar, pitch, etc.) domestically to maintain her large navy and merchant marine. So she was forced to buy them abroad, primarily in the Scandinavian countries. She would have preferred, however, to influence her New England colonies to produce the necessary items. In return, the English were willing to sell the colonists the rope and sails and iron implements which they were unable to make for themselves.

To formalize the above as a differential game, we let $x(t)$ and $y(t)$ denote the stocks of iron tools and of "rope and sails" (one good) present in the colonies at time t, and represent by $z(t)$ the quantity of naval stores then in England. Next we assume that, because the rope and sails are used to power a fishing fleet, the colonists consume fish at a rate $c_1(t)=k_1y(t)$; other goods at the rate $c_2(t)=k_2x(t)$. Similarly, we assume England consumes foreign goods at a rate $\gamma(t)$ proportional to the size of her merchant fleet, which in turn is proportional to the quantity of naval stores available, and write $\gamma(t) = kz(t)$. Also we denote the colonial stock of silver by $s(t)$, and assume that England and her colonies seek to maximize the functionals

$$(1)\quad \begin{aligned} J_1 &= s(t) - s(0) + m \int_0^T c_1(t).c_2(t)dt \\ J_2 &= s(0) - s(T) + \int_0^T \gamma(t)dt \end{aligned}$$

subject to the differential constraints

$$(2) \quad \begin{aligned} \dot{x}(t) &= d_1(t) - r_1 x(t) & \dot{z}(t) &= \delta(t) - \rho z(t) \\ \dot{y}(t) &= d_2(t) - r_2 y(t) & \dot{s}(t) &= p(t)\delta(t) - \pi_1(t) d_1(t) \\ & & &\quad - \pi_2(t) d_2(t) \end{aligned}$$

and the inequalities

$$(3) \quad \begin{aligned} d_1(t) \geq 0 \ , \ d_2(t) &\geq 0, & 0 \leq s(t) \leq S \\ a_1 d_1(t) + a_2 d_2(t) &\leq D, & 0 \leq \delta(t) \leq D \end{aligned}$$

Here S is the (fixed) quantity of silver (measured in pounds Sterling) available for use as a medium of exchange,and $\gamma(t)$ is measured in pounds Sterling (£) per day. Likewise $c_1(t).c_2(t)$ is measured in colonial dollars per day, so that m is an exchange rate in £/\$. Also r_1,r_2 and ρ are depreciation rates for goods X, Y, and Z, $d_1(t)$, $d_2(t)$ and $\delta(t)$ are the corresponding demand rates, and $\pi_1(t)$, $\pi_2(t)$, and $p(t)$ are their prices. Finally D is the maximum rate at which it is possible to transport Z across the Atlantic, measured in tons per day, and a_1 and a_2 are the ratios of the densities of X to Z and Y to Z respectively. The colonies choose $d_1(t)$, $d_2(t)$ and $p(t)$ at each instant t, while England chooses $\pi_1(t)$, $\pi_2(t)$, and $\delta(t)$, in their efforts to maximize the functionals (1) subject to the constraints (2) and (3). We point out that, if the constraint $0 \leq s(t) \leq S$ may be ignored, the functionals (1) may be replaced by

$$(4) \quad \begin{aligned} J_1 &= \int_0^T \{p\delta - \pi_1 d_1 - \pi_2 d_2 + mk_1 k_2 xy\} \, dt \\ J_2 &= \int_0^T \{\pi_1 d_1 + \pi_2 d_2 - p\delta + kz\} \, dt \end{aligned}$$

and the $\dot{s}$ equation eliminated. So if a solution of (2)-(4) does not violate $0 \leq s(t) \leq S$, it will be a solution of (1)-(3) as well.

We shall return to this game (indeed "solve" it) later on. But first we wish to describe a larger class of games which contains this one, and to ask what the nature of a solution for such a game ought to be.

2. The Trading Game

Consider an economy involving N entrepreneurs (players) called 1,2,...,N, and n goods $X_1,\dots,X_n$. We denote by $x_i^j(t)$ the quantity of good X_j possessed by player i at time t, and we call the vector $x_i(t) = (x_i^1(t),\dots,x_i^n(t))$ the "commodity bundle" belonging to player i at time t. For simplicity, we assume there is a fixed stock of silver S in circulation for use as a medium of exchange, and we denote player i's stock of it at time t by $s_i(t)$. As before, silver will be measured in pounds Sterling(£).

We assume that each player i has a production vector $R_i(x_i)$, over which he exercises no control. He also has a price vector $p_i(t) = (p_i^1(t),\dots,p_i^n(t))$ and N demand vectors
$d_{i1}(t) = (d_{i1}^1(t),\dots,d_{i1}^n(t)),\dots,d_{iN}(t) = (d_{iN}^1(t),\dots,d_{iN}^n(t))$.
The component $d_{ij}^k(t)$ is given in units of X_k per month, and represents the rate at which i demands (buys) good X_k from j at time t. Similarly $p_i^k(t)$ is given in £ per unit X_k, and represents the price which i charges for good X_k at time t.

The commodity bundles $x_i(t)$ must therefore satisfy

$$\dot{x}_i(t) = R_i(x_i(t)) + \sum_{j=1}^{N} (d_{ij}(t) - d_{ji}(t)) \tag{5}$$

for each i = 1,...,N and the corresponding flows of silver obeys

$$\dot{s}_i(t) = \sum_{j=1}^{N} < p_i(t),d_{ji}(t) > - < p_j(t),d_{ij}(t) > \tag{6}$$

All vectors must remain non-negative at all times, and may have to obey certain other inequality constraints as well. Finally, we shall assume that the game is to last only from time $t = t_0$ until $t = T$, and that each player has a "utility functional" of the form

$$(7) \qquad J_i = s_i(t) - s_i(0) + \int_0^T L_i(x_i(t))dt ,$$

which he wishes to maximize. The problem of maximizing (7) subject to the differential constraints (5) and (6), and whatever inequality constraints may be present, is a differential game and may be attacked in the usual fashion. It is a reasonable model for international trade if the goods $X_1,\ldots,X_n$ are "capital goods" which need not be consumed (as ropes and sails, iron tools, and naval stores need not) in order to yield their benefits. If "consumption goods" (like food and drink) are to be included, their rates of consumption must be included as control variables which appears both in the stock equations (5) and under the integrals (7). This has been done in a special case [2], but appears to complicate general analyses unduly.

3. On the Nature of Solutions

We observe, at the outset, that this game has an obvious maximin solution. For any player can set all his demand at zero and all his prices at infinity, thus preventing all trade between himself and the other players. The payoff he receives by following this "strategy" is both the largest he can guarantee himself, and the smallest he can ever be forced to accept.

But such a strategy is usually not a good one. For if, for instance, the players $j \neq i$ announce their strategies in advance i can calculate an optimal strategy for trading with them.

And only rarely is his optimal trading strategy to refuse to trade at all.

Of course, it is not likely that all the players but one will announce their strategies in advance. But in a game as familiar as, for instance, international trade has become to international traders, genuine surprises are probably rather rare. So it becomes profitable to try to guess ones opponents' strategies in advance.

Indeed some nations regularly encourage such guessing by announcing key portions of their price strategies (for instance the export prices of automobiles and steel) more than a year in advance. So it seems reasonable, when discussing trading games of the sort described above, to assume that ordinarily the players will proceed first to formulate a guess as to their opponents' likely choice of strategy, and then to choose their own to be roughly optimal against that guess. And what is more, it seems likely that most if not all of the players will afterward feel gratified by their success in both endeavors. Thus we choose, as the object of our study, the outcome of the game when each player is completely successful, both in guessing his opponents' strategies before hand and in calculating his own optimal response thereto.

When each player successfully guesses his opponents' strategies, then responds optimally, the resulting strategy N-tuple is in Nash equilibrium. For each is optimal against the rest. Hence we propose to study the Nash equilibria of the class of trading games. But in what class of strategies should we seek such equilibria ? We must answer this question next, as our conclusions will depend precipitously on the class of strategies chosen.

We interject a word of caution, however, before proceeding. For in times of uncertainty, the players will have less than usual confidence in their guesses. So they will retreat from their optimal responses in the direction of the maximin strategies ; they will charge too much for what they have to sell and they will buy too little. The result will be a "depression" in their trading economy.

Thus our Nash equilibrium analysis, if it is relevant at all to the study of real trading economies, is relevant only in prosperous times. The players will not behave as we have supposed during a recession.

4. The Class of Strategies

A price strategy for player i is a rule which associates with every possible "situation" a price vector p_i appropriate to that situation. And the situation, we shall assume, is described at time t by the values of a finite number of economic "indices" $I_1(t),\dots,I_m(t)$. Thus a price strategy for i is a function P_i from a region in E^m into E^n, and to play it i simply chooses $p_i(t) = P_i(I_1(t),\dots,I_m(t))$.

As indices, we shall allow functions of the state vector $(s(t), x(t)) = (s_1(t),\dots,s_N(t), x_1(t),\dots,x_n(t))$, and certain path integrals thereof as well. The integrals allow us to consider strategies which depend on the previous history of the game.

For instance, in the game described earlier between England and her colonies, one might adopt the quantity

$$I(t) = \int_0^T (\dot{z}(u) + \rho z(u) - \delta_o)\, e^{\alpha u} du \tag{8}$$

as a measure of the past deviations of England's demand from the fixed rate δ_o. By observing $I(t)$, the colonies could "punish" England by refusing to buy English goods whenever $I(t) < 0$.

This new index may be incorporated as a state variable into the earlier formulation of the game. To do this, we consider a fictitious good W, to be stocked in England. Then if $w(t)$ be the quantity of W in England, if $w(0) = 0$, and if $w(t)$ obey the equation

$$\dot{w}(t) = \delta(t) - \delta_o - \alpha w(t)$$

we have immediately $I(t) = e^{\alpha t} w(t)$.

By such devices, a great deal of the history of the general trading game described in § 1 can be incorporated into its formulation in the form of fictitious goods. So we shall assume that all the economic indices known to influence the behaviour of the various players can be, and have been, so incorporated. There is then no further loss of generality in assuming that the price strategies P_i are functions of the state variables alone.

If one has in mind the international trade interpretation of the trading game, this assumption is not too unrealistic. For nations can only be influenced in their behaviour by the economic indices they know, and the common ones (items such as GNP, balance of payments defecits, and inventories of strategic goods) appear regularly in the press almost as soon as they are computed. So these indices really are state variables, as readily observable by one player as another.

Demand strategies, we shall assume, depend on the state variables $(s(t), x(t))$, and on prices as well. For no one, in our experience, ever agrees to any major purchase without knowing its price in advance. Hence, at time t, player i will be charging $p_i(t) = P_i(t,s(t),x(t))$ for the goods he has, and buying at the rates $d_i(t) = D_i(t,s(t),x(t) ; p_1(t),\ldots,p_N(t))$. So a strategy for player i is a pair $\sigma_i = (P_i,D_i)$ consisting of one price strategy and one demand strategy.

It is now reasonable to expect, though the assumptions we have made do not assure it, that every strategy N-tuple $\sigma = (\sigma_1,\ldots,\sigma_N)$ will generate a unique solution $(s_\sigma(t;t_o,s_o,x_o), x_\sigma(t;t_o,s_o,x_o))$ of the kinematic equations (39) and (40), defined on the entire interval $t_o \leqslant t \leqslant T$, such that $s_\sigma(t_o;t_o,s_o,x_o) = s_o$ and $x_\sigma(t_o;t_o,s_o,x_o) = x_o$. And once s_σ and x_σ are known, the corresponding values of the indices $I_1(t),\ldots,I_m(t)$ are known too for $t_o \leqslant t \leqslant T$. So i knows, when he has made his predictions σ_j, $j \neq i$, and his choice σ_i, the index values he may expect to encounter during the course of the game.

Of course i will not always observe exactly the anticipated index values. For in practice all sorts of small errors will inevitably creep into the observations. Only if relatively large deviations from the ancitipated values are observed will he have reason to suspect that some players are behaving otherwise than as expected. As long as he remains convinced, however, that his opponents are playing as he predicted they would, i has no reason to depart from his own intended strategy σ_i. Thus σ_i should be optimal against $(\ldots,\sigma_{i-1}, \sigma_{i+1},\ldots)$ not only along the expected path $(s_\sigma(t;t_o,s_o,x_o), x_\sigma(t;t_o,s_o,x_o))$, but in a neighborhood U_i thereof as well. So σ_i must depend in general on all of the state variables $(t,s_1,\ldots,s_N,x_1,\ldots,x_n)$.

If the moving point $(s(t), x(t))$ should ever leave U_i, i will conclude that his predictions were wrong and that some of the players $j \neq i$ are playing unanticipated strategies. Thus specification of the neighborhood U_i makes precise the notion of "relatively large deviations" introduced above. We shall not attempt to investigate the effect of such deviations. Rather we shall confine our attention to the neighborhood $U = U_1 \cap \ldots \cap U_N$

of the curve $(s_\sigma(t;t_o,s_o,x_o),\ x_\sigma(t;t_o,s_o,x_o))$ in which all N players remain satisfied with their original predictions.

Now suppose that each σ_i is indeed optimal against $(\ldots,\sigma_{i-1},\ \sigma_{i+1},\ldots)$ in U, and that for every point (t_o,s_o,x_o) in U, the point $(t,s_\sigma(t;t_o,s_o,x_o),\ x_\sigma(t;t_o,s_o,x_o))$ also lies in U for every $t_o \leqslant t \leqslant T$. Then the strategy N-tuple σ is in Nash equilibrium not only for the game starting at (t_o,s_o,x_o), but for all the other games as well whose starting points (t,s,x) also lie in U. So if U contained all possible starting points (t_o,s_o,x_o) we could, following Selten (6), call σ a "perfect equilibrium" . But since U will usually be smaller, we shall say only that σ is "locally perfect" .

Our purpose, then, is to calculate locally perfect equilibria for our class of trading games. Clearly this may be done by calculating a perfect equilibrium first, then restricting the result to some appropriate neighborhood U . We shall illustrate the process by analyzing the colonial trade game described in § 1. To do so, we shall need certain results concerning static trading games.

5. Static Trading Games

Consider two isolated islands 1 and 2, which trade only with one another. A native of 1 owns a boat, and sails it to 2 each month. There he meets the local chief, with whom he conducts his business. Both islands use the same currency, which is measured in dollars ($).

The n goods produced in the two-island economy are, as before, $X_1,\ X_2,\ldots,X_n$. Their retail prices on island i are $\hat{P}_i = (\hat{P}_{i1},\ \hat{P}_{i2},\ldots,\hat{P}_{in})$, and their wholesale prices there are $P_i = (P_{i1},\ P_{i2},\ldots,P_{in})$. Naturally $P_{ir} < \hat{P}_{ir}$ so that X_r may be sold at a profit by retailers on i . But it may well be that $P_{2r} > \hat{P}_{1r}$, so that X_r may profitably be bought retail on 1 and sold wholesale on 2. So player 1 (the trader) may offer his goods to player 2 (the chief) at prices $\hat{p}_1 = (\hat{p}_{11}, \hat{p}_{12}, \ldots, \hat{p}_{1n})$ such that $\hat{P}_{1r} < \hat{p}_{1r} < P_{2r}$ if $\hat{P}_{1r} < P_{2r}$ and $\hat{p}_{1r} = \infty$ if $\hat{P}_{1r} \geqslant P_{2r}$. And 2 may sell to 1 at prices $\hat{p}_2 = (\hat{p}_{21}, \hat{p}_{22}, \ldots, \hat{p}_{2n})$ such that $\hat{P}_{2r} < \hat{p}_{2r} < P_{1r}$ if $\hat{P}_{2r} < P_{1r}$ and $\hat{p}_{2r} = \infty$ if $\hat{P}_{2r} \geqslant P_{1r}$.

Next, let Δ_1 be the set of commodity bundles 1's boat can transport from 2 to 1, and let Δ_2 be the set of bundles he can carry from 1 to 2. Thus 1 can deliver any bundle $\hat{d}_2 \in \Delta_2$ to 2 and return with any $\hat{d}_1 \in \Delta_1$.

We shall assume that, in the interest of efficiency, orders are placed each trip for delivery the next. Thus at each meeting 1 presents 2 with his price list $\hat{p}_1$ and asks which bundle $\hat{d}_2 \in \Delta_2$ the latter will have delivered next time. And simultaneously 2 presents $\hat{p}_2$ to 1, who requests that a certain $\hat{d}_1 \in \Delta_1$ await him on his return. The profits on the next trip will then be

$$H_1 = \langle \hat{p}_1 - \hat{P}_1, \hat{d}_2 \rangle + \langle P_1 - \hat{p}_2, \hat{d}_1 \rangle \tag{9}$$

for 1, and

$$H_2 = \langle \hat{p}_2 - \hat{P}_2, \hat{d}_1 \rangle + \langle P_2 - \hat{p}_1, \hat{d}_2 \rangle \tag{10}$$

for 2. The functions H_1 and H_2 are the payoff functions for the players 1 and 2 in a certain game. 1's object in that game is to choose his price vector $\hat{p}_1$ and his demand vector $\hat{d}_1$ so as to maximize his profit H_1. And 2 chooses $(\hat{p}_2, \hat{d}_2)$ to maximize H_2 . Of course both players could do better still if they could arrange to buy wholesale, but we shall ignore this fact.

We have shown in [3] that the game (9)-(10) has ε-optimal Nash equilibrium strategy pairs $(\hat{p}^*_1(\varepsilon), \hat{d}^*_1)$ and $(\hat{p}^*_2(\varepsilon), \hat{d}^*_2)$

such that

$$(11) \qquad \lim_{\varepsilon \to 0} \hat{p}_1^*(\varepsilon) = P_2 \quad \text{and} \quad \lim_{\varepsilon \to 0} \hat{p}_2^*(\varepsilon) = P_1$$

and the well-defined values

$$(12) \qquad H_1^* = \lim_{\varepsilon \to 0} H_1(\hat{p}_1^*(\varepsilon), \hat{d}_1^* \ ; \ \hat{p}_2^*(\varepsilon), \hat{d}_2) = \max_{\hat{d}_2 \in \Delta_2} \langle P_2 - \hat{P}_1, \hat{d}_2 \rangle$$

and

$$(13) \qquad H_2^* = \lim_{\varepsilon \to 0} H_2(\hat{p}_1^*(\varepsilon), \hat{d}_1^* \ ; \ \hat{p}_2^*(\varepsilon), \hat{d}_2) = \max_{\hat{d}_1 \in \Delta_1} \langle P_1 - \hat{P}_2, \hat{d}_1 \rangle$$

to the players 1 and 2, respectively. And since

$$(14) \qquad H_1(\hat{p}_1, \hat{d}_1; \hat{p}_2, \hat{d}_2) + H_2(\hat{p}_1, \hat{d}_1; \hat{p}_2, \hat{d}_2) \leqslant H_1^* + H_2^*$$

for all price vectors $\hat{p}_1$ and $\hat{p}_2$ and for all commodity bundles $\hat{d}_1 \in \Delta_1$ and $\hat{d}_2 \in \Delta_2$, the Nash equilibrium values for the game are Pareto optimal. We shall demonstrate, in the next section, the utility of these facts concerning static trade models for the dynamic theory as well.

6. The Colonial Example

We are ready now to find a perfect Nash equilibrium solution for the colonial trading game described in § 1. We may summarize the description of the game given there as

$$(15) \qquad \begin{aligned} &1 \ \max J_1 = \int_0^T \{ p_1 \delta + p_2 \delta - \pi_1 d_1 - \pi_2 d_2 + mxy \} \, dt \\ &2 \ \max J_2 = \int_0^T \{ \pi_1 d_1 + \pi_2 d_2 - p_1 \delta - p_2 \delta + z \} \, dt \end{aligned}$$

subject to the differential constraints

$$\dot{x}(t) = d_1(t) - rx(t) \qquad \dot{z}(t) = \delta(t) - \rho z(t)$$
$$(16)$$
$$\dot{y}(t) = d_2(t) - ry(t) \qquad \dot{w}(t) = \delta(t) - \delta_o - \alpha w(t)$$

and the physical constraints

$$d_1(t) \geqslant 0 \qquad a_1 d_1(t) + a_2 d_2(t) \leqslant D$$
$$(17)$$
$$d_2(t) \geqslant 0 \qquad 0 \leqslant \delta(t) \leqslant D$$

Here we have specialized the model in an inconsequential fashion by assuming that $k = 1 = k_1 \cdot k_2$ and $r_1 = r_2 = r$, and we have included the fictitious good W in our formulation so that we may later demonstrate its irrelevance. The game (15)-(17) fits the general framework erected in 2 if we define

$$\Delta_1 = \{(d_1, d_2) : d_1 \geqslant 0,\ d_2 \geqslant 0, \quad a_1 d_1 + a_2 d_2 \leqslant D\}$$
$$(18)$$
$$\Delta_2 = \{(\delta_1, \delta_2) : \delta_1 = \delta_2 = \delta, \quad 0 \leqslant \delta \leqslant D\}$$

Incidentally, p_2 is the price 1 charges 2 for W, and it too will be shown to be irrelevant.

The general procedure for solving differential games has been described in (1) and (2). It begins with the writing of the Hamiltonian functions

$$H_1 = p_1\delta + p_2\delta - \pi_1 d_1 - \pi_2 d_2 + mxy$$
$$+ V_x(d_1 - rx) + V_y(d_2 - ry) + V_z(\delta - \rho z) + V_w(\delta - \delta_o - \alpha w)$$
$$= < (p_1, p_2) + (V_z, V_w),\ (\delta, \delta) >$$
$$+ < (V_x, V_y) - (\pi_1, \pi_2),\ (d_1, d_2) >$$
$$(19) \qquad + mxy - rxV_x - ryV_y - \rho zV_z - \alpha wV_w - \delta_o V$$

and

$$H_2 = < (\pi_1,\pi_2) + (W_x,W_y),\ (d_1,d_2) >$$

$$+ < (W_z,W_w) - (p_1,p_2),\ (\delta,\delta) >$$

$$+ z - rxW_x - ryW_y - \rho zW_z - \alpha wW_w - \delta_o W_w \tag{20}$$

$V = V(t,x,y,z,w)$ and $W = W(t,x,y,z,w)$ being the value functions for players 1 and 2 respectively. So if we ignore the parts of H_1 and H_2 which are independent of the control variables $(p_1,p_2,\pi_1,\pi_2,d_1,d_2,\delta)$, and make the identifications

$$\begin{array}{llll} \hat{p}_1 \to (p_1,p_2) & & \hat{p}_2 \to (\pi_1,\pi_2) \\ \hat{d}_1 \to (d_1,d_2) & & \hat{d}_2 \to (\delta,\delta) \\ \hat{P}_1 \to (V_z,V_w) & & \hat{P}_2 \to (W_x,W_y) \\ P_1 \to (V_x,V_y) & & P_2 \to (W_z,W_w) \end{array} \tag{21}$$

the expressions (19) and (20) are seen to have exactly the forms (9) and (10). Therefore when 1 and 2 choose $(\bar{p}_1,\bar{p}_2,\bar{d}_1,\bar{d}_2)$ and $(\bar{\pi}_1,\bar{\pi}_2,\bar{\delta})$ respectively in such a way that

$$\begin{array}{l} H_1(\bar{p}_1,\bar{p}_2,\bar{d}_1,\bar{d}_2;\bar{\pi}_1,\bar{\pi}_2,\bar{\delta}) > H_1(p_1,p_2,d_1,d_2;\bar{\pi}_1,\bar{\pi}_2,\bar{\delta}) \\ H_2(\bar{p}_1,\bar{p}_2,\bar{d}_1,\bar{d}_2;\bar{\pi}_1,\bar{\pi}_2,\bar{\delta}) > H_2(\bar{p}_1,\bar{p}_2,\bar{d}_1,\bar{d}_2;\pi_1,\pi_2,\delta) \end{array} \tag{22}$$

for every $(p_1,p_2,d_1,d_2) \neq (\bar{p}_1,\bar{p}_2,\bar{d}_1,\bar{d}_2)$ and $(\pi_1,\pi_2,\delta)\neq(\bar{\pi}_1,\bar{\pi}_2,\bar{\delta})$, we will have

$$\begin{array}{l} H_1(\bar{p}_1,\bar{p}_2,\bar{d}_1,\bar{d}_2;\bar{\pi}_1,\bar{\pi}_2,\bar{\delta}) = \max\limits_{(d_1,d_2)\in\Delta_1} [(V_x+W_x)d_1+(V_y+W_y)d_2] \\ \qquad + mxy-rxV_x-ryV_y-\rho zV_z-\alpha wV_w-\delta_o V_w \\ \\ H_2(\bar{p}_1,\bar{p}_2,\bar{d}_1,\bar{d}_2;\bar{\pi}_1,\bar{\pi}_2,\bar{\delta}) = \max\limits_{(\delta,\delta)\in\Delta} (V_z+V_w+W_z+W_w)\delta \\ \qquad + z-rxW_x-ryW_y-\rho zW_z-\alpha wW_w-\delta_o W_w \end{array} \tag{23}$$

Therefore, V and W must be the solutions of the Hamilton-Jacobi partial differential equations

$$(24)\quad \begin{aligned} &V_t + \max_{(d_1,d_2)\in\Delta_1} [(V_x+W_x)d_1 + (V_y+W_y)d_2] \\ &\qquad + mxy-rxV_x-ryV_y-\rho zV_z-\alpha wV_w-\delta_o V_w = 0 \\ &W_t + \max_{0\leqslant\delta\leqslant D} (V_z+V_w+W_z+W_w)\delta \\ &\qquad + z - rxW_x-ryW_y-\rho zW_z-\alpha wW_w-\delta_o W_w = 0 \end{aligned}$$

for which $V(T,x,y,z,w) = 0 = W(T,x,y,z,w)$. So the sum $U = V + W$ must be the solution of

$$(25)\quad \begin{aligned} &U_t + \max_{(d_1,d_2)\in\Delta_1} (U_x d_1+U_y d_2) + \max_{0\leqslant\delta\leqslant D} (U_z+U_w)\delta \\ &\qquad + mxy+z-rxU_x-ryU_y-\rho zU_z-\alpha wU_w-\delta_o U_w = 0 \end{aligned}$$

which vanishes identically when $t = T$. And if $U^1(t,x,y)$ and $U^2(t,z,w)$ are the solutions of

$$(26)\quad U^1_t + \max_{(d_1,d_2)\in\Delta_1} (U^1_x d_1+U^1_y d_2) + mxy-rxU^1_x-ryU^1_y = 0$$

and

$$(27)\quad U^2_t + \max_{0\leqslant\delta\leqslant D} (U^2_z+U^2_w)\delta-\rho z-U^2_z-\alpha wU^2_w-\delta_o U^2_w = 0$$

which vanish when $t = T$, then $U = U^1 + U^2$ is the solution of (25) which vanishes then. But (25), (26), and (27) are themselves Hamilton-Jacobi equations. For (26) and (27) arise from the problems

$$\max \int_0^T mxy\, dt$$

(P_1) subject to $\dot{x} = d_1 - rx \qquad \dot{y} = d_2 - ry$

and $d_1 \geqslant 0,\ d_2 \geqslant 0,\ a_1 d_1 + a_2 d_2 \leqslant D$

and
$$\max \int_0^T z dt$$

(P_2) subject to $\dot{z} = \delta - \rho z \qquad \dot{w} = \delta - \delta_o - \alpha w$

and $0 \leqslant \delta \leqslant D$

while (25) arises from the direct sum $(P_1 \oplus P_2)$

The problems (P_1) and (P_2) have the unique optimal strategies

$$(28) \qquad (d_1^*(t,x,y) d_2^*(t,x,y)) \begin{cases} = (0,D) & \text{if } a_1 x > a_2 y \\ = (D,0) & \text{if } a_1 x < a_2 y \\ = (D/2)(1/a_1, 1/a_2) & \text{if } a_1 x = a_2 y \end{cases}$$

and

$$(29) \qquad \delta^*(t,z,w) \equiv D$$

as one may deduce from the standard theorems (see [5]) of optimal control. And clearly (28) and (29) must be optimal for $(P_1 \oplus P_2)$ as well.

But (25) asserts that when the players 1 and 2 play their perfect Nash equilibrium strategies, starting with the initial stocks (x_o, y_o, z_o, w_o) at time t_o, the sum

$$(30) \qquad V(t_o,x_o,y_o,z_o,w_o) + W(t_o,x_o,y_o,z_o,w_o)$$

of their payoffs is as large as $U^1(t_o,x_o,y_o) + U^2(t_o,z_o,w_o)$, which total is attainable only with the demand strategies (28) and (29). So when Nash equilibrium strategies are played by both players, the real goods X, Y, Z must change hands at the rates (28) and (29). But we know nothing, as yet, of the rate at which silver changes hands.

To find out, we must solve the equations (24) for V and W. For the prices which yield (23) are, from (11)

$$(31) \qquad \hat{p}_1 = P_2 = (W_z, W_w) \quad \text{and} \quad \hat{p}_2 = P_1 = (V_x, V_y)$$

Let us try first to find solutions of the form $V = V(t,x,y)$ and $W = W(t,z,w)$. Clearly if such solutions of (24) exist, they will be solutions of (26) and (27) as well. And the only solution of (27) which vanishes when $t = T$ is

$$(32) \qquad W = U^2(t,z) = (D+\rho z)(1-e^{\rho(t-T)})/\rho^2 - D(T-t)/\rho$$

Note that (32) is defined for all t,z, and w, and satisfies (27) everywhere. Similarly the function

$$(33) \qquad V = U^1(t,x,y) = -(m/2r)(e^{2r(t-T)}-1)xy$$

is defined everywhere, but satisfies (26) only on the line $a_1x = a_zy$, where we know $U^1_x d_1 + U^1_y d_2$ must vanish. We could of course, by the method of characteristics, compute $V(t,x,y)$ in the regions $a_1x > a_2y$ and $a_1x < a_2y$ as well. But it seems pointless to do so, for we know from (28) that the stocks $x(t)$ and $y(t)$ of X and Y tend rapidly to the line $a_1x = a_2y$, and stay there. A few of the trajectories $(x(t), y(t))$ which result from the strategy (28) are sketched below in Figure 1.

We note in particular that, from any initial point (x_o, y_o), the paths $(x(t), y(t))$ all proceed first to the line $a_1x = a_2y$, and from there to the point $(D/2r)(1/a_1, 1/a_2)$ on that line. And simultaneously $z(t)$ approaches D/ρ . So after the game has been going on for some time, the point $(x(t), y(t), z(t))$ will be near $\Omega = (D/2)(1/ra_1, 1/ra_2, 2/\rho)$. At Ω, the expressions (32) and (33) for V and W are valid. So there we must have

$$(34)\qquad \begin{aligned} \pi_1^* &= V_x = (mD/4r^2a_2)(1-e^{2r(t-T)}) \\ \pi_2^* &= V_y = (mD/4r^2a_1)(1-e^{2r(t-T)}) \\ p_1^* &= W_z = (1-e^{\rho(t-T)})/\rho \\ p_2^* &= W_w = 0 \end{aligned}$$

Thus, once the point $(x(t), y(t), z(t))$ has reached the vincinity of Ω, silver will be changing hands at approximately the rate

$$\begin{aligned} \dot{S}(t) &= (p_1(t)+p_2(t))\delta(t) - \pi_1(t)d_1(t) - \pi_2(t)d_2(t)) \\ &= (W_z(t)+W_w(t))D - V_x(t)D/2a_1 - V_y(t)D/2a_2 \\ (35)\qquad &= (1-e^{\rho(t-T)})D/\rho - (mD^2/8r^2a_1a_2)(1-e^{2r(t-T)}) \end{aligned}$$

And if T is very large, so that termination is yet in the distant future, the exponentials in (35) are negligable. So we may write

$$(36)\qquad \dot{S}(-\infty) = \frac{D}{\rho} = \frac{mD^2}{8r^2a_1a_2}$$

And if (36) does not vanish, one of the players will require an unlimited quantity of silver to play the game over long periods $[0,T]$. But if the international exchange rate m is given the value

$$(37)\qquad m^* = 8r^2a_1a_2/\rho D$$

the game may be played over arbitrarily long periods with but a limited store of silver. This effect has been observed before, in a slightly different problem [2], so we shall not belabor the point here. Rather, we turn our attention to the question of the relevance of the historical index δ and/or the ficticious good W, for solutions of the type we propose of two-player trading games.

First, we emphasize that the price strategies (34) together with the demand strategies

$$(38)\qquad \begin{aligned} (d_1^*, d_2^*) &= \arg \max_{(d_1,d_2)\in\Delta_1} (\pi_1 d_1 + \pi_2 d_2) \\ \delta^* &= \arg \max_{0 \leqslant \delta \leqslant D} (p_1 + p_2)\delta \end{aligned}$$

constitutes a perfect Nash equilibrium pair for the colonial trading game. A locally perfect equilibrium pair may now be constructed, for the sub-game starting at (x_o, y_o, z_o, w_o) by constructing the path $\Gamma : (x(t), y(t), z(t), w(t)), 0 \leqslant t \leqslant T$ that begins there, restricting the strategies (34), (38) to a neighborhood U of Γ, then defining d_1^*, d_2^*, δ^* and p_1^*, p_2^*, π_1^*, π_2^* in an arbitrary fashion outside of U .

7. The Irrelevance of Fictitious Goods

In [1] it was shown that any pair of strategies (σ_1^*, σ_2^*) in the rather broad class of "switching strategies" there considered gave rise to a pair of "value functions" V and W for players 1 and 2 which were, in an appropriate sense "regular". And if (σ_1^*, σ_2^*) were in Nash equilibrium, then V and W must obey a pair of "Hamilton-Jacobi" partial differential equations which reduce, in the present case to (24). That is, every Nash equilibrium pair of strategies for G yields a pair of solutions of (24) .

In § 6 we exhibited solutions of (24) which were independent of w. Therefore both the price strategies (34) and the demand strategies (38) which together are in Nash equilibrium, are also independent of w and therefore of I. It remains to show that the solutions of (24) are unique, so there can be no other pair of "switching strategies" which is also in Nash equilibrium, and which does depend on I. We shall show this first in the region

$$(39) \qquad R = \{(t,x,y,z,w) : a_1x = a_2y, \quad 0 \leqslant t \leqslant T\}$$

where the equations (24) take an especially simple form.

We showed earlier that (24) implies (28) and (29). Hence $(V_x+W_x)d_1 + (V_y+W_y)d_2$ must vanish in R, for $a_1x = a_2y$ there. Also $a_2V_x = a_1V_y$, etc. in R, so the system (24) may be rewritten

$$(40) \qquad \begin{aligned} &V_t + mxy - 2rxV_x - \rho zV_z - (\delta_o + \alpha w)V_w = 0 \\ &W_t + z + D(V_z+V_w+W_z+W_w) - 2rxW_x - \rho zW_z - (\delta_o+\alpha w)W_w = 0 \end{aligned}$$

there. But the first of these does not contain W. So it has a unique solution V which may be found by the standard method of characteristics (4). Moreover that solution V is independent of both z and w (in fact $V = U^1(t,x,y)$), so the terms V_z and V_w vanish from the second of equations (40). Hence $V = U^1(t,x,y)$ and $W = U^2(t,z)$ are in fact the only solutions of (24) in R. In the region in which $a_1x > a_2y$, the system (24) becomes, again by (28),

$$(41) \qquad \begin{aligned} &V_t + D(V_y+W_y) + mxy - rxV_x - ryV_y - \rho zV_z - (\delta_o+\alpha w)V_w = 0 \\ &W_t + D(V_z+V_w+W_z+W_w) + z - rxW_x - ryW_y - \rho zW_z - (\delta_o+\alpha w)W_w = 0 \end{aligned}$$

Equations (41) are to be solved simultaneously, subject to the boundary conditions $V(T,x,t,z,w) = 0 = W(T,x,t,z,w)$ on $t = T$ and $V = U^1(t,x,y)$, $W = U^2(t,z)$ on R. And since U^1, U^2, and all the coefficients appearing in (41) are analytic functions of (x,t,z,w), we may appeal to the classic theorem of Cauchy and Kowalewski (4) to conclude that, at least in a neighborhood of the hyperplane R in (t,x,y,z,w)-space, the system (41) has a unique analytic solution.

This question could doubtless be pursued further, with the aid of more specific theorems on linear systems. But we shall content ourselves with a few closing observations.

We remark first, that $I(t)$ could have been replaced by a whole vector $I_1(t),\dots,I_\mu(t)$ of measures of the past deviations of England's demand from δ_o. So we could retain the values of

$$(42) \qquad I(\alpha,t) = \int_o^t (\dot{z}(u) - \rho z(u) - \delta_o)e^{\alpha u}du$$

for $\alpha = \alpha_1, \alpha_2,\dots, \alpha_\mu$. And since a knowledge of the function

$$(43) \qquad J(\alpha,t) = \int_o^t f(u)\, e^{\alpha u}du$$

for all values of α and t identifies f uniquely, we could build as much of the history of England's demand for Z as we chose into our formulation of the game by the device of introducing fictitious goods $W_1,\dots,W_\mu$. Indeed, if one chose to extend the theory of nonzero-sum differential games to allow the state vector to lie in a space of infinite dimension, the entire history of England's demand for Z could be so included. And the histories of many other quantities may be included too. But one should not expect such histories to be relevant to the solution of the game. For as long as the Hamilton-Jacobi equations have unique solutions V and W which are independent of the stocks $W_1,\dots,W_\mu$, the perfect Nash equilibrium strategies will be independent of the indices $I_1,\dots,I_\mu$. And although we can demonstrate the existence and uniqueness of such solutions V and W of the Hamilton -

Jacobi equations only in special cases like the one above, we expect our conclusions to hold in much more varied circumstances.

Our program is now complete. We began by defining a special class of differential games (the Trading Games of § 2) for which the reader may be expected to have some intuitive feel. Next we proposed a solution concept (the locally perfect Nash equilibrium of § 4) designed to appeal to the reader's intuition. And finally we solved (in § 6) a simple representative of our class of games and demonstrated (in § 7) the uniqueness of our solution. The key step in our development was the imposition (in § 4) of the requirement of "local perfection" . For as Shapley (7) has shown, differential games may be expected to possess an unfathamable multiplicity of locally imperfect Nash equilibria. The reader should be certain that he understands the meaning of the phrase "locally perfect" and the reasons given in § 4 for its introduction.

We hope to have convinced the reader that (a) Trading Games are an interesting and worthy object of study, (b) locally perfect Nash equilibria in the class of state-dependent switching strategies are an appropriate solution concept for such games, and (c) the task of fingind such solutions is not at all a hopeless one.

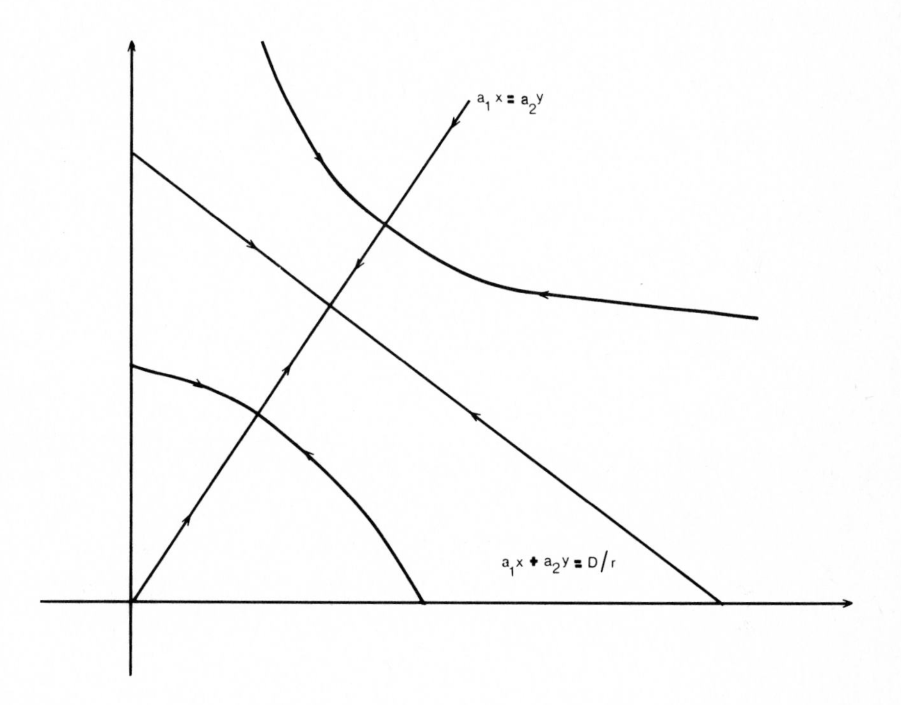

Fig. 1

References

(1) J. CASE, *Towards a Theory of Many Player Differential Games,* SIAM J. Control, Vol.7, n°2 (1969)

(2) J. CASE, *On Ricardo's Problem,* J. Economic Th., Vol.3, n°2, (1971)

(3) J. CASE, *A Class of Games Having Pareto Optimal Nash Equilibrium,* to appear

(4) P.R. GARABEDIAN, *Partial Differential Equations,* J. Wiley, New York (1964)

(5) E.B. LEE, and L. MARKUS, *Foundations of Optimal Control Theory,* J. Wiley, New York (1967)

(6) R. SELTEN, Private Communication

(7) L. SHAPLEY, *Mathematical Concepts of Game Theory - Some Noncooperative Examples,* a lecture delivered at the International Summer School on Mathematical Models of Action and Reaction, held in Varenna, Italy, June 15-27, 1970 .

FIAT MONEY IN AN ECONOMY WITH ONE NONDURABLE GOOD AND NO CREDIT[†] (A Noncooperative Sequential Game)

M. SHUBIK and W. WHITT[††]

1. Introduction

Our purpose is to provide satisfactory connections between current theories in macroeconomics and microeconomics. This study differs from the majority of past efforts in this direction because in a microeconomic context it focuses on money, credit, and financial institutions. From the point of view of monetary theory, the novelty lies in our efforts to connect individual economic behavior with the behavior of the economy as a whole and to apply mathematical models for this purpose. From the point of view of the more closely related mathematical literature on n-person games and competitive equilibrium analysis, the novelty lies in our efforts to explicitly consider money, credit, and financial institutions. We believe that an explicit treatment of money, credit, and financial institutions is necessary for forging links between macroeconomics and microeconomics.

† The research was supported by the Office of Naval Research. The research was also partially supported by a grant from the Ford Foundation. Cowles Foundation Discussion Paper n°355.

†† Yale University, New Haven, USA

The study of money and credit poses many different problems. These concern information, uncertainty, risk, trust, convenience of exchange, and so forth, cf.(12). Our object is to construct and analyze mathematical models which will yield adequate theories to cover these different features. We believe that in appropriate mathematical models the prototypes of institutions familiar to modern economics will naturally emerge. Thus we claim that although particular institutions such as commercial, central, and investment banks, insurance companies, loan societies, and stock markets may reflect particular institutional details pertaining to specific societies, the essential functions that these institutions perform call for the existence of entities to perform these functions in any complex economy. This premise has a significant bearing on our modelling approach. Instead of incorporating as many as possible financial institutions into our models at the outset, we want to see how each of these institutions arises out of necessity. We thus begin with an overly-simplified model, not only because it is easier to analyze, but because we want to explicitly represent economic behavior when important features are missing. We want to establish the fundamental need for the various financial institutions. Furthermore, we are thus able to evaluate the possible forms these institutions can take. For example, in (11) the notion of an optimal bankruptcy law is introduced. (See also (10).)

Having set forth our general objectives, we must say that the results to be reported here constitute only a first step in the overall program. We consider a highly-simplified model which only addresses a few of the problems posed by money and credit. While it is our intent to extend the analysis to more sophisticated models in the future, we believe it is important to stress the virtues of simple models. Among these virtues are :

(i) The simplicity enables us to obtain a rather complete mathematical solution,

(ii) The effect of missing features can be observed in the solution,

(iii) By considering only one or two of the features, the effects of different features are isolated and, hopefully, better understood.

We now give a brief overview. We begin by regarding our simplified economy as a deterministic noncooperative sequential game. The modelling is discussed in Section 2 so we shall not dwell on it here. As a consequence of viewing the economy as a noncooperative sequential game, we confront a system of interdependent dynamic programs, cf. (2.2). We first investigate the case of 2 players and n periods. While we could start by normalizing our game, that is, we could convert it immediately into an equivalent static game, it does not appear to be fruitful to do so. For example, our game is never constant-sum, and even if we converted it to constant-sum, which is possible in some cases, the number of strategies is infinite. Furthermore, the normalization seems to destroy the nice properties in the payoff function. Hence, instead of normalizing, we apply the standard backward recursion of dynamic programming. In this way, we verify under fairly general conditions the existence of a unique noncooperative equilibrium solution. We also describe this solution in detail. These answers were discovered by rather painful analysis of the four-period problem, but of course the proofs here are by induction.

We next consider infinite-horizon extensions of our two-player sequential game. It is not difficult to see that with discounting the infinite-horizon version possesses all the properties of the finite-horizon versions. In fact, the infinite-horizon solution coincides with the n-period solutions for all n sufficiently large because then the (n+1)-period and n-period solutions agree. These answers are obtained by direct argument employing our finite-horizon solutions. Again, the existing (sequential) game theory literature does not appear to be very helpful.

We next extend our model by adding more players. Of course, this is the way in which we intend to relate micro and macro theory. It turns out that the analysis of even a 2-period game for 3-5 players is quite complicated, but great simplicity is achieved when many players are present. If the number of players is sufficiently large with each player sufficiently small in relation to the economy as a whole, then we verify the existence of a unique noncooperative equilibrium solution which has a very simple form. In this solution each player uses his myopic or one-period optimal strategy which dictates spending all his available money each period. It is significant that the nonuniqueness possibilities for two players disappear in the large economies. Furthermore, the complex dynamics in which players exploit their money advantage over several time periods also disappears.

For comparison, we conclude by investigating the set of constrained competitive equilibria. We show for economies of any size that there is always a unique competitive equilibrium solution. This solution also involves each player spending all his money in each period. Thus, for sufficiently large economies, the set of noncooperative equilibrium solutions coincides with the set of competitive equilibrium solutions. This is to be distinguished from recent related work by Aumann (2), Brown and Robinson (3), Hildenbrand (6),(7), and others involving the core and the set of competitive equilibria because here "large" means finite instead of infinite. Our two solution concepts coincide in large finite economies. Hence, there is no need to introduce infinite spaces of players and there is no need to prove a limit theorem (which is not to say that these devices are not very useful in other contexts).

We now turn to Section 2 for a definition of our model. The results appear in Section 3 and the proofs appear in Sections 4-6.

2. A Simple Money Game

2.1 - Goods, Money, and the System of Exchange.

We shall study the distribution of real goods and money over time in a fiat money economy. We assume there is a single nondurable good which goes on the market in constant quantity each of a finite or denumerably infinite number of time periods. (In [12] a model was studied in which players could choose to keep goods off the market, thus causing only part of the real goods to be monetized. This will not be the case here). Each player owns a fraction of this market or, equivalently, each player brings this fraction of the real good to the market. We keep these fractions of ownership fixed over time. Each period all of the commodity on the market is distributed to the players for consumption. By "nondurable", we mean that all of the real good available each period is consumed during that period. Nondurable real goods can not be inventoried.

We also assume the existence of an "institutional stuff" called money whose quantity is fixed. It is neither created nor destroyed over time. Furthermore, each period all the money is initially in the hands of the players. The important point is that we prohibit barter. All real goods must be purchased at the market with money. This money may be thought of as having a physical existence such as poker chips or green pieces of paper; or it may be thought of as a set accounting numbers whose ownership may be transferred. Its value is established by fiat ; i.e., by our assumption here that goods can only be obtained in exchange for money.

We can now describe the exchange over time. Each player begins with certain fractions of the total money supply and of the total ownership of the market. Each period each player must make just one decision : how much money to spend to purchase the real good that period. We assume that the real good is distributed to the players in proportion to the money spent. Furthermore, the money taken in by the market is given back to the players in proportion to their ownership. A player's money at the end of the period thus equals the money he had left over after spending plus his share of the market take. This model has interest because of each player's conflicting desires : to spend more now to get more real goods now or to spend less now to get more real goods later.

2.2 - *Credit.*

We distinguish sharply between money and credit. In this initial model no credit is granted. In particular, this means that there is no market for current money in exchange for claims on future money or goods. In an economy which has neither credit nor barter all exchanges are in the form of a payment of current money for real goods. Obviously, this should make money play a more prominent role than it otherwise would, and this is confirmed by our results. First, players starting with a fraction of money greater than their fraction of ownership are often able to reap a significant advantage in real goods over time which would not be possible with credit. Second, the noncooperative equilibrium solution and the prospective competitive equilibrium allocations turn out not to be Pareto optimal if players have different time discounts. Without credit, the model has difficulty responding to time preferences for goods. Thus, when money is introduced without credit, there exist motivations for introducing credit.

2.3 - *Individual Preferences.*

We assume that individual i has preferences which can be represented by a utility function of the form

$$(1) \qquad U_i(\underset{\sim}{x}) = \sum_{k=0}^{k=n-1} \beta_i^k \varphi_i(x_k)$$

where n $(1 \leqslant n \leqslant \infty)$ is the number of periods, $\underset{\sim}{x} = (x_1,\ldots,x_n)$ is the vector of real goods to be received (and consumed) in successive periods, β_i is the discount factor for individual i, $0 < \beta_i \leqslant 1$, k in β_i^k is an exponent instead of a superscript, and φ_i is the one-period utility function. For the most part, we assume the one-period utility function φ_i is of the special form $\varphi_i(x) = x$, but the principal results for the case of many players hold for quite general φ_i . The important point is that money does not appear in (1). Money is valued only as a means to obtain more real goods.

2.4 - *Solution Concepts.*

2.4.1 - Noncooperative Game. For the most part, we shall view our market as a noncooperative game. Therefore, our object is to identify the set of noncooperative equilibrium state strategy solutions. Such a solution consists of an n-period strategy for each player with the property that no individual acting alone can improve his position by altering his strategy. For specified initial conditions, an equilibrium point is determined by an appropriate schedule of spending for all the players. It is important to distinguish between equilibrium in this game theoretic sense and equilibrium as it may exist in the dynamics of a particular schedule of spending. For example, within our equilibrium solution the situation in which each individual's fraction of money equals his fraction of ownership is a stationary state or an equilibrium point. In other words, if at any period this situation prevails, then it will prevail in every period thereafter if we follow the schedule of spending associated with the noncooperative equilibrium solution.

Since we are viewing the market as an m-person n-period noncooperative game in which the utility functions are separable (as indicated in (1)), the object is to identify the noncooperative solutions to a system of m simultaneous n-period dynamic programs. In particular, individual i has a payoff function which can be defined recursively as

$$U_n^i(p_i,\gamma_i) = \max_{0 \leq x_i \leq p_i+\gamma_i} \{ \varphi_i[K^i(x_1,\ldots,x_m)] + \beta_i U_{n-1}^i(p_i,\gamma_i - x_i + p_i(\sum_{j=1}^m x_j))\} \tag{2}$$

where

$$K^i(x_1,\ldots,x_m) = \begin{cases} \dfrac{x_i G}{\sum_{j=1}^m x_j}, & x_j > 0 \text{ for some } j, \\ 0, & x_1 = \cdots = x_m = 0 \end{cases}$$

p_i is i's fraction of ownership, $0 < p_i < 1$,

$p_i+\gamma_i$ is i's fraction of money, $-p_i < \gamma_i < 1-p_i$,

m is the number of players,

n is the number of periods remaining,

x_j is the amount of money j spends in the first of n periods,

φ_i is i's one-period utility function,

G is the total amount of the real good to be distributed,

and β_i is i's discount factor, $0 \leq \beta_i \leq 1$.

Of course, U_n^i depends not only on p_i and γ_i but also on all the other variables, especially x_j for $j \neq i$. Note that n indicates the number of periods remaining. The representation in (2) is fine as long as $n < \infty$ but would have to be altered with an infinite horizon. With discounting, we would then naturally count time forwards and obtain instead of (2)

$$U^i(p_i,\gamma_i) = \max_{\{x_{ik},k\geqslant 1\}} \sum_{k=1}^{\infty} \beta_i^{k-1} \varphi_i \,[K^i(x_{1k},\ldots,x_{mk})] \tag{3}$$

such that

$$x_{ik} \leqslant p_i + \gamma_{ik}$$

$$\gamma_{i(k+1)} = \gamma_{ik} - x_{ik} + p_i \sum_{j=1}^{m} x_{jk}$$

If we restrict attention to stationary strategies, then we consider the functional equation in (2) where both U_n^i and U_{n-1}^i are placed with U^i . Finally, if there is no discounting in the infinite horizon, then an entirely different criterion is needed, e.g., average return per period.

2.4.2 - State Strategies. In terms of general game theoretic considerations, a strategy is a complete plan of play which might depend in detail upon all aspects of previous history. In particular, a strategy may depend delicately upon information conditions. In oligopolistic financial markets there is ample evidence that information conditions do play a vital role in determining strategies. However, in the models explored here a considerable limitation and simplification of information conditions is assumed. Furthermore, we limite ourselves to an extremely special set of overall strategies which can best be described as simple state strategies where an individual's behavior depends only upon the state and period that he is in and not upon the history of how he arrived in that state. In this particular model the restriction to state strategies does not appear to be particularly binding, but for us it remains an assumption. We start by assuming that the dynamic programming recursion is justified.

2.4.3 - Competitive Equilibrium. More deeply rooted in economics is the concept of a competitive equilibrium. A competitive equilibrium is a set of prices (one price per period here) and a set of allocations (of real goods to each player each period) such that for all i the allocations to individual i are optimal for him alone subject to all budget constraints being met and the outcome being Pareto optimal. Pareto optimality means the players cannot all simultaneously do better by choosing different strategies.

In the customary definition of the competitive equilibrium, *budget constraints* (i.e., limits on an individual's purchasing power in terms of the net worth of an individual's ownership evaluating the goods at the given market prices) but not *cash flow constraints* are active. This is as though either perfect trust exists for the whole trading period or, equivalently, credit is freely available. We consider a modified competitive equilibrium where the individuals have money and its amount is fixed (there is no credit) and trade is carried out in money. In this model the cash flow constraints are of importance.

As before, neither the amount of money in the system nor the amount of real goods put on the market each period changes from period to period, so the prices must be chosen accordingly. The concept of competitive equilibrium is primarily intended for perfect competition where the influence of each individual relative to the market as a whole is negligible. In such a situation it is reasonable to consider the economic behavior in terms of many isolated or decentralized maximization problems taking place simultaneously with each individual ignoring his influence on the market. While the notion of competitive equilibrium is most meaningful when associated with perfect competition including no credit constraints, we can nevertheless identify the set of competitive equilibria given cash flow limitations. In the notation of (2) individual i is confronted with the optimization problem :

$$(4) \qquad \max_{\{x_{ik}\}} \sum_{k=0}^{n-1} \beta^k \varphi_i(x_{ik})$$

subject to :

$$x_{ik} a_k \leqslant M_k^i, \quad 1 \leqslant k \leqslant n,$$

$$M_{k+1}^i = M_k^i - x_{ik} a_k + p_i G a_k, \quad 1 \leqslant k \leqslant n-1,$$

$$M_1^i = (p_i + \gamma_i) M,$$

$$\sum_{i=1}^{m} M_k^i = M, \quad 1 \leqslant k \leqslant n,$$

and $\sum_{i=1}^{m} x_{ik} = G, \quad 1 \leqslant k \leqslant n,$

where M_k^i is the money i has available in period k,

a_k is the price in period k,

and M is the total money in the system.

We shall be interested in comparing the set of (constrained) competitive equilibria with the set of noncooperative equilibria.

2.4.4 - Other Solutions. There are many other solution concepts which we could consider but which we will not. Among these are the cooperative game theory solution concepts such as the core, value, nucleolus, and bargaining set. A rather different approach would be to use a behavioral model in which individuals are assumed to use heuristics or limited optimizations in order to make their decisions. For example, the individuals might respond adaptively over time or spend a random amount each period. Behavioral models differ from more general optimization problems in degree rather in kind, but they are usually characterized by having a relatively simple mechanisms which produces decisions. In terms of the mathematical analysis, the problem is not one of complicated optimization to yield decisions, but one of describing the evolution of the system when the decision-generation mechanism is specified. An example of the behavioral approach to an optimization model is the technique of stationary analysis in inventory theory, cf. Part IV of [1]. Behavioral models have considerable appeal for representing actual human behavior because of limitations in human information processing ability. However, it is difficult to avoid ad hoc modelling. Thus, we will not consider this approach at this time.

3. Results

3.1 - *Noncooperative Game with Two Players and Finite Horizon.*

Consider the system of dynamic programs in (2) with the additional assumptions that $m = 2$, $1 \leqslant n < \infty$, and $\varphi_i(x) = x$, $x \geqslant 0$, $i = 1, 2$. Since the one-period payoff function is now homogeneous of degree zero, we can let $G = 1$ without loss of generality. For notational simplicity, we drop the subscripts on p_1 and γ_1 and use $1-p$ and $-\gamma$ for p_2 and γ_2. Furthermore, we stipulate that $\gamma \geqslant 0$ so that the first player always begins with at least as much money as ownership. It turns out that the money advantage γ rather than the initial money supply $p+\gamma$ determines the strategic character of a player. We thus refer to the first player as the strong player or just I and the second player as the weak player or just II .

To place our game in perspective, first note that it is not constant-sum. It almost is when $\beta_1 = \beta_2$, but it is not because $K(x_1,x_2) = 0$ for $x_1 = x_2 = 0$. If we let $K(0,0) = 1/2$, then we would have a constant-sum game when $\beta_1 = \beta_2$. Even in this special case it does not seem possible to deduce much directly from the existing game theory literature. However, we are able to draw some interesting conclusions from a straight-forward approach.

Theorem 1. *There exists a noncooperative equilibrium solution.*

Theorem 2. *There is only one solution whenever one of the following conditions holds :*

(i) $\beta_2 \leqslant \beta_1$;

(ii) *there are only two periods ;*

(iii) $\beta_1 \leqslant \frac{(1+p)(1-p-\gamma)}{(1-p)}$

and $\beta_2 \leqslant p+\gamma$

(iv) $\beta_2 pA + \beta_1(1-p) \leqslant 1$, *where*

$$A = n^{-1} \sum_{j=0}^{n-1} (\beta_2/\beta_1)$$

and there are n *periods.*

Remark. More refined conditions for uniqueness when $\beta_2 > \beta_1$ are still needed. It is evident that uniqueness holds when $\beta_2 > \beta_1$ in many other situations besides the ones we mentioned. However, it does not always hold. We give a counterexample involving a three-period game in Case 4 of the proof of Theorem 2 in Section 4 .

The following theorems and corollaries describe a (often the) noncooperative equilibrium solution in more detail.

Theorem 3. *The second player always spends all his money each period.*

Remark. Theorem 3 depends critically on the discount factors β_1 and β_2 being less than or equal to one. It is rather contrary to intuition when $\beta_2 > \beta_1$. Evidently the relatively unfavorable initial money distributions still gives II a solution at his boundary. Note, however, that it is precisely when $\beta_2 > \beta_1$ that uniqueness can fail.

Theorem 4. *If* A(n) *represents the first period in which* I *spends all his money, then*

$$(5) \qquad A(n) = \begin{cases} \max\{k : 1 \leqslant k \leqslant n, \ \beta_1^{\frac{k(k-1)}{2}} > c\}, & \beta_1 > c \text{ and } n \geqslant 2, \\ 1 & \beta_1 \leqslant c \text{ or } n = 1 \end{cases}$$

where

$$(6) \qquad c = \frac{1-p-\gamma}{1-p}$$

Corollary 5.

(a) *If* $\beta_1 = 1$ *and* $\gamma > 0$, *then* I *never spends all his money until the last period* .

(b) *If* $\beta_1 < 1$, *then* I *spends all before the last period if the horizon is sufficiently long.*

(c) *If* $\beta_1 < c$, *then both players spend all their money in the first period.*

(d) *If* $A(n) = k$, *then both players spend all their money each period from* k *to* n.

(e) *If* $A(n) = k < n$, *then* $A(n+m) = k$ *for all* $m \geqslant 0$.

Remark. As a consequence of Corollary 5, we refer to the state in which $\gamma = 0$ or, equivalently, the state in which both players spend all their money as the equilibrium or stationary state within the noncooperative equilibrium solution.

Theorem 6. *If* $A(n) = k$, *then*

(a) *in the first period* I *spends*

$$\hat{x}_{11} = c^{\frac{k-1}{k}} \beta_1^{-\frac{(k-1)}{2}} - (1-p-\gamma)$$

(b) *in period* j $(2 \leq j \leq k)$ *the total money spent is*

$$\hat{x}_{11} + \hat{x}_{21} = c^{\frac{k-j}{k}} \beta_1^{-(k-j)}$$

(c) *the discounted value of the real good consumed by* I *in period* j *is*

$$\frac{\beta_1^{j-1} \hat{x}_{1j}}{\hat{x}_{1j}+\hat{x}_{2j}} = \begin{cases} \beta_1^{j-1} - \beta_1^{\frac{k-1}{2}} (1-p)^{\frac{k-1}{k}} (1-p-\gamma)^{\frac{1}{k}}, & j \leq k \\ \beta_1^{j-1} p & , k < j \leq n \end{cases}$$

(d) *the discounted value of the real good consumed by* II *in period* j *is*

$$\frac{\beta_2^{j-1} \hat{x}_{2j}}{\hat{x}_{1j}+\hat{x}_{2j}} = \begin{cases} \beta_2^{j-1} \beta_1^{\frac{k-2j+1}{2}} (1-p)^{\frac{k-1}{k}} (1-p-\gamma)^{\frac{1}{k}}, & j \leq k \\ \beta_2^{j-1} (1-p) & , k < j \leq n \end{cases}$$

(e) *The n-period utilities* $U_n^1(p,\gamma)$ *and* $U_n^2(1-p,\gamma)$ *are continuously differentiable strictly increasing functions of* γ, *which are convex in* $(0, 1-p)$ *and* $(0,p)$ *respectively.*

Virtually all aspects of this solution are now easy to describe. We list a few additional properties.

Corollary 7. *If* $A(n) < n$, *then the* $(n+m)$*-period solution coincides with the* n*-period solution during the first* n *periods for all* $m \geq 1$. *At the end of period* n, $\gamma = 0$, *so that both players spend all thereafter.*

Corollary 8. *If* $A(n) = k$, *then* II's *utilities from consumption during the first* k *periods increases or decreases according to whether* $\beta_2 > \beta_1$ *or* $\beta_2 < \beta_1$. *If* $\beta_2 = \beta_1$, *then* II's *utility from consumption in period* j *is :*

$$\frac{\beta_1^{j-1}\hat{x}_{2j}}{\hat{x}_{1j}+\hat{x}_{2j}} = \begin{cases} \beta_1^{\frac{k-1}{2}}(1-p)^{\frac{k-1}{k}}(1-p-\gamma)^{\frac{1}{k}} & , \quad 1 \leqslant j \leqslant k \\ \beta_1^{j-1}(1-p) & , \quad k+1 \leqslant j \leqslant n \end{cases}$$

Remark. Corollary 8 gives an interesting characterization of the solution. If I spends all his money for the first time in period k, then his spending is such that II's discounted utility from consumption would be constant over the first k periods if he used I's discount factor β_1 .

Corollary 9. *If* $A(n) = k$ *and* $\beta_1 < 1$, *then both* I's *actual consumption and his utilities decrease from period to period during the first* k *periods. If* $\beta_1 = 1$, *then* $A(n) = n$ *and* I's *actual consumption as well as utility each period is*

$$\frac{\hat{x}_{1j}}{\hat{x}_{1j}+\hat{x}_{2j}} = 1-(1-p)^{\frac{n-1}{n}}(1-p-\gamma)^{\frac{1}{n}}$$

Let V_n represent the excess utility I receives beyong what he would receive if he received goods equal to his ownership every period. In other words, let

$$V_n^1(p,\gamma) = U_n^1(p,\gamma) - p\sum_{k=0}^{n-1}\beta_1^k \tag{7}$$

Corollory 10.

(a) *If* $A(n) = k$, *then* $V_j = V_k$ *for* $j \geqslant k$ *and* $V_{j+1} > V_j$ *for* $1 \leqslant j \leqslant k-1$.

(b) *If* $\beta_1 = 1$, *then*

$$V_n(p,\gamma) = n(1-p)(1-c^{1/n})$$

$$\lim_{n\to\infty} V_n(p,\gamma) = -(1-p)\log(1-\frac{\gamma}{1-p})$$

and

$$\lim_{\gamma\uparrow(1-p)} \lim_{n\to\infty} V_n(p,\gamma) = \infty$$

where c *is defined in* (6).

Corollary 11. *Players* II's *amount of money is strictly increasing from* $1-p-\gamma$ *until* $(1-p)$ *is reached. If* $A(n) = k$, *then* II's *amount of money increases from period to period by a factor of* $c^{-1/k}\beta_1^{-1} \geq 1$ *for* $3 \leq j \leq k-1$.

Theorem 12. *If* $\beta_1 = \beta_2$, *then every noncooperative equilibrium solution is Pareto optimal, but if* $\beta_1 \neq \beta_2$, *then none is.*

3.2 - *Infinite Horizon.*

We now consider infinite-horizon versions of our two-person sequential game. We still assume that $\varphi_i(x) = x$, $x \geq 0$.

3.2.1 - $\beta_i < 1$. If the discount factors are both less than one, then we can use the discounted utility functions in (3). If we restrict attention to stationary strategies, then we can use the functional equations in (2) with both U_n^i and U_{n-1}^i replaced by U^i , namely,

$$(8) \qquad U^i(p_i,\gamma_i) = \sup_{0\leq x_i\leq p_i+\gamma_i} \{K^i(x_1,x_2) + \beta_i U^i(p_i,\gamma_i - x_i + p_i(x_1+x_2))\}$$

We use supremum instead of maximum because the maximum might not be attained.

If $\beta_1 = \beta_2$, our game is almost, but not quite, strictly competitive or constant-sum. It is not constant sum because of the one-period distribution of goods at (0,0) spending. If we redefine the one-period distributions at (0,0) to be

$K^1(0,0) = K^2(0,0) = 1/2$, then our game would be constant-sum. The resulting model is then similar to the deterministic version of Shapley's (9) stochastic game. It is not quite the same , though, because Shapley only considered finite sets of states and actions. However, it is known that the theory also applies to various infinite sets under additional assumptions. In particular, this is evident from the fundamental paper by Denardo (5). Another treatment in the spirit of Blackwell (3) has recently been provided by Maitra and Parthasarathy (8). Unfortunately, none of these papers appears to be directly applicable here, although the flexible framework provided by Denardo (5) is promising. In the more general context of nonconstant-sum sequential games, the recent paper by Sobel (13) is also related but again not directly applicable. The discontinuity of K in (2) at (0,0) is the principal source of difficulty.

Although the game theory literature does not appear very helpful, we can easily apply our previous finite-horizon results to this infinite-horizon model. For this purpose, define $A(\infty)$ just as in (5). Since $\beta_i < 1$, $A(\infty) < \infty$. On the basis of Corollary 7, it is obvious that the infinite-horizon model has the same properties as the n-period problem for $n > A(\infty)$. Let

$$W_n^i(p_i,\gamma_i) = U_n^i(p_i,\gamma_i) + p_i \sum_{k=n}^{\infty} \beta_i^k \tag{9}$$

Obviously W_n^i represents the infinite-horizon utility to individual i when both players follow the n-period solution for the first n periods and are in the steady-state thereafter.

Corollary 13. *If* $A(\infty) = k$ *and* $n \geq k$, *then* (W_n^1, W_n^2) *defined in* (9) *is a solution to the infinite-horizon functional equations in* (8). *Moreover, the allocations and spending agree with the* n-*period solution for the first* n *periods and are in the steady-state thereafter.*

It seems that the solution to (8) above should be the only one which is bounded below by 0 and above by $(1-\beta_i)^{-1}$. It also seems that nothing would change if we allowed non-stationary strategies. However, we do not yet have proofs.

3.2.2 - $\beta_1 = \beta_2 = 1$. When there is no discounting, the criterion is usually changed to average return per period, but we shall not consider this criterion here. Instead, we shall consider the infinite-horizon functional equations related to the excess utilities V_n^i in (7). Corollary 10 (b) suggests such a system might have a finite solution. In particular, we shall consider the functional equations

$$(10) \qquad V^i(p_i,\gamma_i) = \max_{0 \leq x_i \leq p_i+\gamma_i} \{K^i(x_1,x_2)-p_i + V^i(p_i,\gamma-x_i + p(x_1+x_2))\}$$

for i = 1, 2 and K^i in (2).

Theorem 14. *The limit in Corollary 10 (b) is a solution to* (10) *and the limit in Theorem 6 (a) is the amount* I *spends initially, i.e.,*

$$V^1(p,\gamma) = - V^2(1-p-\gamma) = - (1-p) \log(1- \frac{\gamma}{1-p})$$

and

$$\hat{x}_{11} = c - K = \frac{p(1-p-\gamma)}{1-p}$$

Remarks. Since the proof of Theorem 14 is easy, we omit it. It is easy to see that the same result holds for I is $\beta_2 < 1$. Then β_2 must be included in the functional equation involving V^2. Of course, II spends all every period. Again we have not yet resolved the issue of uniqueness.

Corollary 15. *The solution above is not realized by* I *following the indicated strategy. Consumption each period equals ownership and* I *carries his extra money* γ *into the future.*

Since the optimal value cannot actually be attained, we look for strategies which can come arbitrarily close.

Corollary 16. *An* ε*-optimal (stationary) strategy for* I *can be obtained by using an appropriate optimal finite-horizon strategy.*

3.3 - *Many Players.*

We now consider our sequential money game with more than two players. In the beginning we still assume that $\varphi_i(x) = x$, $x \geq 0$, but later we show how this can be generalized.

When there are more than two players, each player can still think of himself being in a two-person game because he can lump all the other players together, but as the number of players increases, the number of cases and the complexity of the analysis increases. Even a two-period version with ten players would present a formidable task. The m-person game does reduce to a two-person game in some special cases however. In particular, this occurs if all the players or all but one player would be at the boundary (spending all their money in the first period) in the associated two-person game in which all other players are lumped together. The two-person results in Sections 3.1 and 3.2 thus immediately imply the following two corollaries.

Corollary 17. *(No Big Strong Player). If* $\gamma_i \leq 0$ *or* $\beta_i \leq (1-p_i-\gamma_i)/(1-p_i)$ *for each individual* i, *then everyone spending all their money in every period yields a noncooperative equilibrium solution.*

Corollary 18. *(One Big Strong Player). If* $\gamma_1 > 0$ *and* $\beta_1 > (1-p_1-\gamma_1)/(1-p_1)$ *while either* $\gamma_i \leq 0$ *or* $\beta_i \leq (1-p_i-\gamma_i)/(1-p_i)$

for all $i \geq 2$, *then there exists a noncooperative equilibrium solution in which individual* 1 *follows the two-person strategy for player* I *in Sections 3.1 and 3.2 and the others spend all each period.*

Remark. We have not yet successfully characterized uniqueness in Corollaries 17 and 18, but it is easy to verify for a finite horizon in special cases of interest as we illustrate below.

We now apply Corollary 17 to investigate large economies in a state approaching perfect competition. We obtain a "law of large numbers" comparable to the classical probability theorem with that name. Just as in the probabilistic setting, we need to require not only that the number of individuals be large but also that each one be asymptotically negligible in relation to the whole. Since p_i and $p_i+\gamma_i$ represent the fractions of total ownership and money respectively, it suffices for these to be small.

Theorem 19. *(Perfect Competition) If* $p_i \leq n^{-1}$, $\gamma_i \leq n^{-1}$, *and* $\beta_i \leq 1-\delta < 1$ *for each individual* i, *then for sufficiently large* n *there exists an equilibrium solution in which each individual spends all his money every period. If the horizon is finite, then this is the only solution.*

A similar result corresponding to Corollary 18 also holds.

Theorem 20. *(One Fat Cat) Let* p_1 *and* $\gamma_1 > 0$ *be independent of* n . *If* $p_i \leq n^{-1}$, $\beta_i \leq n^{-1}$, *and* $\beta_i \leq 1-\delta < 1$ *for each individual* $i \geq 2$, *then for sufficiently large* n *there exists an equilibrium solution in which each individual* i, *for* $i \geq 2$, *spends all his money every period. If the horizon is finite, then this is the only solution.*

Remark. Theorems 19 and 20 go beyond Corollaries 17 and 18 by providing uniqueness. Unfortunately, our uniqueness proofs in Section 5 only apply to arbitrary finite horizons. It seems intuitively obvious that uniqueness should also hold for the model with an unbounded horizon but a proof eludes us. Since the actual horizon out there in the world appears to be unbounded, it is natural to question the value of our finite-horizon results. However, our finite-horizon results do have a natural interpretation in the infinite horizon. You can think of each player using a rolling strategy ; that is, each period each player looks a specified finite number of periods into the future and selects his strategy assuming the world or his interest in the world terminates at the end of those periods. This process is repeated each period so that the players are always making their decisions based on the present plus a specified number of periods of the future. With the time discounts, each player knows, in our model at least, that the part of the future he is failing to consider is negligible. It is significant that in the setting of Theorems 19 and 20 the solution in succeeding periods after the first is the same (everyone spends all) using a rolling strategy or the second period strategy from a fixed finite-horizon strategy. We also expect the right argument will yield uniqueness in the infinite horizon, which would imply that the rolling strategy solution coincides with the infinite-horizon solution.

Remark. The results in this section obviously hold for quite general one-period utility functions φ_i, cf. (2) . When φ_i is changed, we must specify the total amount G of the good to be distributed. It is then natural to let G grow linearly in m as m, the number of players, increases. Suppose that φ_i is twice-differentiable with $\varphi_i' > 0$ and $\varphi_i'' < 0$. Then it is easy to see that Corollary 17 still holds if, for each i ,

$$\frac{\varphi_i'\left(\frac{Gx_i}{x_i + 1-p_i-\gamma_i}\right)}{\varphi_i'(G\,[p_i + (1-p_i)(p_i+\gamma_i-x_i)\,])} \geqslant \frac{\beta_i(1-p_i)}{G(1-p_i-\gamma_i)} \tag{11}$$

for $0 \leqslant x_i \leqslant p_i+\gamma_i$. To get (11), it suffices to have

$$\frac{\varphi_i'(G\,[p_i+\gamma_i\,])}{\varphi_i'(Gp_i)} \geqslant \frac{\beta_i(1-p_i)}{G(1-p_i-\gamma_i)} \tag{12}$$

where the right side is less than G^{-1} . The uniqueness in Theorems 19 and 20 also obviously carries over to more general φ_i , but we have no nice conditions.

3.4 - *Constrained Competitive Equilibria.*

For comparison, we now investigate the set of competitive equilibria. In particular, assume each player is confronted with the optimization problem in (4). We look for a set of prices and allocations such that the allocations are optimal for each individual at those prices, the constraints are met, and the overall solution is Pareto optimal.

Theorem 21. *If* $\beta_i = \beta \leqslant 1$ *for all* i, *then there is a unique competitive equilibrium solution. The price is* M/G *every period and the players spend all their money every period. If* $\beta_i \neq \beta_j$ *for some* i *and* j, *then Pareto optimality is lost.*

Corollary 22. *Under the conditions of Theorem 19, the set of noncooperative equilibrium solutions coincides with the set of competitive equilibrium solutions for sufficiently large* n .

Remark. Corollary 22 is strictly correct only when $\beta_i = \beta$ for all i, but is true more generally if we relax the requirement of Pareto optimality in the definition of a competitive equilibrium.

As we have noted before, the current models without credit do not cope with uneven time preferences for goods. This has been illustrated here with different interest rates. It can also be illustrated by allowing ownership to change over time.

4. Proofs for the Two-Person Finite-Horizon Games.

Proof of Theorem 1 : Existence.

Our proof will be by induction. We will thus want to combine our existence proof (Theorem 1) with the description proofs (Theorems 3, 4 and 6 plus associated corollaries). All the results are trivial for one period since each player is clearly motivated to spend all his money regardless of what the other does. Hence, we shall verify that the solution described in Section 3.1 is in fact a solution for n+1 periods assuming that it is for k periods for each k, $1 \leqslant k \leqslant n$. To simplify expressions, we shall let $K = (1-p-\gamma)$ and $c = K/(1-p)$.

Case 1 : Player I's Optimization when $A(n+1) = k < n+1$.

We first look at I, assuming that II spends all his money in the first period as well as every period thereafter. Suppose $A(n+1) < n+1$, where $A(n)$ is defined in (5). Player I's optimal value is thus

$$(13) \qquad U^1_{n+1}(p,\gamma) = \max_{0 \leqslant x_1 \leqslant p+\gamma} \left\{ \frac{x_1}{x_1+K} + \beta_1 U^1_n(p,\ (1-p)(p+\gamma-x_1)) \right\}$$

where $K = 1-p-\gamma$. We will drop the subscript on β_1 since the other discount β_2 will be of no concern to I. Using Theorem 4, Corollary 7, and the induction hypothesis, we have

$$(14) \qquad U^1_n(p,\ (1-p)(p+\gamma-x_1)) = U^1_{n-1}(p,\ (1-p)(p+\gamma-x_1)) + p\beta^{n-1}$$

if $\beta^{\frac{n(n-1)}{2}} \leqslant c_n$, where we let $c_n = (1-p-\gamma_n)/(1-p)$ with γ_n

denoting I's excess money at the beginning of the second period (with n of n+1 periods to go). Here

$$(15) \qquad c_n = \frac{1-p-(1-p)(p+\gamma-x_1)}{1-p} = 1-p-\gamma+x_1 .$$

Hence, for $x_1 \geqslant \beta^{\frac{n(n-1)}{2}} - K$, we can substitute $U^1_{n-1}+p\beta^{n-1}$ for U^1_n inside (13) which means $U^1_{n+1}(p,\gamma) = U^1_n(p,\gamma) + \beta^n p$. Then, by Theorem 6 (a),

$$(16) \qquad \hat{x}_{11} = c^{\frac{k-1}{k}} \beta^{-\frac{(k-1)}{2}} - K$$

$$\geqslant \beta^{\frac{k(k-1)}{2}} - K$$

so that $\hat{x}_{11} \geqslant \beta^{\frac{n(n-1)}{2}} - K$ as required. If $k = n$, then we use the hypothesis that $A(n+1) < n+1$ to get $\beta^{\frac{n(n+1)}{2}} \leqslant c$, which implies that $\hat{x}_{11} \geqslant \beta^{\frac{n(n-1)}{2}} - K$.

It remains to show that I does not want x_1 for which $\beta^{\frac{n(n-1)}{2}} > c_n$ or, equivalently, for which $x_1 < \beta^{\frac{n(n-1)}{2}} - K$.

If $\beta^{\frac{n(n-1)}{2}} > c_n$, then by Theorem 6 (c),

$$(17) \qquad U^1_n(p,\gamma_n) = U^1_n(p,\ (1-p)(p+\gamma-x_1))$$

$$= -n\beta^{\frac{n-1}{2}}(1-p)^{\frac{n-1}{n}}(1-p-\gamma_n)^{\frac{1}{n}} + \sum_{j=0}^{n-1} \beta^j$$

$$= -n\beta^{\frac{n-1}{2}}(1-p)(1-p-\gamma+x_1)^{\frac{1}{n}} + \sum_{j=0}^{n-1} \beta^j$$

Next we differentiate (13) using (17) to get

$$(18)\qquad f'(x_1) = \frac{df(x_1)}{dx_1} = \frac{K}{(x_1+K)^2} - \beta^{\frac{n+1}{2}}(1-p)(x_1+K)^{-\frac{(n-1)}{n}}$$

and

$$(19)\qquad f''(x_1) = \frac{-2K}{(x_1+K)^3} + \beta^{\frac{n+1}{2}}(1-p)\left(\frac{n-1}{n}\right)(x_1+K)^{-\frac{(2n-1)}{n}}$$

where $f(x)$ denotes the expression inside the maximum in (13) when (17) is used. If we set $f'(x_1) = 0$, then we get the unique solution

$$(20)\qquad \bar{x}_{11} = c^{\frac{n}{n+1}}\beta^{-\frac{n}{2}} - K$$

We now must verify that $f''(\bar{x}_{11}) < 0$ to show that (20) gives a maximum. This is not immediately obvious because (19) contains one positive term as well as one negative term. However,

$$(21)\qquad f''(\bar{x}_{11}) = (\bar{x}_{11}+K)^{-3}\left[-2K+(1-p)\beta^{\frac{n+1}{2}}\left(\frac{n-1}{n}\right)(\bar{x}_{11}+K)^{\frac{n+1}{n}}\right]$$

$$= -(\bar{x}_{11}+K)^{-3}(1-p)^{-1}\beta^{\frac{n+1}{2}} \times$$

$$\left[\left(\frac{2n}{n-1}\right)c\beta^{-\frac{(n+1)}{2}} - (\bar{x}_{11}+K)^{\frac{n+1}{n}}\right]$$

where

$$\left(\frac{2n}{n-1}\right)c\beta^{-\frac{(n+1)}{2}} > c\beta^{-\frac{(n+1)}{2}} = \left(c^{\frac{n}{n+1}}\beta^{-\frac{n}{2}}\right)^{\frac{n+1}{n}} = (\bar{x}_{11}+K)^{\frac{n+1}{n}}$$

Now note that $\bar{x}_{11}$ in (20) exceeds $\beta^{\frac{n(n-1)}{2}} - K$ because $\beta^{\frac{n(n+1)}{2}} \leqslant c$. Hence, the maximum value possible for $x_1 \leqslant \beta^{\frac{n(n-1)}{2}} - K$ occurs at the boundary, i.e.,

$$(22)\qquad \tilde{x}_1 = \beta^{\frac{n(n-1)}{2}} - K$$

We now must verify that I prefers (16) to (22). Applying the induction hypothesis with (16), we have

$$(23)\qquad f(\hat{x}_{11}) \geqslant \sum_{j=0}^{n-1} \beta^{j} - n\beta^{\frac{n-1}{2}}(1-p)^{\frac{n-1}{n}}(1-p-\gamma)^{\frac{1}{n}} + p\beta^{n}$$

$$= \sum_{j=0}^{n} \beta^{j} - n\beta^{\frac{n-1}{2}}(1-p)c^{\frac{1}{n}} - (1-p)\beta^{n}$$

where the inequality is due to our using the solution generated by $A(n) = n$. If $A(n) < n$, then (23) is a strict inequality. Corresponding to (22), we have

$$(24)\qquad f(\tilde{x}_{11}) = \frac{\beta^{\frac{n(n-1)}{2}} - K}{\beta^{\frac{n(n-1)}{2}}}$$

$$+ \beta\left[\sum_{j=0}^{n-1} \beta^{j} - n\beta^{\frac{n-1}{2}}(1-p)^{\frac{n-1}{n}}\left(1-p-(1-p)\left(1-\beta^{\frac{n(n-1)}{2}}\right)\right)\right]$$

$$= \sum_{j=0}^{n} \beta^{j} - K\beta^{-\frac{n(n-1)}{2}} - n(1-p)\beta^{n}$$

Comparing (23) and (24), we see that

$$(1-p)^{-1}[f(\hat{x}_{11})-f(\tilde{x}_{11})] \geqslant c\beta^{-\frac{n(n-1)}{2}} + n\beta^{n} - n\beta^{\frac{n-1}{2}}c^{\frac{1}{n}} - \beta^{n}$$

$$= \left(\frac{c^{1/n}}{\beta^{(n-1)/2}}\right)^{n} - \beta^{n} - n\beta^{n-1}\left(\frac{c^{1/n}}{\beta^{(n-1)/2}} - \beta\right)$$

$$= (\alpha^{n} - \beta^{n}) - n\beta^{n-1}(\alpha-\beta)$$

$$= (\alpha-\beta)\left(\sum_{k=0}^{n-1} \alpha^{k}\beta^{n-k-1} - n\beta^{n-1}\right)$$

$$(25)\qquad \geqslant 0$$

where $K = 1-p-\gamma$ and $c = K(1-p)^{-1}$ as before, $\alpha = c^{\frac{1}{n}}\beta^{-\frac{(n-1)}{2}} \geqslant \beta$ $\alpha^{k}\beta^{n-k-1} \geqslant \beta^{n}$. We have thus shown that if $A(n+1) < n+1$, then the $(n+1)$-period solution for I is the same as the n-period solution except I gets $\beta^n p$ more in the $(n+1)^{st}$ period. We have in fact shown a bit more, namely, that the prospective strategy in (22) is dominated by the strategy in (16) with $k = n$, which in turn is dominated by the optimal strategy.

Case 2 : Players I's Optimization when $A(n+1) = n+1$.

Now we assume that $\beta^{\frac{n(n+1)}{2}} > c$. As in Case 1, I will spend less than all his money each period until the last period, beginning in the second period, if $x_1 \leqslant \beta^{\frac{n(n-1)}{2}} - K$. Assuming this to be the case, we get the solution generated by $\overline{x}_{11}$ in (20) . However, since $\beta^{\frac{n(n+1)}{2}} > c$ now, $\overline{x}_{11} < \beta^{\frac{n(n-1)}{2}} - K$, so $\overline{x}_{11}$ is the natural candidate for the initial spending which generates the optimal solution. The return using $\overline{x}_{11}$ in (20) is easily calculated :

$$f(\overline{x}_{11}) = \frac{\overline{x}_{11}}{\overline{x}_{11}+K} + \beta U_n^1 \left(p,\ (1-p)\left(1 - c^{\frac{n}{n+1}} \beta^{-\frac{n}{2}}\right)\right)$$

$$= 1 - \frac{K}{c^{\frac{n}{n+1}} \beta^{-\frac{n}{2}}} + \beta \left[\sum_{j=0}^{n-1} \beta^j - n\beta^{\frac{n-1}{2}} (1-p)^{\frac{n-1}{n}} (1-p-\gamma_n)^{\frac{1}{n}}\right]$$

$$= \sum_{j=0}^{n} \beta^j - Kc^{-\frac{n}{(n+1)}} \beta^{\frac{n}{2}}$$

$$- n\beta^{\frac{n+1}{2}} (1-p)^{\frac{n-1}{n}} \left[1-p-(1-p)\left(1-c^{\frac{n}{n+1}} \beta^{-\frac{n}{2}}\right)\right]^{\frac{1}{n}}$$

$$= \sum_{j=0}^{n} \beta^j - (n+1)\ \beta^{\frac{n}{2}} (1-p)^{\frac{n}{n+1}} (1-p-\gamma)^{\frac{1}{n+1}}$$

(26)

The second step in (26) is justified because

$$(27) \quad \beta^{\frac{n(n-1)}{2}} > c_n = \frac{1-p-\gamma_n}{1-p} = c^{\frac{n}{n+1}} \beta^{-\frac{n}{2}}$$

since $\beta^{\frac{n(n+1)}{2}} > c$ by assumption.

It remains to rule out $x_1 \geqslant \beta^{\frac{n(n-1)}{2}} - K$. As in (13)-(16) of Case 1, we would obtain $U^1_{n+1}(p,\gamma) = U^1_n(p,\gamma) + \beta^n p$ if x_1 is constrained to be bigger than $\beta^{\frac{n(n-1)}{2}} - K$. The propsective solution would then be generated by $\hat{x}_{11}$ in (16) for some $k \leqslant n$. If $\hat{x}_{11} \geqslant \beta^{\frac{n(n-1)}{2}} - K$, then $\hat{x}_{11}$ is a legitimate candidate. For this case it suffices to show that $f(\bar{x}_{11}) > f(\tilde{x}_{11})$ with $f(\bar{x}_{11})$ in (26) and

$$(28) \quad f(\hat{x}_{11}) = \sum_{j=0}^{n} \beta^j - k\beta^{\frac{k-1}{2}}(1-p)c^{\frac{1}{k}} - (1-p)\sum_{j=k}^{n}\beta^j ,$$

for any $k \leqslant n$. In fact, it is easier to prove a stronger result. We shall show that (28) is strictly increasing in k. For $k \leqslant n$,

$$(1-p)^{-1}[f(\hat{x}_{11})_{k+1} - f(\hat{x}_{11})_k]$$

$$= k\beta^{\frac{k-1}{2}} c^{\frac{1}{k}} + \beta^k - (k+1)\beta^{\frac{k}{2}} c^{\frac{1}{k+1}}$$

$$= \left(\beta^k - \beta^{\frac{k}{2}} c^{\frac{1}{k+1}}\right) - k\left(\beta^{\frac{k}{2}} c^{\frac{1}{k+1}} - \beta^{\frac{k-1}{2}} c^{\frac{1}{k}}\right)$$

$$= \beta^{\frac{k}{2}}\left[\left(\beta^{\frac{1}{2}}\right)^k - \left(c^{\frac{1}{k(k+1)}}\right)^k\right] - k\beta^{\frac{k-1}{2}} c^{\frac{1}{k+1}}\left[\beta^{\frac{1}{2}} - c^{\frac{1}{k(k+1)}}\right]$$

$$= \beta^{\frac{k}{2}}\left(\beta^{\frac{1}{2}} - c^{\frac{1}{k(k+1)}}\right)\left[\sum_{j=0}^{k-1}\left(\beta^{\frac{1}{2}}\right)^{(k-j-1)}\left(c^{\frac{1}{k(k+1)}}\right)^j - k\beta^{-\frac{1}{2}} c^{\frac{1}{k+1}}\right]$$

$$(29) \quad = \beta^{\frac{k-1}{2}}\left(\beta^{\frac{1}{2}} - c^{\frac{1}{k(k+1)}}\right)\sum_{j=0}^{k-1}\left[\beta^{\frac{k-j}{2}} c^{\frac{j}{k(k+1)}} - c^{\frac{k}{k+1}}\right]$$

which is positive because

$$\beta^{\frac{1}{2}} - c^{\frac{1}{k(k+1)}} > 0$$

and

$$\beta^{\frac{k-j}{2}} c^{\frac{j}{k(k+1)}} - c^{\frac{k}{k(k+1)}} > 0$$

since

$$\beta^{\frac{k(k+1)}{2}} \geqslant \beta^{\frac{n(n+1)}{2}} > c$$

Furthermore, the argument above shows that $f(\hat{x}_{11})_k = f(\hat{x}_{11})_{k+1}$ if $\beta^{\frac{k(k+1)}{2}} = c$.

There is still one point more to dispose of in this case. It could happen that $\hat{x}_{11} < \beta^{\frac{n(n-1)}{2}} - K$. Applying (19) and (21) for $k \leqslant n$, we see that $\hat{x}_{11}$ is a maximum and $f(x_1)$ decreases as x_1 moves away from $\hat{x}_{11}$. Therefore, the candidate to be considered is $\tilde{x}_{11}$ in (22) instead of $\hat{x}_{11}$ in (16). Thus, it remains to show that $f(\overline{x}_{11}) > f(\tilde{x}_{11})$ for $f(\overline{x}_{11})$ in (26) and $f(\tilde{x}_{11})$ in (24), but

$$\beta^{-\frac{n}{2}} (1-p)^{-1} \left[f(\overline{x}_{11})-f(\tilde{x}_{11})\right]$$

$$= \beta^{-\frac{n}{2}} \left[c\beta^{-\frac{n(n-1)}{2}} + n\beta^{n} - (n+1)\, \beta^{\frac{n}{2}} c^{\frac{1}{n+1}}\right]$$

$$= n \left(\beta^{\frac{n}{2}} - c^{\frac{1}{n+1}}\right) \left(c^{\frac{1}{n+1}} - c\beta^{-\frac{n^2}{2}}\right)$$

$$= \left(\beta^{\frac{n}{2}} - c^{\frac{1}{n+1}}\right)\left[n - c^{\frac{1}{n+1}} \beta^{-\frac{n^2}{2}} \left(\beta^{\frac{n^2}{2}} - c^{\frac{n}{n+1}}\right)\left(\beta^{\frac{n}{2}} - c^{\frac{n}{n+1}}\right)\right]$$

$$= \left(\beta^{\frac{n}{2}} - c^{\frac{1}{n+1}}\right)\left[n - c^{\frac{1}{n+1}} \beta^{-\frac{n^2}{2}} \sum_{j=0}^{n-1} \left(\beta^{\frac{n}{2}}\right)^{j} \left(c^{\frac{1}{n+1}}\right)^{n-j-1}\right] , \tag{30}$$

which is positive because

$$\beta^{\frac{n}{2}} - c^{\frac{1}{n+1}} > 0$$

and

$$c^{\frac{1}{n+1}} \beta^{-\frac{n^2}{2}} \beta^{\frac{nj}{2}} c^{\frac{n-j-1}{n+1}} = c^{\frac{n-j}{n+1}} \beta^{-\frac{n(n-j)}{2}} < 1,$$

since

$$\beta^{\frac{n(n+1)}{2}} > c .$$

This completes the proof for I.

Case 3 : Possible Advantage for Player II.

We now show that the weak player spends all his money every period. Continuing the induction proof, we assume II spends all his money every period in each k-period problem for $k \leqslant n$ and we use the strategies just verified for I.

We begin by assuming $\beta_1 < c$ so that I spends all his money in the first period. Then I loses his money advantage in the second period. The new advantage to I with n periods to go becomes

$$\begin{aligned} \gamma_n &= \gamma + px_2 - (1-p)x_1 \\ &= -\,p(1-p-\gamma-x_2) \leqslant 0, \end{aligned} \tag{31}$$

where x_2 is II's initial spending. Hence, II's optimization problem is

$$U^2_{n+1}(1-p,\ -\gamma) = \max_{0 \leqslant x_2 \leqslant 1-p-\gamma} g(x_2), \tag{32}$$

where

$$\begin{aligned} g(x_2) &= \frac{x_2}{x_1 + x_2} + \beta_2 U^2_n(1-p,\ +\gamma_n(x_2)) \\ &= \frac{x_2}{x_2+p+\gamma} + \beta_2 U^2_n(1-p,\ p(1-p-\gamma-x_2)). \end{aligned} \tag{33}$$

Note that $U^2_n(1-p,\ +\gamma)$ in (33) corresponds to I's strategy in period n with p and 1-p switched, obtained via the induction hypothesis, because now II has the money advantage. We know the second term in (33) increases as x_2 decreases and as β_2 increases, so let $\beta_2 = 1$. This will only make $\hat{x}_{21}$, the initial spending for II, smaller. Then

$$(34)\qquad U_n^2(1-p,\ +\gamma_n) = U_n^2(1-p,\ p(1-p-\gamma-x_2))$$

$$= np^{\frac{n-1}{n}}\,(p-\gamma_n)^{\frac{1}{n}} = np(p+\gamma+x_2)^{\frac{1}{n}}$$

by virtue of (26) or Theorem 6 (c). Then

$$(35)\qquad g'(x_2) = \frac{p+\gamma}{(x_2+p+\gamma)^2} - \frac{p}{(x_2+p+\gamma)^{\frac{n-1}{n}}}$$

and

$$(36)\qquad g''(x_2) = \frac{-2(p+\gamma)}{(x_2+p+\gamma)^3} + \frac{p(n-1)}{n(x_2+p+\gamma)^{\frac{2n-1}{n}}}$$

for g in (33). Setting $g'(x_2) = 0$, we get the unique solution

$$(37)\qquad \bar{x}_{21} = \left(\frac{p+\gamma}{p}\right)^{n/(n+1)} - (p+\gamma)$$

which is obviously strictly greater than $K = 1-p-\gamma$ for $\gamma > 0$. Also $g''(\bar{x}_{21}) > 0$ because

$$(38)\qquad \frac{2(p+\gamma)}{(\bar{x}_{21}+p+\gamma)^3} > \frac{p(n-1)}{n(\bar{x}_{21}+p+\gamma)^{\frac{2n-1}{n}}}$$

or, equivalently,

$$(39)\qquad \frac{2n}{n-1}\,\frac{p+\gamma}{p} > (\bar{x}_{21}+p+\gamma)^{\frac{n+1}{n}} = \frac{p+\gamma}{p}$$

This means $\bar{x}_{21}$ is indeed a maximum. Since $\bar{x}_{21} > 1-p-\gamma$, II's initial spending should be

$$(40)\qquad \hat{x}_{21} = 1-p-\gamma\ .$$

The present case demonstrates that II will not spend less to capitalize on a money advantage he could obtain over I. We have shown this for $x_1 = p+\gamma$, but it also applies to all other possible x_1, i.e., $c-K = \frac{p(1-p-\gamma)}{1-p} \leqslant x_1 \leqslant p+\gamma$. Note from (34) that

$$(41) \qquad \frac{dU_n^2(1-p, \gamma)}{d\gamma} = \left(\frac{p}{p-\gamma} \right)^{(n-1)/n}$$

The largest money advantage II can obtain at the end of the first period occurs when I spends all his money in the first period. Hence,

$$(42) \qquad \frac{dU_n^2(1-p, \gamma_n)}{d\gamma_n} \frac{d\gamma_n}{dx_2} \leqslant \left(\frac{p}{p+\gamma+x_2} \right)^{(n-1)/n} p \leqslant p \ .$$

On the other hand, the least marginal advantage II can obtain from spending in the first period occurs at the point where he spends all his money. This rate is

$$(43) \qquad \frac{x_1}{(x_1+1-p-\gamma)^2} = 1 - (1-p-\gamma)^{1-a} (1-p)^a \geqslant p \ ,$$

for $x_1 = c^a - K$, $0 \leqslant a \leqslant 1$, which covers all x_1, $c-K \leqslant x_1 \leqslant p+\gamma$. Comparing (42) and (43), we see that the conclusion in this case applies to all x_1 which I might select as optimal following the strategies previously determined for him.

Case 4 : The Possibility of II Reducing I's Advantage.

The critical difference between this case and the one before is that now II is considering spending less so that I will have a smaller money advantage in the second period instead of spending less so that II can achieve an actual money advantage for himself. We shall see that II still spends all his money.

We have treated $A(n+1) = 1$ in the last case. Now suppose $A(n+1) = k \geqslant 2$ with I's advantage

$$\gamma_n = \gamma + px_2 - (1-p)\left[c^{\frac{k-1}{k}}\beta_1^{-\frac{k-1}{2}} - K\right] \geq 0 \tag{44}$$

Then, instead of (33), we have

$$g(x_2) = \frac{x_2}{x_1+x_2} + \beta_2 U_n^2(1-p, -\gamma_n(x)). \tag{45}$$

If $x_2 = (1-p-\gamma)$, then

$$g(x_2) = \frac{1-p-\gamma}{c^{\frac{k-1}{k}}\beta_1^{-\frac{k-1}{2}}} + \beta_2\beta_1^{\frac{k-2}{2}}(1-p)^{\frac{k-2}{k-1}}(1-p-\gamma_n)^{\frac{1}{k-1}}\sum_{j=0}^{k-2}(\beta_2/\beta_1)^j$$

$$= \frac{K}{c^{\frac{k-1}{k}}\beta_1^{-\frac{k-1}{2}}} + \beta_2\beta_1^{\frac{k-2}{2}}(1-p)^{\frac{k-2}{k-1}}\left[(1-p)c^{\frac{k-1}{k}}\beta_1^{-\frac{k-1}{2}}\right]^{\frac{1}{k-1}}\sum_{j=0}^{k-2}(\beta_2/\beta_1)^j \tag{46}$$

by virtue of the induction hypothesis. If $\beta_1 = \beta_2$, then $U_n^2(1-p, -\gamma_n(x)) = \sum_{j=0}^{n-1}\beta_1^j - U_n^1(p, \gamma_n(x))$, but if $\beta_1 \neq \beta_2$, then we must include the last sum in (46).

It is evident that both terms of the derivative in (45) are decreasing in x_2. Hence, it suffices to consider the derivative evaluated at the largest possible value of x_2. In other words,

$g'(x_2) \geq g'(1-p-\gamma) =$

$$\begin{cases} \dfrac{z-K}{z^2} - p\beta_2, & k = 2 \\ \dfrac{z-K}{z^2} - p\beta_2\beta_1^{\frac{k-2}{2}}(k-1)^{-1}z^{-\frac{k-2}{k-1}}\displaystyle\sum_{j=0}^{k-2}(\beta_2/\beta_1)^j, & k \geq 3 \end{cases} \tag{47}$$

where

$$z = c^{\frac{k-1}{k}} \beta^{-\frac{k-1}{2}} . \tag{48}$$

If A(n+1) = 2, then (47) reduces to

$$g'(1-p-\gamma) = \frac{c^{1/2} \beta_1^{-1/2} - K}{c\beta_1^{-1}} - p\beta_2$$

$$= c^{-1/2} \beta_1^{1/2} - [\beta_1(1-p) + p\beta_2]$$

$$\geqslant 1 - 1 = 0 \tag{49}$$

because $\beta_1 > c$. Hence, II will spend all his money every period if A(n+1) = 2.

For $A(n+1) = k \geqslant 3$, it suffices to let $\beta_2 = 1$. This can only make the negative term in (47) larger in absolute value.Then

$$g'(1-p-\gamma) \geqslant \frac{z-K}{z^2} - \frac{p\beta_1^{-(k-2)/2}}{z^{(k-2)/(k-1)}}$$

$$= z^{-\frac{k-2}{k-1}} \left[\frac{1}{z^{1/(k-1)}} - \frac{(1-p)c}{z^{k/(k-1)}} - p\beta_1^{-\frac{k-2}{2}} \right]$$

$$\geqslant \beta_1^{-(k-2)/2} - (1-p) - p\beta_1^{-(k-2)/2}$$

$$\geqslant (1-p) (\beta_1^{-(k-2)/2} - 1) \geqslant 0 \tag{50}$$

Hence, the weaker player always spends all his money.

Related Theorems and Corollaries.

The proof of Theorem 1 has employed Theorems 3, 4 and 6. Therefore, to properly complete the proof of Theorem 1, we must verify that these descriptions prevail in the (n+1)-period problem. The only parts remaining are (b)-(e) in Theorem 6.

Theorem 6 (b) is easy to verify directly for $j = 2$ and by induction for $j > 2$. Using (b), it is easy to compute the money each player has in the beginning of period j . Then the next consumption is easy to determine, from which (c) and (d) follow easily. Part (e) is evident for a given strategy. The argument in (29) shows there is no difficulty at the transition points.

The remaining corollaries in Section 3.1 follows easily from Theorems 1, 3, 4 and 6. In Corollary 10 (b) the first limit can be obtained by applying Taylor's Theorem to $c^{n^{-1}} = e^{n^{-1}\log c}$.

Proof of Theorem 2 : Uniqueness

We now consider whether the solution just obtained is the only solution. We show that it is if either $\beta_2 \leqslant \beta_1$ or $\beta_1 \leqslant c_n$, where

$$c_n = \frac{1-p-\gamma_n}{1-p} = \frac{1-p-\gamma-px_2+(1-p)x_1}{1-p}$$

However, we do not obtain uniqueness when $\beta_1 > c_n$, that is, when the strong player is not motivated to spend all his money in the second period. After proving uniqueness where it holds, we give a counterexample to uniqueness when $\beta_2 > \beta_1$ and $\beta_1 > c_n$.

Again our proof will be by induction. Just as with the existence, the uniqueness is obvious for one period. Therefore, we shall show it is true for n+1 periods given that is true for k periods for each $k \leqslant n$.

Case 1 : Boundary Values .

Let (x_1, x_2) denote the initial spending by I and II for a prospective second solution with n+1 periods. First, note that $0 < x_1 < p+\gamma$ and $0 < x_2 < 1-p-\gamma$. If $x_2 = 1-p-\gamma$, then the previous solution would be obtained. Furthermore, we have seen in Case 3 of the last proof that II spends all whenever I does. Hence, $x_1 < p+\gamma$. Finally, zero spending for either player is obviously not an equilibrium point. The other player is then motivated to select an arbitrarily small positive initial spending in order to get all the goods in the first period without significantly jeopardizing his position in the second period. Of course, this does not yield a well-defined strategy, but even if it did, the player who had planned to spend nothing would then himself be motivated to spend a small positive value. In other words, we can begin by considering solutions in the interior of the possible spending intervals.

Case 2 : $\beta_1 \geqslant \beta_2$

In the interior of the possible spending intervals all possible solutions are solutions to the pair of equations obtained by taking derivatives of the return functions. We allow arbitrary spending in the first period but we use the previous solution thereafter. In other words, we apply the induction hypothesis. If I's money advantage in the second period is nonnegative, that is, if

(51) $$\gamma_n = \gamma + px_2 - (1-p)x_1 \geqslant 0,$$

and if $A_2(n) = k \geqslant 2$, where $A_2(n)$ is the number of the period beginning with the second when I first spends all his money, then the two equations are :

(52)
$$\frac{x_2}{(x_1+x_2)^2} = \frac{\beta_1(1-p)(1-p)^{(k-1)/k}\ \beta_1^{(k-1)/2}}{(1-p-\gamma-px_2+(1-p)x_1)^{(k-1)/k}}$$

$$\frac{x_1}{(x_1+x_2)^2} = \frac{\beta_2 pA(1-p)^{(k-1)/k}\ \beta_1^{(k-1)/2}}{(1-p-\gamma-px_2+(1-p)x_1)^{(k-1)/k}}$$

where

(53) $$A = k^{-1} \sum_{j=0}^{k-1} (\beta_2/\beta_1)^j$$

The equations in (52) come from (c) and (d) of Theorem 6 or (26) and (46). As an immediate consequence, we get a relationship between x_1 and x_2 namely,

(54) $$x_1 = \frac{\beta_2 pA}{\beta_1(1-p)} x_2$$

Substituting (54) into (51) we get

(55) $$\gamma_n = \gamma + (1-p)x_1 \, [(\beta_1/\beta_2 A)-1]$$

so that $\gamma_n \geqslant \gamma$ if and only if $\beta_2 \leqslant \beta_1$. If $\beta_2 \leqslant \beta_1$, then

$$x_2 = \frac{\beta_1(1-p)\left[1-p-\gamma-px_2\,[1-(A\beta_2/\beta_1)]\right]^{(k-1)/k}}{[\beta_2 pA+\beta_1(1-p)]^2(1-p)^{(k-1)/k}\beta_1^{(k-1)/2}}$$

$$\geqslant \frac{(1-p)^{1/k}\left[(1-p-\gamma)\left[1-p\,[1-(A\beta_2/\beta_1)]\right]\right]^{(k-1)/k}}{[\beta_2 pA + \beta_1(1-p)]^2\beta_1^{(k-3)/2}}$$

$$= \frac{(1-p)^{1/k}\,(1-p-\gamma)^{(k-1)/k}}{\left[\beta_2 pA+\beta_1(1-p)\right]^{\frac{k+1}{k}}\,\beta_1^{(\frac{k-3}{2}+\frac{k-1}{k})}}$$

$$\geqslant (1-p)^{1/k}\,(1-p-\gamma)^{(k-1)/k}$$

$$\geqslant 1-p-\gamma.$$

Hence, we have a contradiction with our requirement that $0 < x_2 < 1-p-\gamma$. The only solution with $\beta_1 \geqslant \beta_2$ is the one previously determined. In the second step of (56) we have made the right side smaller by replacing x_2 with $(1-p-\gamma)$.

Case 3 : $A_2(n) = 1$

Suppose now that $\beta_2 > \beta_1$. If

$$\beta_1 \leqslant \frac{(1+p)(1-p-\gamma)}{(1-p)} \qquad \text{and} \qquad \beta_2 \leqslant p+\gamma\ ,$$

then, for all x_1 and x_2, $\beta_1 \leq c_n$ if $\gamma_n \geq 0$ and $\beta_2 \leq \bar{c}_n$ if $\gamma_n \leq 0$, where

$$\bar{c}_n = \frac{p - (-\gamma_n)}{p} = \frac{p + \gamma_n}{p}$$

cf. Theorem 2 (iii). Note that this is always true for two periods, cf. Theorem 2 (ii). In this case the equations in (52) or (68) reduce to

$$\frac{x_2}{(x_1+x_2)^2} = \beta_1(1-p) \tag{57}$$

and

$$\frac{x_1}{(x_1+x_2)^2} = \beta_2 p .$$

We again get (54) but with A = 1. Substituting (54) into (57), we get

$$x_2 = \frac{\beta_1(1-p)}{[\beta_1(1-p) + \beta_2 p]^2} \tag{58}$$

and

$$x_1 + x_2 = [\beta_1(1-p) + \beta_2 p]^{-1}$$

so that $x_1+x_2 > 1$ unless $\beta_1 = \beta_2 = 1$. However, if $\beta_1 = \beta_2 = 1$, then $x_2 = 1-p \geq 1-p-\gamma$. Hence, there is no other solution for k = 1. This case completes the uniqueness proof for the two-period problem.

Case 4 : A Counterexample to Uniqueness if $\beta_2 > \beta_1$

Suppose $\beta_2 > \beta_1$, $A_2(n) > 1$, and $\gamma_n > 0$. Then we still have (52) and the first equation in (56). If

$$\beta_2 pA + \beta_1(1-p) \leq 1, \tag{59}$$

which is not necessarily to be expected now, then

$$
\begin{aligned}
x_2 &\geqslant \beta_1(1-p)\, c_n^{(k-1)/k}\ \beta_1^{-(k-1)/2} \\
&\geqslant (1-p)\, c_n^{(k-1)/k} \\
&\geqslant (1-p)\, c^{(k-1)/k} = (1-p)^{1/k}\,(1-p-\gamma)^{(k-1)/k} \\
&\geqslant 1-p-\gamma\ .
\end{aligned}
\tag{60}
$$

However, without (59), uniqueness can be lost. We demonstrate this for $A_2(n) = 2$, which corresponds to a three-period game. We then have

$$
x_2 = \frac{\beta_1^{1/2}(1-p)^{1/2}\left[1-p-\gamma+px_2\,[\,(A\beta_2/\beta_1)-1]\right]^{1/2}}{[\,\beta_2 pA + \beta_2(1-p)]^2}
\tag{61}
$$

or

$$
ax_2^2 - bx_2 - K = 0,
\tag{62}
$$

where $a > 1$, $b > 0$, and $K = 1-p-\gamma$. In particular,

$$
a = \frac{[\,\beta_2 pA + \beta_1(1-p)]^4}{\beta_1(1-p)}
\tag{63}
$$

and

$$
b = p\,[\,(A\beta_2/\beta_1) - 1\,]
$$

It is evident that (62) has exactly one positive real root, namely,

$$
x_2 = \frac{b + (b^2 + 4aK)^{1/2}}{2a}
\tag{64}
$$

Now we show that it is possible to select a, b, and K appropriately so that $0 < x_1 < p+\gamma = 1-K$ and $0 < x_2 < 1-p-\gamma = K$. It turns out that haphazard selections of p, γ, β_1, and β_2 will not do. It is important to make A and β_1^{-1} very large. For example, think of β_1^{-1} as 10^{100} although this may be a bit bigger than necessary ; we shall just let $\beta_1^{-1} = N$ with the understanding that N is big. Let $p = 1-p = 1/2$; let $\beta_2 = 1$ and let $K = 1/2$, which means that $\gamma = 1/4$. We shall express x_2 approximatively using N and generic constants c_i , $i \geqslant 1$.

We get $A = c_1 N$, $a = c_2 N^5$, $b = c_3 N^2$, and

$$x_2 = \frac{c_3 N^2 + (c_3^2 N^4 + 4Kc_2 N^5)^{1/2}}{2c_2 N^5} \tag{65}$$

$$= c_4 N^{-5/2} < K$$

Moreover,

$$x_1 = \frac{\beta_2 pA}{\beta_1 (1-p)} \; x_2 = c_5 N^2 x_2 = c_6 N^{-1/2} < 1-K \tag{66}$$

and

$$\gamma_n = \gamma - px_2 \, [\, (A\beta_2/\beta_1) - 1]$$

$$= \gamma - pc_4 N^{-4} c_7 N^2$$

$$= \gamma - c_8 N^{-2} \geqslant 0 \tag{67}$$

Case 5 : $\gamma_n \leqslant 0$

We have not discussed the case in which $\gamma_n \leqslant 0$, that is, when the advantage shifts to II. Instead of (52), we have

$$\frac{x_2}{(x_1+x_2)^2} = \frac{\beta_1 (1-p) p^{(k-1)/k} \; \beta_2^{(k-1)/2}}{(p+\gamma+px_2 - (1-p)x_1)^{(k-1)/k}} \tag{68}$$

and
$$\frac{x_1}{(x_1+x_2)^2} = \frac{\beta_2 pAp^{(k-1)/k} \; \beta_2^{(k-1)/2}}{(p+\gamma+px_2 - (1-p)x_1)^{(k-1)/k}} \; .$$

Again we obtain (54). Substituting it into the first equation of (68), we get

$$x_2 = \frac{\beta_1(1-p)\left[p+\gamma+(1-p)x_1\,[(\beta_1/A\beta_2)-1]\right]^{(k-1)/k}}{[\beta_1(1-p)+\beta_2 Ap]^2\, p^{(k-1)/k}\,\beta_2^{(k-1)/2}}$$

$$\geqslant \frac{\beta_1(1-p)\left[p+\gamma+(1-p)(p+\gamma)[(\beta_1/A\beta_2)-1]\right]^{(k-1)/k}}{[\beta_1(1-p)+\beta_2 Ap]^2\, p^{(k-1)/k}\,\beta_2^{(k-1)/2}}$$

(69)

$$= \frac{\beta_1(1-p)(p+\gamma)^{(k-1)/k}\,[\beta_1(1-p)+\beta_2 Ap]^{(k-1)/k}\beta_2^{-(k-1)/k}}{[\beta_1(1-p)+\beta_2 Ap]^2\, p^{(k-1)/k}\,\beta_2^{(k-1)/2}}$$

$$= \beta_1(1-p)\left(\frac{p+\gamma}{p}\right)^{(k-1)/k} \times$$

$$[\beta_1(1-p)+\beta_2 Ap]^{-(k+1)/k}\,\beta_2^{-[(k-1)(k+2)/2]}$$

where the second step involves replacing x_1 by $p+\gamma$ on the right. Since β_1 can be arbitrarily small, (69) is rather difficult to work with directly. If we assume (59), then we can apply (54) plus (69) to obtain

$$x_1+x_2 \geqslant [\beta_1(1-p)+\beta_2 Ap]\left(\frac{p+\gamma}{p}\right)^{(k-1)/k} \times$$

$$[\beta_1(1-p)+\beta_2 Ap]^{-(k+1)/k}\,\beta_2^{-[(k-1)(k+1)/2]}$$

(70)

$$= \left(\frac{p+\gamma}{p}\right)^{(k-1)/k}[\beta_1(1-p)+\beta_2 Ap]^{-1/k}\,\beta_2^{-[(k-1)(k+1)/2]}$$

$$> 1,$$

which contradicts the basic spending constraints : $x_1 \leqslant p+\gamma$ and $x_2 \leqslant 1-p-\gamma$. However, (70) is not possible without (59) . A counterexample for $A_2(n) = 2$ is easily constructed here just as in (61)-(67). This completes our discussion of uniqueness. In particular, we shall neither attempt to provide more detailed conditions for uniqueness when $\beta_2 > \beta_1$ nor attempt to describe the other noncooperative equilibrium solutions at this time.

Proof of Theorem 12 : Pareto Optimality

If $\beta_1 = \beta_2$, then our game is constant-sum except for the case in which both players spend nothing. Since such spending never occurs in any solution, cf. Case 1 of the Proof of Theorem 2, any solution is Pareto optimal when $\beta_1 = \beta_2$. However, if $\beta_1 \neq \beta_2$, then both players could simultaneously do better by judiciously (cooperatively) spending less initially. Since both players spend strictly positive amounts in any solution, both players are always free to spend less. The returns to I and II using positive x_1 and x_2 initially and the optimal strategies thereafter are

$$(71)\qquad R^1_{n+1}(p,\gamma) = \frac{x_1}{x_1+x_2} + \beta_1 U^1_n(p,\ \gamma+px_2 - (1-p)x_1)$$

and

$$R^2_{n+1}(1-p,\ -\gamma) = \frac{x_2}{x_1+x_2} + \beta_2 U^2_n(1-p),\ \gamma+px_2-(1-p)x_1) .$$

For $\beta_1 = \beta_2$, it is easy to see that

$$(72)\qquad -\frac{dU^2_n(1-p,\ -\gamma)}{d\gamma} = \frac{dU^1_n(p,\gamma)}{d\gamma} \geqslant \frac{dU^1_n(p,0)}{d\gamma} = 1$$

If $\beta_1 > \beta_2$, then it is easy to verify that

$$(73)\qquad -\frac{dU^2_n(1-p,\ -\gamma)}{d\gamma} < \frac{dU^1_n(p,\gamma)}{d\gamma}$$

Hence, $R^1_{n+1}(p,\gamma)$ can be kept constant by decreasing both x_1 and x_2 a small amount so that I's first period loss equals his future gain. This can be done in many ways. Then II's first period gain coincides with I's first period loss, but II's second period loss is less than I's second period gain because $\beta_2 < \beta_1$. Hence, II is better off while I is indifferent. A parallel argument applies to $\beta_1 < \beta_2$.

5. Proofs for Many Players

Proof of Theorem 19 : Perfect Competition

The bounds on p_i and γ_i imply that

(74) $$c_i = \frac{1 - p_i - \gamma_i}{1 - p_i} \geqslant \frac{n-2}{n}$$

so that $\beta_i \leqslant 1-\delta \leqslant c_i$ for sufficiently large n. Hence, the condition of Corollary 17 is satisfied. We now investigate uniqueness beginning with two periods. Suppose some player, say I, does not spend all his available money in the first period. This means that I, faced with the problem

(75) $$\max_{0 \leqslant x_1 \leqslant p_1+\gamma_1} \left\{ \frac{x_1}{x_1+y} + \beta_1(p_1+\gamma_1-x_1+p(x_1+y)) \right\}$$

where $y = \sum_{j=1}^{m} x_j - x_1$, elects to spend

(76) $$\hat{x}_{11} = \left[\frac{y}{\beta_1(1-p)} \right]^{1/2} - y < p_1+\gamma_1 \leqslant 2n^{-1}$$

which he obtains by differentiating in (75) as in Section 4. It is easy to see from (76) that we must have $y < n^{-1}$. First, $\hat{x}_{11}$ is increasing in y for $0 \leqslant y \leqslant [4\beta(1-p)]^{-1}$ and decreasing in y for $[4\beta(1-p)]^{-1} \leqslant y \leqslant 1-p_1-\gamma_1$. If $y = 1-p_1-\gamma_1$, then $\hat{x}_{11} > p_1+\gamma_1$ since $\beta_1 \leqslant c_1$. If $y = n^{-1}$, then

(77) $$\hat{x}_{11} = \beta_1^{-1/2}(1-p)^{-1/2}n^{-1/2}-n^{-1} > (n^{1/2}-1)n^{-1} .$$

Since $y < n^{-1}$, one of the remaining players must spend less than $[n(n-1)]^{-1}$ because $y = \sum_{j=1}^{m} x_j - x_1$ and $m \geqslant n$. These two arguments can be repeated to show that $y < (n-1)^{-k}$ and $\hat{x}_{11} < (n-1)^{-(k+1)}$ for all $k \geqslant 1$, where of course the distinguished player changes each time. This argument, easily made precise by induction, demonstrates that there is no spending at all in the first period, which is a contradiction because zero spending is obviously not an equilibrium solution. Hence, there is no second solution with some players spending less than all their money in the first period. This proof was for two periods. It is extended to any finite number of periods by induction. For sufficiently large n, the position in the second of several periods will correspond to the initial position with two periods, which we have just analyzed in detail. For example, player i's money supply in the second period can be no greater than $2p_i+\gamma_i$. Hence, $c_i \geqslant (n-3)/n$ in the second period and the spending is bounded above by $3n^{-1}$. Thus, by virtue of the induction hypothesis, I is again faced with (75) and the same argument applies.

Proof of Theorem 20 : One Fat Cat

The proof above applies to show that none of the small players will spend less than all their money each period. This guarantees uniqueness.

6. Constrained Competitive Equilibrium

Proof of Theorem 21

First, consider the one-period problem. Everyone is clearly motivated to spend all available money at any price. Since $\sum_{i=1}^{m} x_i = G$, the price must be M/G. Next, assume the theorem to be true for n periods and consider the (n+1)-period optimization problem in (4) . By virtue of the induction hypothesis, it reduces to the following two period problem for each individual :

$$\max_{\{x_i\}} \{x_1 + \beta x_2\} \tag{78}$$

subject to

$$x_1 a_1 \leqslant (p_1 + \gamma_1)M$$

$$x_2 \frac{M}{G} = (p_1 + \gamma_1)M - x_1 a_1 + p_1 G a_1 \ .$$

If $\frac{M}{G} < a_1\beta$, then each player wants x_2 instead of x_1 . This would lead to zero spending in the first period. If $\frac{M}{G} = a_1\beta$, then each player is indifferent between x_1 and x_2 . If $\beta < 1$, then both cases can be ruled out because the price in the first period must be less than or equal to $\frac{M}{G}$ since no more than M will be spent for the total goods G. If $\beta = 1$, then $a_1 = \frac{M}{G}$ is only possible if each player spends all his money in the first period. If $\frac{M}{G} > a_1\beta$, then each player prefers x_1 to x_2 . Hence, each player will spend all his available money in the first period. The associated price is then $\frac{M}{G}$.

If $\beta_i = \beta$ for all i, then the allocations are constant-sum and thus obviously Pareto optimal. If $\beta_i > \beta_j$ for some i, j, then both players could simultaneously do better if j gave i some goods in the future in exchange for some goods in the present.

References

(1) K.J. ARROW, S. KARLIN, and H. SCARF, *Study in the Mathematical Theory of Inventory and Production*, Stanford University Press, Stanford, California, 1958

(2) R.J. AUMANN, Markets with a continuum of traders. *Econometrica*, vol.32, 1964, pp.39-50

(3) D. BLACKWELL, Discounted dynamic programming. *Ann. Math. Statist.*, vol.36, 1965, pp.226-235

(4) D.J. BROWN and A. ROBINSON, A limit theorem on the cores of large standard exchange economies. *Proc. Nat. Acad. Sci. USA*, vol.69, 1972, pp.1258-1260

(5) E.V. DENARDO, Contraction mappings in the theory underlying dynamic programming. *SIAM Review*, vol.9, 1967, pp.165-177

(6) W. HILDENBRAND,Measure spaces of economic agents. *Proceedings of Sixth Berk. Symp. Prob. Stat.*, II, 1972, pp.81-96

(7) W. HILDENBRAND,*Core and Equilibria of a Large Economy*. 1973. Forthcoming.

(8) A. MAITRA and T. PARTHASARATHY, On stochastic games. *J. of Opt. Theory and Appl.*, vol.5, 1970, pp.289-300

(9) L. SHAPLEY, Stochastic games. *Proc. Nat. Acad. Sci. USA*, vol.39, 1953, pp.1095-1100

(10) L. SHAPLEY and M. SHUBIK, A Theory of Money and Financial Institutions, Part VI. The Rate of Interest, Noncooperative Equilibrium and Bankruptcy. Cowles Foundation Discussion Paper 334, April 5, 1972

(11) M. SHUBIK, Commodity money, oligopoly, credit and bankruptcy in a general equilibrium model. A theory of money and financial institutions. Part III. Cowles Foundation Discussion Paper 324, 1971. (revised February 1973, forthcoming in *Western Economic Journal*).

(12) M. SHUBIK, A theory of money and financial institutions. Part IV, Fiat money and noncooperative equilibrium in a closed economy. Cowles Foundation Discussion Paper 330, 1972

(13) M. J. SOBEL, Continuous stochastic games. *J. Appl. Probability*, 1973. To appear.

AUTHOR INDEX